# ADVANCES IN ENVIRONMENTAL BIOLOGY

# ADVANCES IN ENVIRONMENTAL BIOLOGY

G. Tripathi
B.M. Sharma
T.K. Ghosh

Oxford Book Company
Jaipur ▪ New Delhi

*Published By*

Oxford Book Company

**Regd. Off.** "Bony Residency", Gate No. 2
Opp. Tilak Public School, Vishweswariya Nagar,
Gopalpura Road, Jaipur - 302018, India
Ph: 0141 - 2761280, Telefax: 2761381
email - oxfordbookjpr@gmail.com
website - oxfordbookcompany.com

**Delhi Office:** 102, 1st Floor, Satyam House,
4327/3, Ansari Road, Darya Ganj,
New Delhi - 110002, India
Ph: 011-45652440

*Distributed by*
**KEYA INTERNATIONAL PRESS LLP**
E-mail: keyainternationalpressllp@gmail.com

Published in India, 2025



ISBN : 9788189473773



Typeset by Shivangi Computers, Jaipur

Printed and bound in India by Replika Press Pvt. Ltd.

# Preface

The book deals with the latest research on environmental biology right from ecofriendly management of waste to biochemical effects of pesticides. There are 17 chapters written by nationally and internationally reputed scientists from India and abroad. Billions of tons of agricultural wastes are produced around the world and impose environmental problems. Hence chapter 1 dictates an integrated approach to the recycle of organic waste from agro-industrial system. Chapter 2 explains an ecofriendly approach for treatment and disposal of pulp and paper mill wastewater. For safe disposal of treated coloured pulp and paper mill wastewater on land, high rate transpiration system has been developed. Chapter 3 describes biological consequences of increased $CO_2$ concentration. The information on this aspect is useful for conservation of biological diversity and sustainable development of any ecosystem in multifaceted activity comprising various abiotic and biotic components. Chapter 4 throws light upon monitoring of colour removal capacity of soil. Chapter 5 illustrates agro-economic evaluation of wastewater.

Chapter 6 discusses the importance of resource conservation through land treatment of municipal wastewater. Chapter 7 enumerates profiles of toxic metals and particulate matter in urban areas. Chapter 8 gives highlights of conservation of endangered nitrogen fixing plants of cold desert. Chapter 9 elucidates role of botanical insecticides in pest management. Chapter 10 presents an interesting aspect on fish pheromones and their importance in management of aquaculture environment. Chapter 11 summarises drug metabolism and detoxification in organisms, especially the role of cytochrome P450 and aryl hydrocarbon receptor. Chapter 12 explains the mechanism of oxidative stress and diseases due to such stress. Chapter 13 deals with biology of cancer as a consequence of environmental pollution. Chapter 14 describes the biochemical effects of pesticides in fish. Chapter 15 dictates the hazards of arthropod borne diseases in changing environment. Chapter 16 discusses impact of steel industry wastewater on the lentic water bodies. Chapter 17 explains the impact cooling water discharges on primary producers of a reservoir.

The book will be useful to the persons engaged in learning, teaching and working in the area of environmental biology. It may also be used for environmental consultancy and counseling.

**G. Tripathi**
**B.M. Sharma**
**T.K. Ghosh**

# List of Contributors

**A.A. Juwarkar**
National Environmental Engineering Research Institute (NEERI), Nehru Marg, Nagpur-440020, India.

**A.K. Pandey**
National Burev of Fish Genetic Resources (NFGR), Lucknow, India.

**A.V. Kulkarni**
National Environmental Engineering Research Institute (NEERI), Nehru Marg, Nagpur-440020, India.

**B. Deepa**
Holy Cross College, Trichy, India.

**B.M. Sharma**
NIMS University, Shobha Nagar, Jaipur - 303 001.

**D.G. Gajghate**
National Environmental Engineering Research Institute (NEERI), Nehru Marg, Nagpur-440020, India.

**G. Tripathi**
Deptt. of Zoology, J.N.V. University, Jodhpur-342 033, India.

**H. Singh**
Deptt. Of Zoology, J.N.V. University, Jodhpur-342 033, India.

**J. Shasmal**
Deptt. Of Zoology, J.N.V. University, Jodhpur-342 033, India.

**K. Muthu**
Department of Biochemistry J.J. College of Arts and Science, Pudukkottai-622 404, India.

**K. Ramlingam**
PG and Research Department of Zoology, Govt. College, Nandanam, Chennai-35, India.

**Monika Tyagi**
Plant Physiology Discipline, Botany Division, Forest Research Institute, Dehradun, India.

**N. Bandooni**
Deptt. of Zoology, J.N.V. University, Jodhpur-342 033, India.

**P. Krishnamoorthy**
Department of Biochemistry J.J. College of Arts and Science, Pudukkottai-622 404, India.

**P. Kumar**
National Environmental Engineering
Research Institute (NEERI),
Nehru Marg, Nagpur-440020, India.

**P. Nema**
National Environmental Engineering
Research Institute (NEERI),
Nehru Marg, Nagpur-440020, India.

**P. Rapana**
National Council of Research-Instituto di
Biologia Agroambientale e Forestale
Area della Ricerca di Roma 1 Via Salaria,
km 29.300-00016
Monterotondo scalo (Roma), Italy.

**P. R. Chaudhari**
National Environmental Engineering
Research Institute, (NEERI),
Nehru Marg, Nagpur 440020 (India).

**P.R. Thawale**
National Environmental Engineering
Research Institute (NEERI),
Nehru Marg, Nagpur-440020, India.

**P.S. Rao**
National Environmental Engineering
Research Institute (NEERI),
Nehru Marg, Nagpur-440020, India.

**Pankaj Kumar**
Plant Physiology Discipline, Botany
Division, Forest Research Institute,
Dehradun, India.

**S. Arun**
Division of Chlor-Alkali Central
Electrochemical Research Institute,
Karaikudi-630 006, India.

**S.K. Singh**
National Environmental Engineering
Research Institute (NEERI),
Nehru Marg, Nagpur-440020, India.

**S.R. Verma**
National Environmental Engineering
Research Institute, (NEERI),
Nehru Marg, Nagpur 440020 (India).

**S. R. Wate**
National Environmental Engineering
Research Institute, (NEERI),
Nehru Marg, Nagpur 440020
(India).

**S. Zargar**
Environmental Biotechnology Division,
National Environmental Engineering
Research Institute, Nehru Marg,
Nagpur – 440 020.

**T.C. Pokhriyal**
Plant Physiology Discipline,
Botany Division, Forest Research Institute,
Dehradun, India.

**T.K. Ghosh**
National Environmental Engineering
Research Institute (NEERI),
Nehru Marg, Nagpur-440020,
India.

**U. Tomati**
National Council of Research-Instituto di
Biologia Agroambientale e Forestale
Area della Ricerca di Roma 1 Via Salaria
km 29.300-00016
Monterotondo scalo (Roma), Italy.

**V. Sharma**
National Environmental Engineering
Research Institute (NEERI),
Nehru Marg, Nagpur-440020,
India.

**V.R. Sadevra**
National Environmental Engineering
Research Institute (NEERI), Nehru Marg,
Nagpur-440020,
India.

# Contents

# 1

# An Integrated Approach to the Recycle of Organic Wastes from Agro-industrial System

*Umberto Tomati and Pompilio Rapanà*

People make only a marginal use of what the earth produces through photosynthesis by green plants. More of 70% of agricultural and forest materials are discarded as waste and/or excess of production. Billions of tons of agricultural waste are yearly produced around the world as residues of high-yield crops. Moreover, agricultural processing yields many by-products that have significant value. They are mainly destroyed by burning, burying or uncontrolled disposed. Consequently, a great part of what is produced by plants is getting to be irremediably lost and, if incorrectly used or disposed, results in environmental problems. In the last decades an increasing attention has been posed on integrated or alternative solutions involving recycling as the most positive approach that provide for a beneficial use of waste. The challenge is now to consider all waste produced by anthropic activity as a resource and evaluate how they can be used for further productions. Agriculture and agro-industries produce about the 50% of the total waste. In contrast with industrial and urban refuses, agricultural/agro-industrial waste is an "healthy" organic matrix generally lacking of harmful and hazardous compounds. This huge biomass could be usefully recycled in agriculture as safe soil conditioners following the simple recipe: "What comes from soil must be returned to the soil" (Sequi, 1990). Moreover, it could be transformed into food or feed or used as a starting materials for new high value products. That tunes up with the exhortation: "Now time has come not to expect the earth to produce more, but to expect man to do more with what the earth produces" (Pauli, 1996).

Waste is not an indication of an inherent property of any material. It only describes a situation, namely the transition of a material or product from being "useful" to being "useless" ( Van Belle, 1996). Therefore, waste can be either considered as a polluting stream or a resource to be recycled through appropriate processes or biotechnologies (Fig 1).

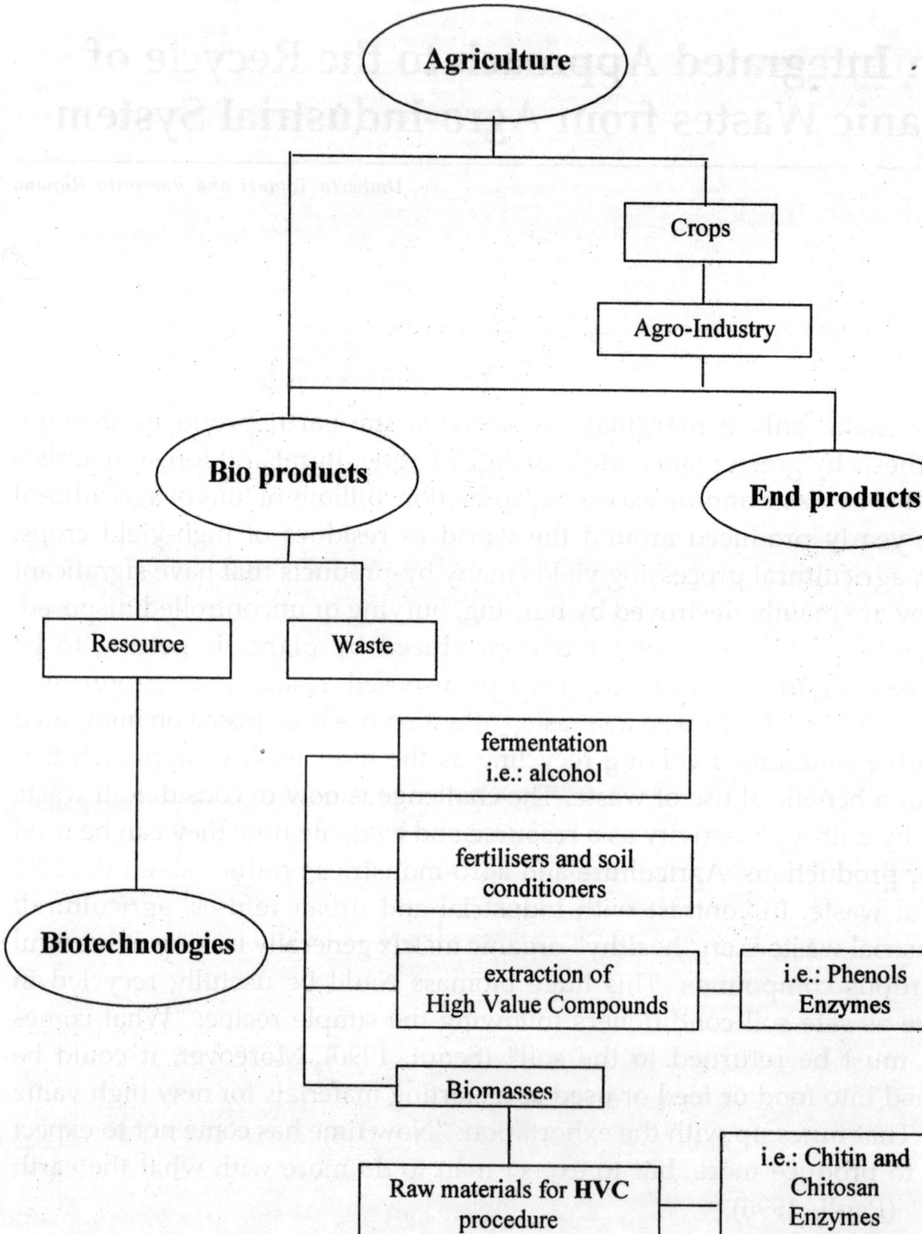


**Fig 1.**

The opportunities of converting agro-industrial waste into raw materials for new activities are unlimited, although sometimes this task is not easily to do. Many Countries lack the national capacity to handle and manage waste, primarily due to inadequate infrastructures. This includes inadequate facilities, lack of trained personnel, lack of information and monitoring systems, inadequate regulatory programmes and insufficient financing. The establishment of an effective waste management programme involves building institutions, training, development of human resources, research programmes aimed at individuating new technologies and, in general, building the capacity to control and manage waste. Difficulties are often due to the economic and/or legal aspects of the problem rather than to technical solutions proposed from time to time to face the arising problems. Waste has a cost for its disposal and/or has a limited value at its place of origin. In order to effectively carry out recycling, managers involved in treatment programmes must operate in a business-like manner as raw materials and energy suppliers. They must treat the users of their products as costumers. This means that they must offer products that meet the costumer's request at a competitive price.

Especially in developed Countries – but it should be desirable in everywhere – regulations have been stated to control the national plans of restoration. The legal aspects mainly consider the possibility of planning the management of waste having in mind some the main objectives aimed at:

1. preserving, protecting and improving the quality of the environment;
2. contributing towards protecting human health;
3. ensuring a prudent and rational usage of natural resources.

The establishment of national regulatory control programmes with appropriate legislation, regulations, ordinances and licenses is an extremely important step for a rational policy.

The waste problems occur when, either because of ignorance or irresponsible behaviour, there is negligence in the handling materials. The status-of-the-art of waste treatment and disposal has advanced rapidly in recent years: today, technologies are available to treat and dispose of waste in an environmentally sounding manner. The three main strategies available to dispose or recovery waste are incineration, sanitary landfill and recycling. These three options can co-exist; how to use them is a political choice that must firstly take into consideration the environment safe-guard. At least at a first glance, incineration and landfill seem to be clean and fascinating technologies since incineration allows energy recovery and landfill permits waste disposal without apparent difficulties in appropriately selected areas. As a matter of facts, benefits are more apparent than actual. There is a risk of pollution from emission into atmosphere when waste are burnt whereas the recovery of energy remain often a dream. The drawbacks due to landfill mainly arise from the relative low availability of suitable land surfaces, from the transformation of disposed materials, from the emission of gases into the atmosphere, and harmful percolates in the medium and long term. Recycling is in general the most difficult practice to face the problem. It requires the highest degree of technical and scientific professional competences including microbiology, chemistry, biochemistry and engineering. Recycling is the technology which prevents the existence itself of waste. Any material, which actually could become waste if their owner should get rid them, can be transformed in useful materials or commercial products. Agricultural and agro-industrial waste, because of their organic matrix, are particularly suitable to be submitted to recycling technologies. Among them, composting is the most important biotechnology from the quantitative point of view, also considering the vast quantity of residues to be disposed. A wise policy in matter of waste should promote recycling technologies and subsidise the formation of adequate competences.

Many Countries have defined far reaching objectives for waste management, and have implemented sophisticated legislative, technological and logistic systems to reach these objectives. Goals mainly include protection of man and the environment, and conservation of resources. The main question is if the planned waste management systems are the most efficient means to reach the objectives set.

## HOW TO FACE THE PROBLEM: A PROPOSAL

Criteria describing the different scenarios have to be defined in ways which allow to calculate the relevant variables quantitatively and to fulfil the requirement of an integrative evaluation of each scenario. In this paper, a methodology and a case of study is proposed. To gather the most of information, the research programmes should consider:

### 1. The Existing Situation with Regard to the Place of Production, Quality and Quantity, Waste Characterisation, Economical Aspect of Their Disposal or Recovery and Regulation.

Generally, there is a lack of reliable data concerning the territorial aspects of waste production. This first step is aimed at providing Agencies involved in waste management with every information suitable to draw rational disposal plans. A correct policy which permits an integrated approach to define suitable strategies to waste managing needs a good knowledge of

- Areas of production
- Quantity and quality of waste
- Technologies of production and disposal
- Existing laws
- Economic and social aspects of the problem

## 2. The Environmental Impact

The determination of the hydro-pedo-morphological, climatic and agronomic aspects concerning the areas proposed for landspreading is of a capital importance. Especially in the presence of liquid wastes, the above reported characteristics must be carefully considered. Because of their organic matrix, agricultural and agro-industrial waste are often disposed onto soil either before or after biostabilisation. The main inconvenient of the agro-industrial wastes is due to the high organic load (COD and $BOD_5$) and, sometimes, the presence of several organic compounds (i.e. phenols, pesticides) which can be resistant to biodegradation. Therefore, when disposed into environment, they can provoke undesirable effects on soil and aquatic biota. A massive waste disposal can include potential leaching to groundwater or runoff to the surface waters from waste application sites. Leaching into groundwater is related to the solubility of the organic compounds and the sorption and desorption of the compounds from colloidal surfaces such as humic materials and clay. Soil characteristics are of a great importance to avoid such undesirable effects so it is necessary to evaluate the pedological, hydrological, agronomical and climatic aspects of disposal sites to avoid environmental pollution. Their knowledge permits to acquire the information necessary to prepare maps for a correct disposal/agronomic use of wastes, prepare rational plans for waste managing, and propose suitable regulations.

## 3. Waste Stabilisation and/or Reduction of Volume and/or Mass

The waste stabilisation and/or reduction of volume and/or mass are two important aspects of waste recycle. For that the primary aim of a waste legislation is the prevention of waste generation. Waste prevention refers to three types of practical actions, i.e. strict avoidance, reduction at source, and product re-use (Pohjola & Pongracz, 2002; Pongracz, 2003;). Considering the waste management options, at the top of the hierarchy stands waste minimisation that includes (Riemer & Kristoffersen, 1999):

– waste prevention i.e. reduction of waste by application of more efficient production technologies;

- internal recycling of production waste;
- source-oriented improvement of waste quality, e.g. substitution of hazardous substances;
- re-use of products or parts of products, for the same or other purpose.

The organic matrix of agricultural and agro-industrial waste easily undergoes alteration due to fermentative processes. Transforming this potentially putrescible organic matter into a stable state and produce a material that may be used for soil improvement or other beneficial purposes is an important goal to be achieved. Reducing waste must be broadly taken into consideration by industries, enterprises and governmental agencies. Getting to the bottom of the problem, waste production can be reduced through raw material substitution and redesign of products and processes. Governments should encourage this policy (Pongracz & Pohjola, 2004).

## 4. The Waste as Raw Material for High Added Value Compound Recovery

Proposing technologies for high value compounds extraction/biosynthesis is a promising trend for the scientists involved in agricultural and agro-industrial waste recovery. To prevent or reduce pollution, there is a growing interest for technologies aimed at using renewable and sustainable commodities such as those from agriculture in alternative to mineral and synthetic ones. This new approach should develop strategies economically sustainable for designing environmentally friendly processes for the production of safe products. In this contest, agricultural waste represents an important source of renewable products of a great economic and social value that can considered as a starting material for the extraction of natural compounds and a growth substrate for micro-organisms.

Plants are attractive sources of natural compounds so their extraction from vegetable materials could be a further way to take advantages of waste. It has been established that by-products of agro-industry are suitable sources to obtaining, directly or indirectly, a great variety of high value compounds, especially polyphenols which are the major plant compounds with antioxidant activity (Moure *et al.* 2001). The growing interest in the substitution of synthetic antioxidants for food, cosmetics, pharmaceuticals by natural ones has fostered research on vegetable sources and the screening of raw materials for identifying new molecules. Special attention has been focused on their extraction from inexpensive or residual sources from agricultural industry (Table 1a and 1b).

Natural phenols showing a powerful antioxidant activity can be regarded as a cheap and suitable source of natural non toxic antioxidants for use in food industry. The antioxidant activity of monomeric and polymeric compounds has been object of investigation, and there is evidence that such compounds can be recovered for their use as food preservative to preserve flavour, colour and avoid vitamin degradation as well as to prolong shelf live. They could substitute the synthetic antioxidants, used to preserve food (i.e. butylated hydroxytoluene - BHT and butylated hydroxyanisole -BHA) (Camire &Dougherty, 1998; Zandi & Gordon, 1999; Che Man& Tan,1999).

The organic matrix of waste and its dressing of nutrients constitute an optimal substrate for the growth of micro-organisms. These biomasses are in turn an important raw material for new compounds to be used for many industrial purposes. Algae and photosynthetic bacteria are among the most widely used micro-organisms. Because of their metabolism, photosynthetic micro-organisms transform the most of organic carbon present in the growth substrate into active biomasses (about 1.2-1.3 kg per kg of $BOD_5$) that can be usefully used to obtain soil conditioners or high value compounds. Many of them are already used as commercial products.

Many polysaccharides have been obtained from bacterial biomasses. The number of potential polysaccharide structure is almost limitless although currently only a small number ob biopolymers are commercially produced. Among them, xanthan from *Xanthomonas campestris* (Bilanovich *et al.*, 1994), succinoglycan from *Rhizobium* spp. and other species, gellan and a range of structurally related polysaccharides from *Sphingomonas paucimobilis* (Sutherland, 1996), bacterial alginates from *Pseudomonas* spp., *Azotobacter vinelandii* and *A. chroococcum* (Gonzales Lopez *et al.*, 1996), pullulan from *Aureobasidium pullulans* (Sutherland, 1996).

The possibility of producing mycelial biomasses from agricultural wastes is another attractive perspective which permits to recycle wastes into "nutraceutical" products (Coulibaly *et al.*2003).

In view of recent advances in experimental mycology and mushroom biotechnology, interest in higher Basidiomycetes as a natural source of valuable biologically active metabolites has significantly increased. The U.S. National Cancer Institute classified many of them as antitumor chemicals with immuno-modulating activity. The therapeutic properties are mainly related to polysaccharides and protein-bound polysaccharides (Bohn & Bemiller, 1995). Several antitumoral polysaccharides with different chemical structures such as a- glucan, a-heterogalactan, and a-glucanprotein have been isolated from different mushrooms (Furue 1986; Yadomae & Ohno, 1992; Bohn & Bemiller, 1995; Wasser & Weiss, 1999)

## Table 1a

| Residue | Solvent | Solubles yield (% dry weight) or T.E.P. (as equivalent) | Identified compounds | Reference |
|---|---|---|---|---|
| 1 | 2 | 3 | 4 | 5 |
| Durum wheat bran | Ethanol | 12.1 (% dry weight) 2.769 (HPLC) | PA, pBA, GA, CaA, VA, CA, SA, pCA, FA | Onyeneho and Hettiarachchy (1992) |
| Fraxinus ornus bark | Ethanol | 14.5 (dry weight) | Hydroxycoumarin (Es, Est, Fx, Fxt) | Marinova et al. (1994) |
| Corn bran hemicellulose | NaOH | 3.259 (FA) | FA; pCA; dFA | Ohta et al. (1994) |
| Potato peel extract | Water (100 C) | 0.048 (HPLC) | CA, GA, PA, CCA | Rodrìguez de Sotillo et al. (1994a) |
| | Methanol (4 C) | 0.041 (HPLC) | CA, GA, PA, CCA | |
| Shrimp shell waste | 95% Ethanol | | 1,2 diamino-1(o-hydroxyphenyl)propene | Seymour et al. (1996) |
| Red grape pomace peels | Methanol water (1,1) | 4.3 (tannic acid) | | Larrauri et al. (1997) |
| | Acetone,water (7,3) | c. t., 27.0 (") | | |
| Apple pomace | 70% acetone | | EC, CA, 3- hPhlz, Phl-2-x, Phlz, Q-3-gal, Q-3-glu, Q-3-xyl, Q-3-ara, Q-3-rha | Lu and Foo (1997) |
| Buckwheat hulls | Ethanol | 23.8 (HPLC) | PA, 3,4-dihydroxybcenzaldehyde, Hy, Ru, Q, Vi, iVi | Watanabe et al. (1997) |
| Oat hulls | Methanol | 0.035 (GC-MS) | FA, pCA, VA, pBA, V, 4-PhA, catechol, o-coumaric acid, SA, SaA | Xing and White (1997) |
| Lemon seeds | Methanol | 0.2333 (1) | CaA, pCA (*cis* and *trans*), FA, SA, Eri, Nar, Neh | Bocco et al. (1998) |
| Sweetorange seed | | 0.0544 | CaA, pCA (*cis* and *trans*), FA, SA, Nat, Nar, Hes | |
| Sour orange peel | | 2.526 | CaA, pCA (*cis* and *trans*), FA, SA, Ner, Nat, Nar, Hes | |
| Bergamot peel | | 1.359 | CaA, pCA (*cis* and *trans*), FA, SA, Ner, Nar, Neh | |
| Lentil seed coat (brown) | Water | 22.3 (% dry weight) | L, Q, K, PC, PD, FA, PA, CaA | Muanza et al. (1998) |
| (green) | | 8.7 (% dry weight) | | |
| Grape marc | Ethyl acetate | 0.224 (HPLC) | K, Q, iRh3glu; K-3-glu; Q-3-glu; Q-3-gal; EC; C; VA; PA; GA | Bonilla et al. (1999) |
| | Water | 0.021 (HPLC) | M-3(6-a)glu; Po-3(6-a)glu; M-3-glu; Po-3-glu; Pt-3-glu; Cy-3-glu; D-3-glu; PC | |
| Grape pomace | 80% Ethanol | 42% (dry weight) | GA; GA 3-bglup; GA 4-bglup; trans-CfA; cis and trans CtA; 2-h-5(2he)pb-D-glp; C; EC; PCB1; Q-3glup; Q-3-glurcp; K-3-glup; K.3-galp; Ey; At; En | Lu and Foo (1999) |
| Lemon peel | Methanol-Water | 0.02% (dry weight) | coumarins (8-geranyloxypsolaren, 5-geranyloxypsolaren, 5 geranyloxy-7 methoxycoumarin) | Miyake et al. (1999) |

| 1 | 2 | 3 | 4 | 5 |
|---|---|---|---|---|
| Grape seeds | 95% Ethanol | | pCA; GA; CAA; C,-EC, procyanidins (B1-B8) | Saint-Cricq de Gaulejac et al. (1999) |
| Non-volatile residue from orange essential oil | | | a-toc; 3,30,40,5,6,7 hexamethoxyflavone; 3,30,40,5,6,7,8,-heptamethoxyflavone; 40,5,6,7,8 pentamethoxyflavone | Vargas-Arispuro et al. 1998) |
| Olive mill waste waters | Water:Ethanol | 8.11 (% dry weight) | hT, T, EA, oleuropein derivatives, Lu-7-glu, Q, CnA derivatives, hT derivatives | Visioli et al. (1999) |
| Grape seed extract | Ethyl acetate 20% Ethanol | 22.78 (% dry weight) | proanthocianidins, monomeric flavonols | Yamaguchi et al. (1999) |

a e.p., extractable polyphenols; n.e., non extractable; UP; unidentified polyphenols; c.t., condensed tannins; (1)glycosylated flavanone+phenolic acids content; GC-MS, gas chromatography-mass spectro-scopy. Phenolic acids; CA, chlorogenic; CaA, caffeic; Cf A, caftaric; CnA, cinamic; Ct A, Coutaric; EA, elenoic; FA, ferulic; dFA, diferulic; GA, gallic; KA, kiwic; QA, quinic; VA, vanillic; SA, syringic; CnA, cinnamic; PA, protocatechuic; pBA, p-hydroxybenzoic; GeA, gentisic; pCA, p-coumaric; VA, vanillic; At, Astilbin; Ct, catechol; EC, Epicatechin; C, Catechin; Cy-3-glu, Cyanidin-3-glucoside; D-3-glu, Del-phinidin-3-glucoside, En, Engeletin; Eri, eriocitrin; Es, Esculin; Est, Esculetin; Ey, Eucryphin: Fx, Fraxin; Fxt, Fraxetin; GA 3-bglup, Gallic acid 3-b-glucopyranoside; GA 4-bglup, Gallic acid 4-b-glucopyr-anoside; 2-h-5(2he)pb-D-glp, 2-hydroxy-5-(2-hydroxyethyl(phenyl-b-D-glucopyranoside; Hes, hesperidin; Hy, Hyperin; K, Kaempferol; K-3-glu, Kaempferol-3-glucoside; K-3-glup: Kaempferol 3-b-D-glucopyranoside; K-3-galp: Kaempferol 3-b-D-galactopiranoside; Lu-7-glu, Lu-7-glucoside M-3(6-a)glu,Malvidin-3-(6-acetyl)-glucoside; M-3-glu,Malvidin-3-glucoside; Nar, naringin; Nat, narirutin; Neh, neo-hesperidin; Ner, neoeriocitrin; PC, Procyanidin; PD,Prodelphinidin; 4-PhA, 4-hydroxyphenilacetic acid; 3,4-PhE, 3,4-hydroxyphenilethanol; 3,4-dB, 3,4-dihydroxybenzaldehyde; 3-hPhlz, 3-Hydroxyphloridzin; Phl-2-x,Phloretin-20- xyloglucoside; Phlz,Phloridzin; Po-3(6-a)glu,Peonidin-3-(6-acetyl)-glucoside; Pt-3-glu,Petunidin-3-glucoside; Q,Quercetin; Q-3-gal, Quercetin-3-galactoside; Q-3-glu, Quercetin-3-glucoside; Q-3-glucr, Quercetin-3-glucuronide; Q-3-glurcp: Quercetin 3-b-D glucuropyranoside; Q-3-glup: Quercetin 3-b-D glucopyranoside; Q-3-xyl, Quercetin-3-xyloside; Q-3-ara, Quercetin-3-arabinoside; Q-3-rha, Quercetin-3-rhamnoside; iRh3Glu, isoRhamnetin-3-glucoside; Ru, Rutin; SaA, Salicilyc acid; T, tyrosol; hT, hydroxytyrosol; V,Vanillin Vi, Vitexin; iVi, isovitexin

**Table 1b**

**High value commercial compounds obtained from phothosynthetic micro-organisms**

| Item | Use |
|---|---|
| Health foods<br>Vitamins (E, $B_{12}$)<br>Ficobiliproteins | Dietary supplement |
| Astaxantin<br>Polysaccharides | Dyes for food and feed<br>Many industrial applications |
| β-carotene | Dyes |
| Poly-unsaturated fatty acids | Food and feed |
| Microbial biomasses<br>Water cleaning | Soil conditioners |
| Ion exchangers | Metal biosorption |
| Biological active substances | Antibiotics |

## Recycling Olive Oil Mill Industry By-Products: A Representative Example of Waste Recovery

(References: see appendix)

The disposal of olive oil waste is a typical problem in the Mediterranean area where the cultivation of olive trees has been an established feature since more than 6000 years. Olive trees grow in fact in rather arid areas, where protect soils against desertification and erosion, and help to regulate hydric cycle. The extraction and use of olive oil is linked to the Mediterranean culture and history (Civantos, 1995; Tartaguila *et al*, 1996). Besides its use in cosmetic, olive oil has always been an essential component of Mediterranean diet for its high dietetic and nutritional quality.

Nowadays, the olive oil industry yearly produce about 30 million tons of waste in the short and rainy harvesting time. The production of olives can greatly vary from one year to other; several factors, mainly related to weather and plagues, can affect the trees and olives. The rough mean data of olive production is reported in Table 2. Although they can greatly vary, they show the importance of the olive oil sector and the magnitude of the problem related with the huge amounts of wastes and by-products generated by olive mill industry. On the basis of FAOSTAT data, Figure 2, (FAOSTAT 2000), the three main producers are Spain, Italy and Greece, with about 80 % of world production, followed by Tunisia and Turkey. The contribution of other Mediterranean countries is much lower.

**Table 2**

**Mean data of production (ton/year).**

| | EU | Other | World |
|---|---|---|---|
| Olive yield | 7 700 000 | 2 000 000 | 9 700 000 |
| Olive oil | 1 450 000 | 375 000 | 1 825 000 |
| Waste material | 5 800 000 | 1 500 000 | 7 300 000 |

## Waste Production

The olive industry produces large amounts of by-products; the quantity and, in a minor extent, the quality depend on the extraction technology. On average every 100 kg of olives originate 35 kg of solid waste and 55 – 120 l of liquid waste. Speaking about olive by-products, olive-mill extraction waste are mainly considered.

The olive consists in pulp (75 – 85 % by weight), stone (13 – 23 %) and seed (2 – 3 %) (Maestro 1985). The chemical composition is reported in table 3. Apart from the oil, all the other components constitute olive mill waste.

**Table 3**

**Composition of olives**

| CONSTITUENTS | PULP | STONE | SEED |
|---|---|---|---|
| Water | 50 – 60 | 9.3 | 30 |
| Oil | 15 – 30 | 0.7 | 27.3 |
| Constituents containing nitrogen | 2 – 5 | 3.4 | 10.2 |
| Sugar | 3 – 7.5 | 41 | 26.6 |
| Cellulose | 3 – 6 | 38 | 1.9 |
| Minerals | 1 – 2 | 4.1 | 1.5 |
| Polyphenols (aromatic substances) | 2 – 2.25 | 0.1 | 0.5 – 1 |
| Others | - | 3.4 | 2.4 |
| Values in percent by weight [%] | | | |

The traditional methods of extraction and the continuous three-phase decanting generate one stream of olive oil and two stream of wastes (liquid and wet solid). Nowadays, a new low water consuming water extraction method (two-phase decanting process), extensively used in Spain and in expansion in Italy and Greece, produces one stream of oil and one stream of a wet solid waste (Fig. 3).

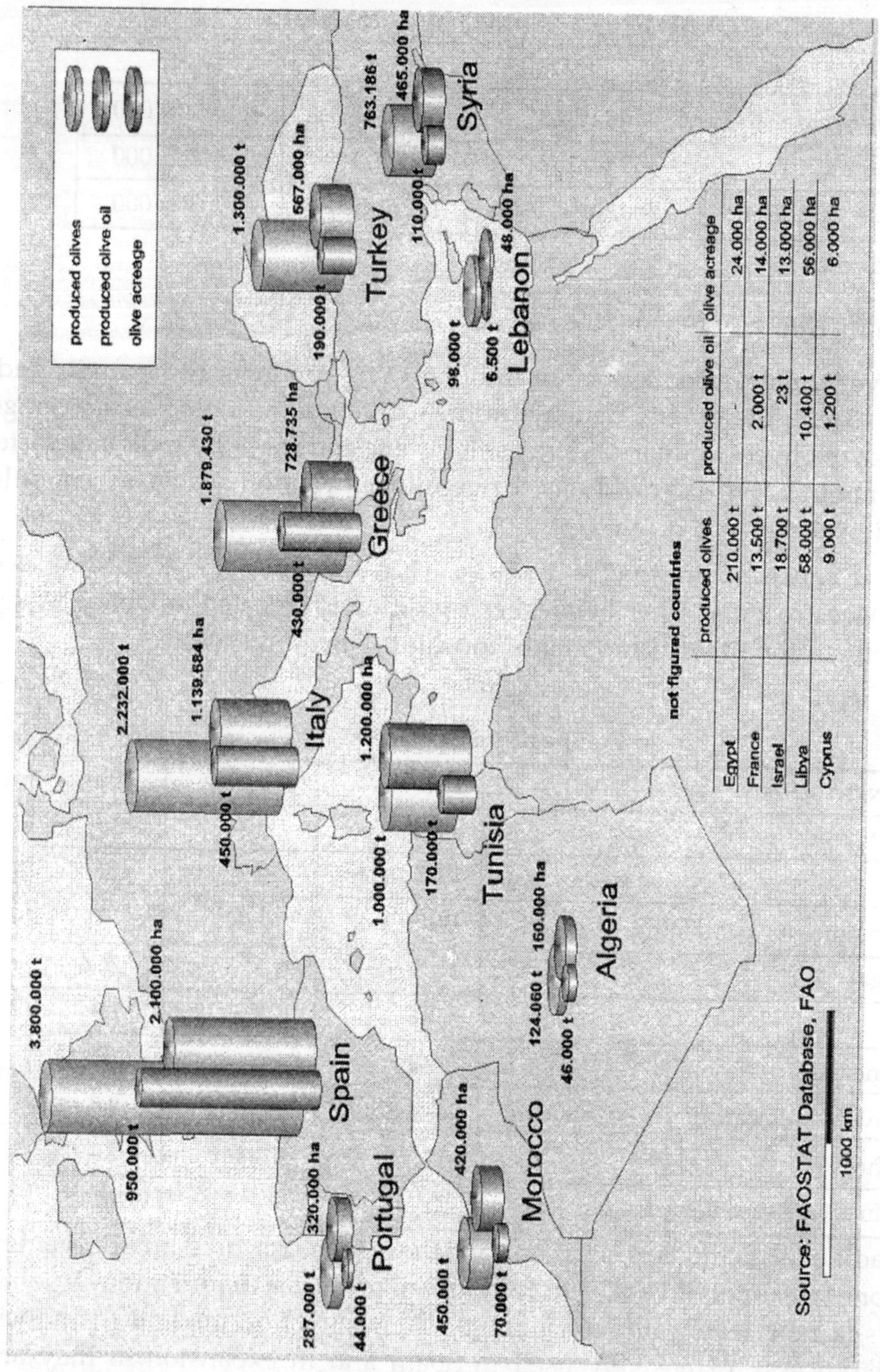


**Figure 2. Olives and olive oil production in the Mediterranean area during 1998.**

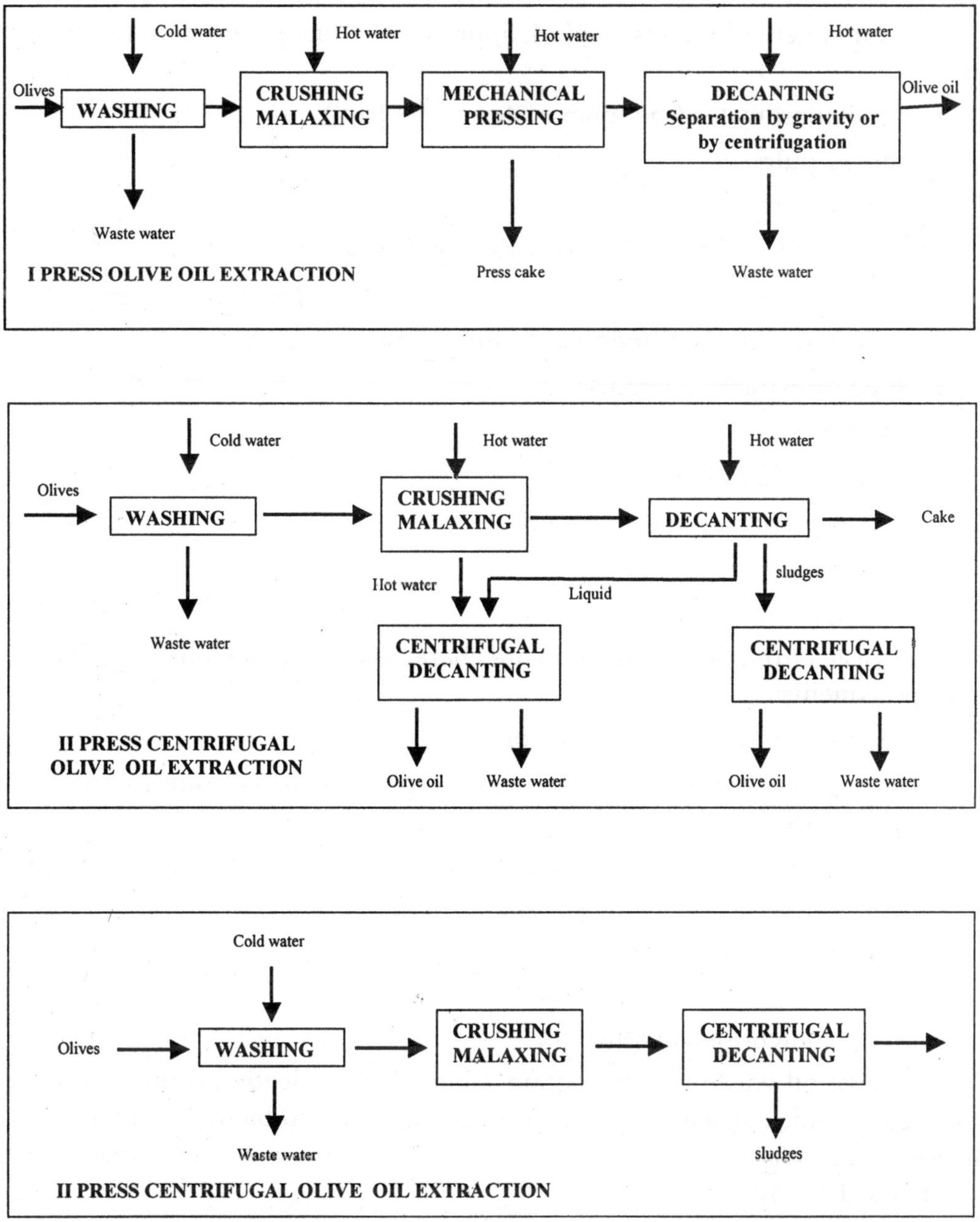


**Fig 3. Present olive oil production systems (Vlyssides et al., 1998)**

The component of wastes (Table 4, 5) produced during the extraction process are:

- vegetation water (the inner water of the fruit)
- milled pulp
- ground stones
- water added during the extraction process.

The waste quality is influenced by the following factors:

- type of extraction process
- cultivar
- area of cultivation
- use of pesticides and fertilisers
- harvesting time, stage of maturity
- climate/weather conditions.

The wastes from olive extraction are characterised by the following features and components:

- intensive violet-dark brown colour
- strong specific olive oil smell that rapidly turn bad due to anaerobic processes
- high concentration of hardly degradable organic compounds (COD up to 220 g/l and a COD/$BOD_5$ ratio between 2.5 and 5)
- pH between 3.0 and 5.9
- high organic matter load and salt
- high content in polyphenols.

The olive oil waste waters contain a high organic load and many complex coloured organic substances which are very resistant to biodegradation. When disposed into environment, they can provoke dangerous effects that can be summarised as follows:

### *Discolouring of Natural Waters*

This is the most visible effects of the pollution so often observed in the Mediterranean countries (Fung, 1997) and it is due to tannins that come from the olive skin and remain in the waste water.

**Table 4**

**Composition of olive mill wastewaters**

| | Pressure Extraction Process | | | Continuous extraction process | | |
|---|---|---|---|---|---|---|
| | Min | Average | Max | Min | Average | Max |
| pH | 5.1 | 5.4 | 5.8 | 4.7 | 5.4 | 5.5 |
| Water % | 79.85 | 86.4 | 91.7 | 90.4 | 93.5 | 96.5 |
| Organic compounds % | 7.22 | 12.0 | 18.3 | 2.6 | 5.2 | 8.0 |
| Lipids % | 0.02 | 0.5 | 1.0 | 0.5 | 10 | 2.3 |
| Nitrogen compounds % | 1.2 | 1.8 | 2.4 | 0.17 | 0.3 | 0.4 |
| Sugars % | 2.0 | 4.5 | 8.0 | 0.5 | 1.5 | 2.6 |
| Organic acids % | 0.5 | 0.9 | 1.5 | - | trace | - |
| Polyalcohols % | 1.0 | 1.1 | 1.5 | 0.9 | 1.1 | 1.4 |
| Pectins, tannins % | 1.3 | 1.5 | 1.7 | 0.23 | 0.37 | 0.5 |
| Glycosides % | - | trace | - | - | trace | - |
| Polyphenols % | 1.2 | 1.7 | 2.4 | 0.3 | 0.63 | 0.8 |
| Minerals at 550°C | 1.0 | 1.5 | 1.7 | 0.2 | 0.4 | 0.5 |
| $P_2O_5$ (%) | 0.14 | 0.21 | 0.23 | 0.03 | 0.06 | 0.07 |
| $CO_2$ (%) | 0.2 | 0.3 | 0.35 | 0.04 | 0.08 | 0.1 |
| $SO_3$,$SiO_2$ ,FeO, MgO (%) | 0.06 | 0.09 | 0.1 | trace | 0.02 | 0.03 |
| CaO (%) | 0.06 | 0.09 | 0.01 | 0.01 | 0.02 | 0.03 |
| $K_2O$ (%) | 0.47 | 0.71 | 0.81 | 0.11 | 0.19 | 0.24 |
| $Na_2O$ (%) | 0.07 | 0.1 | 0.11 | 0.01 | 0.03 | 0.03 |
| Suspended solids % | 0.08 | 0.1 | 0.15 | 0.7 | 0.9 | 1.1 |
| Dry matter at 105°C | 8.3 | 13.6 | 2015 | 3.5 | 6.5 | 9.6 |
| COD (g/l) | 54.1 | 208.0 | 318.0 | 28.9 | 49.5 | 7901 |
| $BOD_5$(g/l) | 19.2 | 90.2 | 134.8 | 17.0 | 28.7 | 41.2 |

### *Threat to the Aquatic Life*

The waste water are characterised by a high COD and $BOD_5$ and a considerable quantity of reduced sugar. When discharged directly into natural waters, they increase the number of micro-organisms that would use the sugar as a source of substrate. The greater consumption of oxygen dissolved in the water reduce the share available for other living organisms. This may cause an imbalance of the whole ecosystem. A similar process can result from the high phosphorus content

**Table 5**

**Chemical characterisation of wet solid wastes from two-phase process (SW)**

| | Feb. 97 – Jun. 98 | | | Jun. 98 – Dec. 98 | |
|---|---|---|---|---|---|
| | SW | Stonefree SW | Oil reduced Stonefree SW | SW | Dried SW(400 °C) |
| pH | 5.30 | 4.87 | 5.00 | 5.80 | 5.80 |
| Ash (%w/w) | 7.10 | 7.65 | 9.12 | * | * |
| Lipids (%w/w) | 4.34 | 7.18 | 6.38 | 7.46 | 12.48 |
| Proteins (%w/w) | 13.56 | 9.44 | 8.65 | 14.80 | 15.96 |
| Sugars (%w/w) | 2.31 | 1.48 | 1.21 | 1.30 | 1.87 |
| Tannins (%w/w) | 2.70 | 2.18 | 2.61 | 1.25 | 1.33 |
| Nitrogen (%w/w) | 2.48 | 2.10 | 1.96 | 3.16 | 3.08 |
| LHV** (kcal/kg) | 27.61 | 15.04 | 22.45 | * | * |

* Not determined **Low heating value

that encourages and accelerates the growth of algae and increases the eutrophisation, destroying the whole ecological balance in natural waters. The presence of the large quantity of nutrients supplied with the waste water can also provide a growth medium for pathogens with severe consequences to the local aquatic life. Moreover, the lipids in the waste water may form an impenetrable film on the surface of rivers, worsening the aeration conditions (Fung, 1997).

### *Surface and Ground Water Pollution*

The runoff of wastes when disposed on sloping land can provoke pollution of surface water. Ground water can be endangered by water leaching when soil pedomorphology is not suitable to accept them.

### *Soil Quality and Plant Growth*

The uncontrolled waste disposal can provoke inconvenient to soil properties (decrease of the cation exchange capacity, increase of salinity, alteration of microbial population, alteration of the soil-air- and the air-water balance) (Steegmans, 1992). Moreover, it can affect seed germination and plant growth by the phytotoxicity due to the fresh organic matter and some compounds i.e. phenols and organic acids (Fung, 1997).

### *Odours*

They are due to the anaerobic fermentation of the waste water. This leads to a considerable pollution by odours even in greater distance especially when such a waste is stored in ponds (Fung, 1997).

**************

Like the most of agro-industrial by-products, the wastes produced by the olive mill industry can be either considered as a polluting stream or a resource to be recycled. This double aspect of olive waste generates antagonism among the various political groups, mainly agriculture and environment ones, due to their different point of view on this topic. The organic matter and nutrient dressing are undoubtedly a resource for agricultural and industrial applications. However, because of the possible dangerous effects, Authorities in every country tried to develop regulation that often vary greatly among them with a consequent non uniform application of even generally accepted guidelines. Due to the insufficient regulation, a normative that impose a common behaviour among the oil producer countries should be highly desirable. Such a regulation is particularly important for European countries in view of an efficient safeguard of the Mediterranean area. Due to the development of olive tree cultivation, a regulation that involves both recycling and disposal of olive waste is getting to be a worldwide problem.

## Management of Olive Oil Industry By-products

The problem posed by Olive mill waste has been faced through several European project aimed at evaluating sectional aspects of waste disposal or recycle. In Italy, where the extraction activity of olive mills produces about 2 million tons of waste/year, the "olive mill question" has long been under discussion from the scientific, economic and regulatory point of view. At the end of '90s, the Italian National Council of Research, in collaboration with the Ministry for University and Research, promoted the integrated project "Recycling the waste of the agro-industrial system: olive mill and animal waste" in order to define and develop new technologies transferable to the industrial sector. The project involved 60 research units belonging to Universities, National Council of Research and the Ministry of Agricultural Policies as well as Industry Partners and manufacturing sector. Based on the philosophy that it is better to recover rather than destroy, the "Olive mill waste subproject" was divided into four main topics aimed at evaluating the various problems that olive mill industry poses for the community, and proposing suitable solutions for the various territorial contests having in account the environmental safeguard and the recovery of a re-usable resource. The topics taken into consideration are summarised in Figure 4.

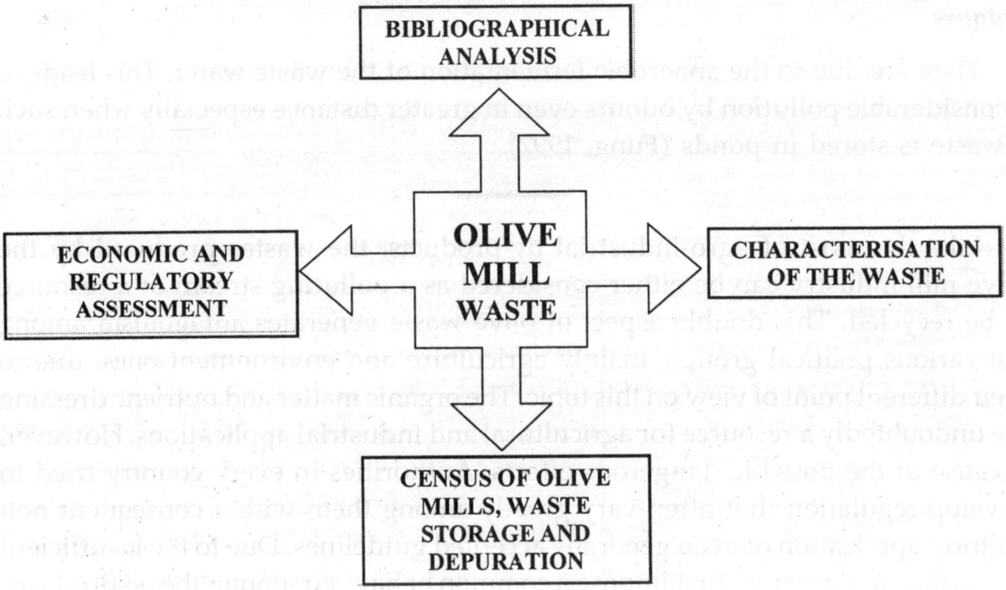
BIBLIOGRAPHICAL ANALYSIS
ECONOMIC AND REGULATORY ASSESSMENT
OLIVE MILL WASTE
CHARACTERISATION OF THE WASTE
CENSUS OF OLIVE MILLS, WASTE STORAGE AND DEPURATION

Fig 4a

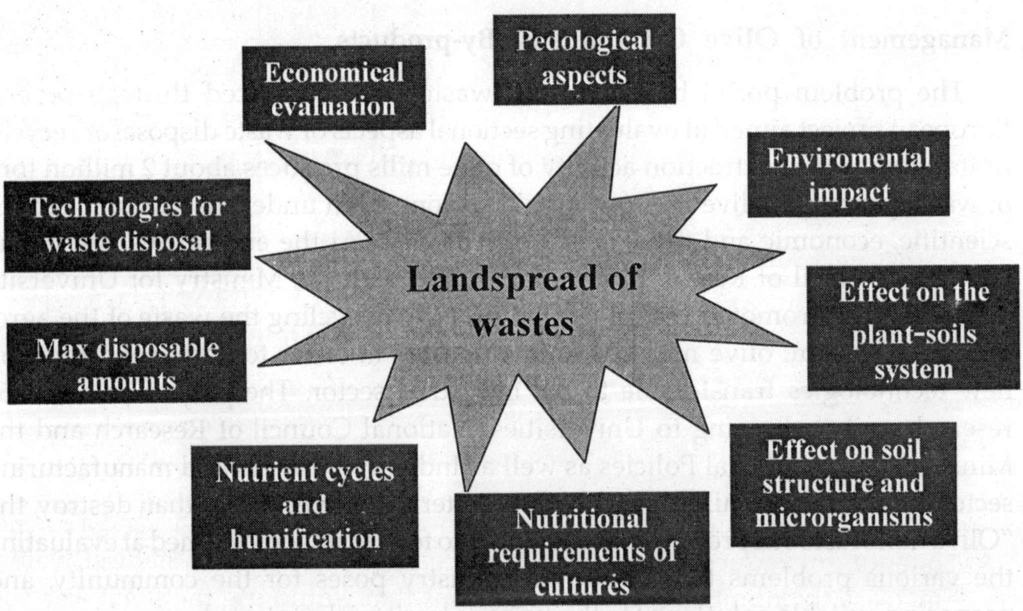
Economical evaluation
Pedological aspects
Enviromental impact
Technologies for waste disposal
Landspread of wastes
Effect on the plant-soils system
Max disposable amounts
Effect on soil structure and microrganisms
Nutrient cycles and humification
Nutritional requirements of cultures

Fig. 4b

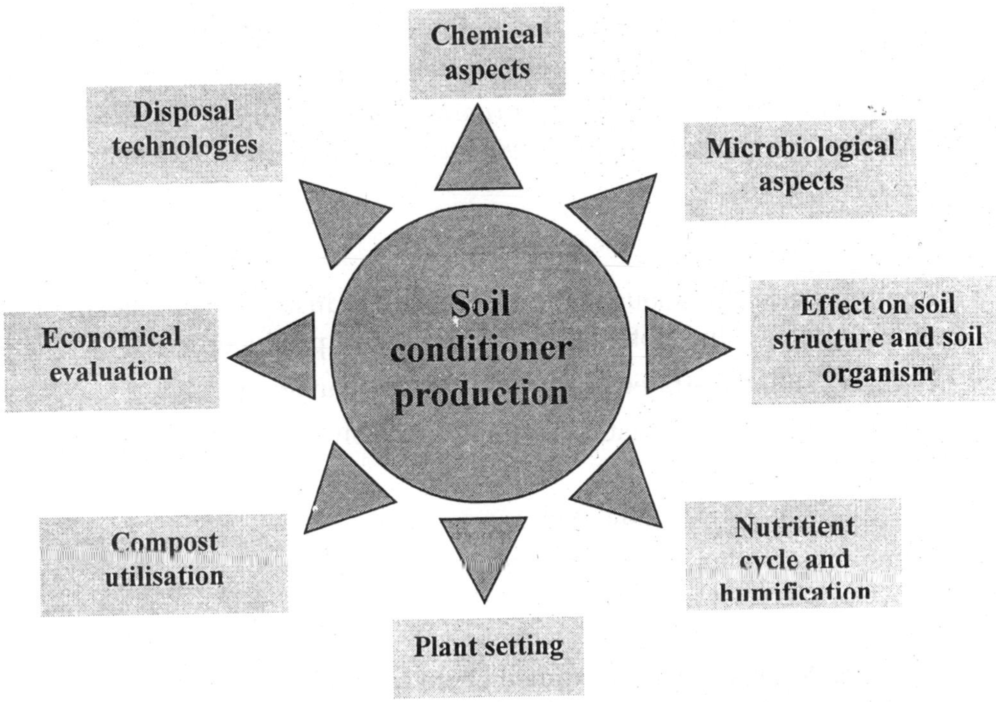
Chemical aspects
Disposal technologies
Microbiological aspects
Soil conditioner production
Economical evaluation
Effect on soil structure and soil organism
Compost utilisation
Nutritient cycle and humification
Plant setting

Fig. 4c

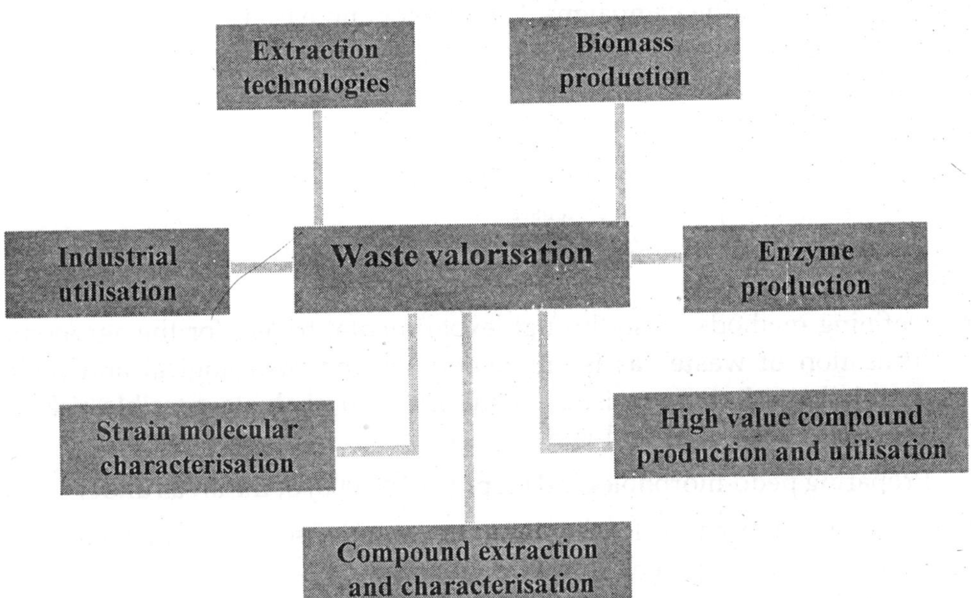
Extraction technologies
Biomass production
Industrial utilisation
Waste valorisation
Enzyme production
Strain molecular characterisation
High value compound production and utilisation
Compound extraction and characterisation

Fig. 4d

In every topic, both scientific and technological aspects were taken into consideration, Legal and economical implications were also evaluated.

**Topic 1:** *"Review of the existing situation with regard to the production, characterisation and spreading of olive mill wastes; extracting technologies and operating plants"*

*Objectives*

- Providing a census about all the aspect of oil industry (i.e. number of olive mills, typology and potentiality, oil and waste production).
- Defining suitable strategies aimed at waste managing.
- Providing useful information to Agencies involved in drawing up disposal plans.
- Promulgating new regulations.
- Economical evaluation of the proposed technologies.

*Results*

- A complete census was performed about the Italian olivemill industry on the whole national territory. Data about oil producers, oil production, number of olive trees, number of olive mills, their potentiality and extraction process, quality and quantity of waste were collected considering the last harvesting campaigns. For a better and easier understanding, the data were linked within MapInfo Programme, a software GIS (Geographical Information System) that permit to consult data directly on geographical maps: figure 5.

**Topic 2:** *"Determination of the hydro-pedo-morphological, climatic and agronomic aspects concerning the spreading of wastes"*

*Objectives*

- Defining methods - also through experimental trials - for the agronomic utilisation of waste "as is" in relation to the hydrological and pedo-morphological characteristics of the areas to evaluate possible risks of contamination of ground and surface waters.
- Preparing pedo-morphological maps (<1:100.000) of the areas under study.
- Acquiring information for the correct agronomic use of waste, safeguarding or increasing the potential fertility of soils.
- Developing technological packages that take into account the different territorial situations.

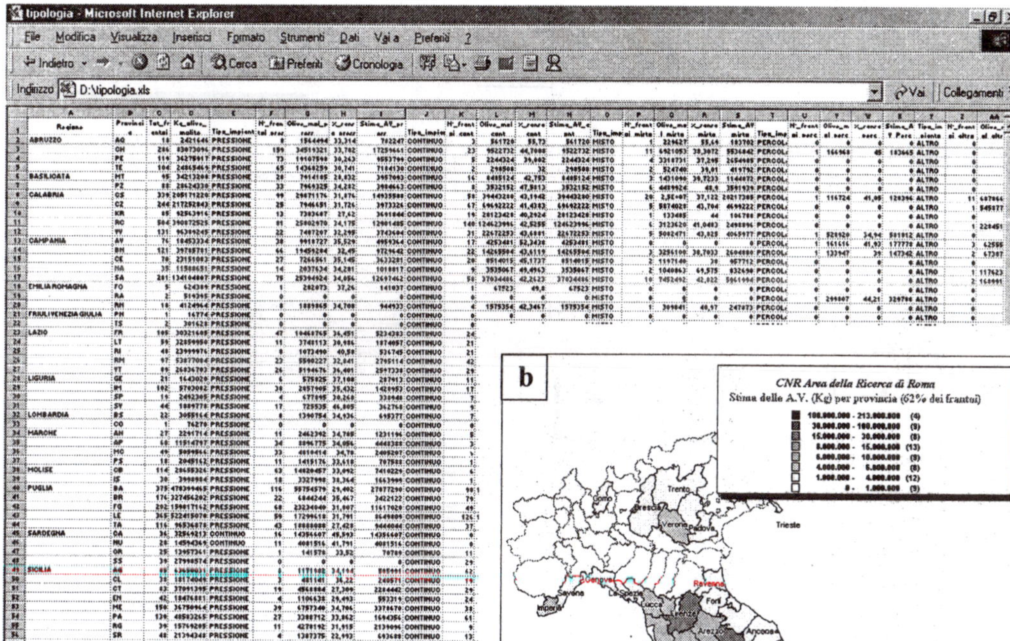


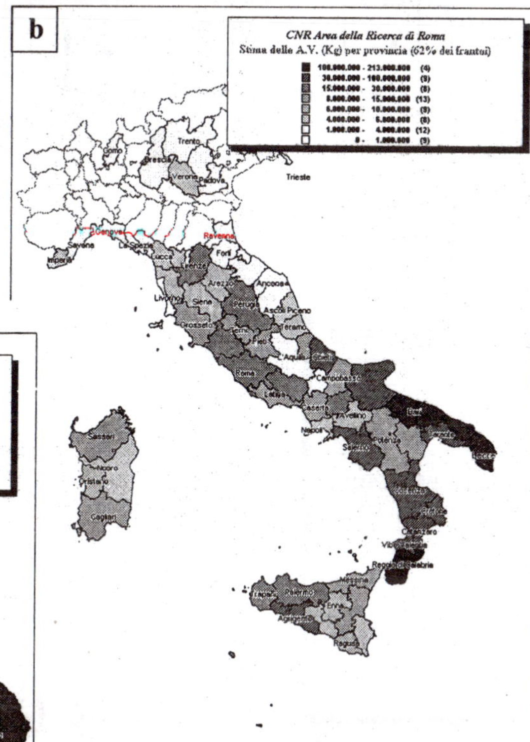


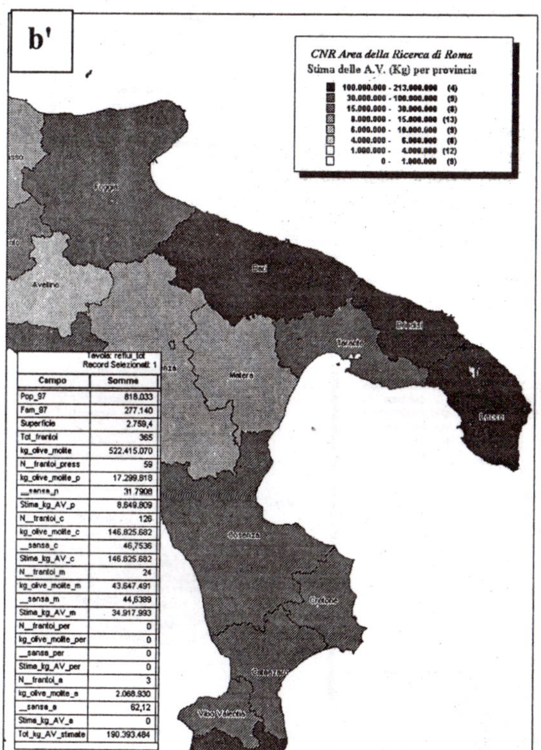


**Fig. 5: Graphical elaboration of olive oil industry trough MapInfo:**

(a) raw data on an electronic sheet

(b) evaluation of OMW product in Italy

(b′) evaluation of OMW product in a restrict area.(Puglia).

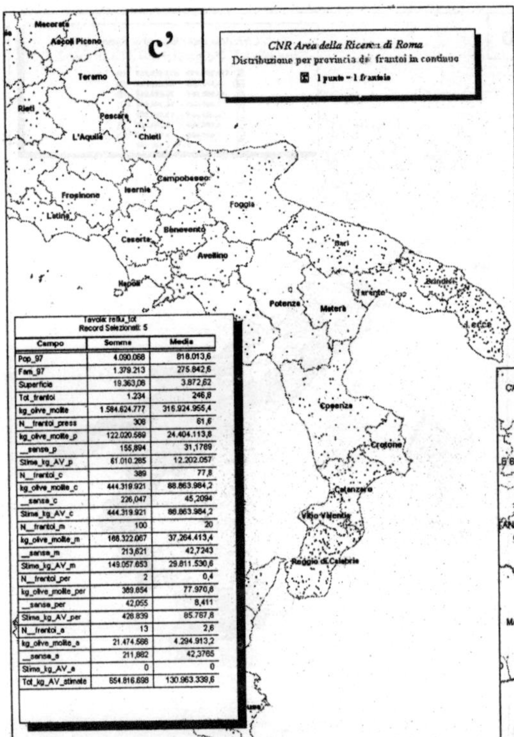

(c) Olive Mills distribution on the whole national area of Italy.

(c′) Olive Mills distribution on a restricted area (Puglia); all data concerning OM industry are showed in the window.

(c″) Olive Mills distribution at the country level (Melendugno); all data concerning OM industry are showed in the window.

- Developing suitable technologies for the waste spreading

*Results*

- Surveying of the hydro-pedo-morphological, climatic and agronomical situations was performed in the main producing regions. Contemporary, waste production was also mapped. All the data are kept available in a graphic form (figure 6). The waste disposal/supply onto agricultural soils was performed taking in account the need of nutrients (N, P, K) of cultures and the characteristics of soils. The suitability of soil to receive olive waste strictly depends on the percentage of the soil components. Clay soil are particularly suitable to receive waste because of clay ability to adsorb waste components. A dangerous leaching of waste compounds was observed for sandy soils. A mathematical simulation of phenol leaching in the different soil typology was also performed.

**Topic 3**: *"Biotransformation, stabilisation of waste and reduction of its volume/ mass"*

*Objectives*

- Developing appropriate treatment to stabilise the waste and making it suitable for a year-round agronomic use
- Obtaining soil correctives that can be used "as is" or as bioremediated organo-mineral soil conditioners to be used for field or protected cultures
- Acquiring scientific and technological know-how for the industrial production of "quality" composts

*Results*

- A computerised composting plant in a pre-industrial scale. Two main objectives were reached at the same time: first, a consistent reduction of waste volume, and then the production of a soil conditioner that can be used to improve soil fertility because of its high content of nutrients, humified organic matter, and micro-organisms. With little modifications, this plant can be used for every kind of agro-industrial refuse. The re-use of the organic substance coming from agrofood by-products or/and from vegetal scrap using a controlled bio-oxidative process to produce "fertilisers" is an useful, if not essential, practice to manage a modern agriculture in the respect of the environment
- Technologies able to reduce waste production are very suitable to face the environmental problem. The first and still now used extraction technology

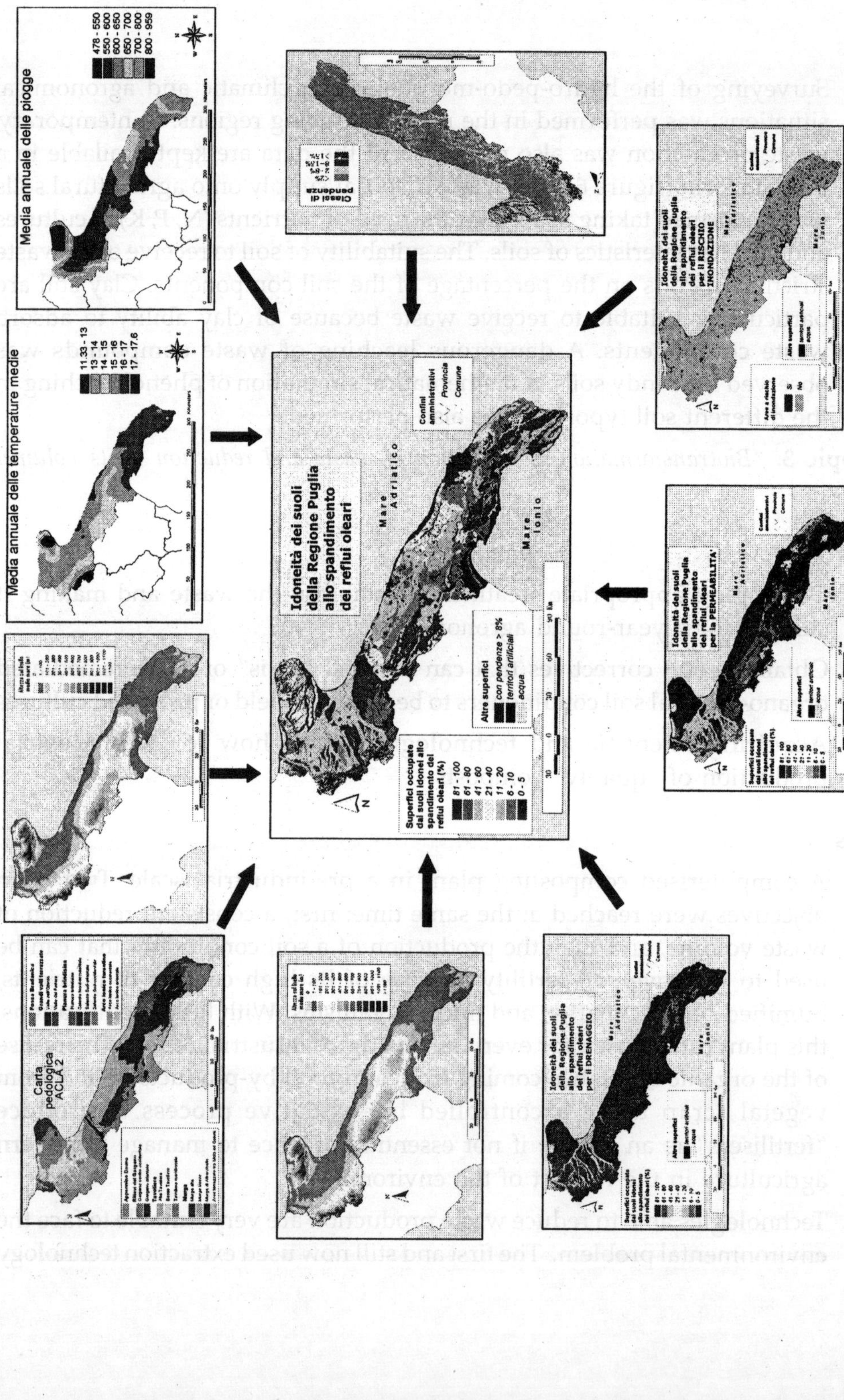


**Fig. 6: Soil suitability for waste disposal.**

*(Soil suitability was stated based on hydro-pedo-morphological, climatic and agronomical conditions)*

is the so called 3-phase extraction that produces liquid waste amounts lying between 80- 120 l of waste/100 kg of olives. A new low water consuming technology was subsequently introduced that produces about 60 kg of a wet waste per 100 kg of olives and only a limited flow of liquid waste. A multipurpose decanter VDP –variable differential pressure - that permits to improve oil quality strongly reducing waste stream was realised.

- A prototype of an *ad hoc* waste water spreading machine able to operate in olive groves, conforming to the law. Based on soil characteristics, a software permits to regulate the speed and evaluate the right amount of waste to spread

**Topic 4:** *"Utilisation of the waste as a growth substrate for microbial biomasses and /or as a raw material for recovering high value compounds"*

*Objectives*

- Developing extractive technologies for compounds to be used in different industrial sectors (food, pharmaceutical, cosmetics...)
- Developing biotechnologies for biomass production
- Characterising and certifying the most suitable microbial strains for targered fermentation

*Results*

- Extractive technologies to recover high value products such as phenols, tocopherols, squalene, extracellular enzymes (polygalacturonase, lipase, laccase, Mn-dependent peroxidase)
- Fertiliser production from algal biomasses
- Fungal biomasses as a raw material for b-glucan, chitin and chitosan, and enzyme (polygalacturonase, lipase, laccase, Mn-dependent peroxidase) extraction.

## Legal Aspects

According to UE directives, every Country established regulation, guidelines and standards aimed at evaluating the quality of industrial and agricultural waters in view of the safeguard of the environment. In this optic, the quality of solid and liquid waste to be released into environment is evaluated based on pollutants that are grouped and listed according their toxicity, and the organic load. The selection of pollutants is based on the following criteria: human exposure and health effect, plant uptake of pollutants, phytotoxicity, effect on soils and terrestrial ecosystems, effect on domestic animals and wildlife, effect on aquatic ecosystems, bioaccumulation.

The main UE directives are:

Related to the pollution caused by certain dangerous substances. It is aimed at eliminating or reducing the contamination caused in the waters by dangerous substances grouped in two main lists according to their toxicity.

The organic load, expressed as COD and $BOD_5$, is among the quality parameters taken into consideration. Because of their organic load, thousand times greater than legal limits, many organic waste must be cleaned before their disposal into environment.

As for as olive mill by-products, UE directives have been modified taking in to consideration scientific results and technological application.

The problem of olive mill waste arose in Italy in 1976, as a consequence of the law 319/76 "Water protection against pollution" that classified olive mills as industrial areas. Because of the organic load, expressed as COD and $BOD_5$, was thousand times greater than the legal limits (COD = 500) olive wastewaters had to be cleaned before their release. Due to the difficulty of reaching such a limit, and the high cost of cleaning, there was a crisis for the olive mill industry that, before the law, could release wastewaters after a rough depuration. To avoid both the economical and social problems due to olive mill industry crisis and the very restrictive regulation in force, several provisory legislative and ministerial decrees were promulgated. According many studies and research performed especially in Spain and Italy that demonstrated the high fertilising value of olive oil waste and the possibility of using them as a soil conditioner, a regulation (law 574/96: "New regulation about the agronomical re-use of olive mill wastewaters") about olive waste disposal and/or recycle was promulgated.

This law and its subsequent modification (d.l.152/1999 and d.l. 258/2000) encourages the agronomical use of olive waste as is or bioremediated.

As a general remark there is a need of a normative that imposes a common behaviour among all producer countries to safeguard the Mediterranean sea since oliviculture is a general feature in the Mediterranean basin.

**Economical Aspect**

Evaluating the economical aspect of the recycle of any waste is a very complex problem since many implications must be contemporary considered. Legal, technical, scientific, social and territorial aspects can greatly influence the cost of the operation. As for as olive mill waste, the cost of recovery has been calculated starting from the census of olive mill industry and hydro-pedo-morphological maps that permitted to know the dislocation of mills, the qualitative and quantitative production of waste, the suitable areas for waste disposal. Moreover, a study about the cost of different typology of composting plants has been performed. The marketing of compost has been also investigated. Based on that, a data bank has been built up that permit to evaluate from time to time how face the waste recycle in the most convenient way.

## Conclusions

The opportunities of converting agro-industrial waste into raw materials for new activities are unlimited, although sometimes this task is not easily to do. The establishment of an effective waste management programme involves building institutions, training, development of human resources, research programmes aimed at individuating new technologies and, in general, building the capacity to control and manage waste.

Waste minimisation and waste recycle optimisation are the two main requirements to correctly approach the management of waste. To achieve these goals, there is a need of a very close collaboration among the forces of the research, industry and administration. Providing and encouraging communication between researchers and practitioners is the way to individuate from time to time the best solution to face the problems posed by the different typologies of waste.

## References

Bilanovic D., Shelef G., & Green M. (1994). Xanthan fermentation of citrus waste *Bioresource Technol.* 48:169-172.

Bocco A., Cuvelier M.E., Richard H., & Berset, C. (1998). Anti-oxidant activity and phenolic composition of citrus peel and seed extracts. *Journal of Agricultural and Food Chemistry,* 46: 2123-2129.

Bohn J.A., Bemiller J.H. (1995) (1à3)-â -D-glucans as biological response modifiers: a review of structure – functional activity relationships. *Carbohydrate Polymers* 28: 3-14.

Bonilla F., Mayen M., Merida J., & Medina M. (1999). Extraction of phenolic compounds from red grape marc for use as food lipid antioxidants. *Food Chemistry,* 66: 209-215.

Camire M.E. & Dougherty M.P. (1988). Added phenolic compounds enhance stability in extruded corn. *Journal of Agricultural and Food Chemistry,* 63: 516-518.

Che Man Y.B. & Tan C.P. (1999). Effect of synthetic and natural antioxidants on changes of refined, bleached, and deodorized palm olein during deep-fat frying of potato chips. *Journal of the American oil Chemists' Society* 76: 331-339.

Coulibaly L., Gourene G, Agathos N.S. (2003). Utilization of fungi for biotreatment of raw wastewaters. A review. *African Journal of Biotechnology* 2 (12): 620-630

Furue H. (1986). Biological characteristic of schyzophyllan (SPG) and its clinical effect. *Med Immun* 12: 65-77.

Gonzales Lopez J., Pozo C, Martinez Toledo M.V., Rodelas B. & Salmeron V. (1996) Production of polyhydroxyalkaonates by *Azotobacter chroococcum* H23 in wastewater from olive oil mills. *Int Biodeterioration & Biodegradation*. 38 (3-4): 271-276.

Larrauri J.A., RupeÂrez P. & Saura-Calixto F. (1997). Effect of drying temperature on the stability of polyphenols and antioxidant activity of red grape pomace peels. *Journal of Agricultural and Food Chemistry,* 44: 1390-1393.

Lu Y., & Foo L.Y. (1997). Identification and quantification of mayor polyphenols in apple pomace. *Food Chemistry,* 59: 187-194.

Lu Y., & Foo L.Y. (1999). The polyphenol constituents of grape pomace. *Food Chemistry,* 65: 1-8.

Márinova E.M., Yanishlieva N.VI., & Kostova I.N. (1994). Anti-oxidative action of the ethanolic extract and some hydro-xycoumarins of *Fraxinus ornus* bark. *Food Chemistry,* 51: 125-132.

Miyake Y., Murakami A., Sugiyama Y., Isobe M., Koshimizu K. & Ohigashi H. (1999). Identification of coumarins from lemon fruit (*Citrus limon*) as inhibitors of in vitro tumour promotion and super-oxide and nitric oxide generation. *Journal of Agricultural and Food Chemistry,* 47: 3151-3157.

Moure A., Cruz J.M., Franco D., Dominguez J.M., Sineiro J., Dominguez H., Nunez M.J. & Parajo J.C. (2001). Review: Natural antioxidants from residual sources. *Food Chemistry:* 72: 145-171

Muanza D., Robert R. & Sparks W. (1998). Antioxidant derived from lentil and its preparation and uses. US Patent. US5762936.

Ohta T., Yamasaki S., Egashira Y., & Sanada H. (1994). Anti-oxidative activity of corn bran hemicellulose fragments. *Journal of Agricultural and Food Chemistry,* 42: 653-656.

Onyeneho S.N. & Hettiarachchy N.S. (1992). Antioxidant activity of durum wheat bran. *Journal of Agricultural and Food Chemistry,* 40: 1496-1500.

Pauli G. (1996) Breakthroughs: What business can offer Society. *Epsilon Press,* Surrey, UK. Pp241-256. In: (De Bertoldi M., Sequi P., Lemmes B., Papi T. eds.) *The Science of composting*. Black Academic and professional. Glasgow

Pohjola V.J. & Pongrácz E. (2002) Approach to the formal theory of waste management. *Resources Conservation & Recycling* 35 (1-2): 17-29.

Pongrácz E. (2003). Industrial Ecology and Waste Management: From Theories to Applications. Accepted for publication in *Progress in Industrial Ecology*.

Pongrácz E. & Pohjola V.J. (2004). Re-defining waste, the concept of ownership and the roles of waste management. *Resources Conservation & Recycling*. 40 (2): 141-153.

Reimer J. & Kristoffersen M. (1999). Information on waste management practices. A proposed framework. European Agency; Copenhagen, Denmark

Rodriguez de Sotillo D., Hadley, M., & Holm, E. T. (1994a). Phenolics in aqueous potato peel extract, extraction, identification and degradation. *Journal of Food Science,* 59, 649-651.

Saint-Cricq de Gaulejac N., Provost C. & Vivas N. (1999). Comparative study of polyphenol scavenging activities assessed by different methods. *J. Agric.Food Chemistry* 47:425-431

Sanchez Villasclaras S., Martinez Sancho M.E., Espejo Caballero M.T. & Delgado Perez A. (1996). Production of microalgae from olive oil wastewaters. *Int. Biodeterioration & Biodegradation* 38 (3-4):245-248.

Sequi P. (1996). The role of composting in Sustainable Agriculture. In: (De Bertoldi M., Sequi P., Lemmes B., Papi.T.eds.) *The Science of composting*. Black Academic and professional. Glascow pp.23-29.

Steegmanns R. (1992). Optimierung der anaeroben Verfahrenstechnik zur Reinigung von organischen hochverschmutzten Abnwässern aus der Olivenölgewinnung. Forschungsinstitut für Wassertechnologie an der RWTH Aachen (Hrsg.), *Forschungsbericht AZ* 101/81 der Oswald.

Tardáguila J., Montero F., Olmeda M., Alba J. y Bernabéu R. (1996). Análisis del Sector del Aceite de Oliva, *Alimentación, Equipos y Tecnología*, Abr., 22: 245-248

Van Belle W.A (1996). Waste management and Legislation. In: (De Bertoldi M., Sequi P., Lemmes B., Papi.T.eds.) *The Science of composting*. Black Academic and professional. Glascow pp 40-46.

Vargas-Arispuro, I., Sanz, B.I., Martinez-Tellez, M.A., & Primo-Yufera, E. (1998). Actividad antioxidante de compuestos aislados del residuo no-volatil del aceite esencial de naranja. *Grasas y Aceites*, 49: 159-164.

Visioli, F., Romani, A., Mulinacci, N., Zarini, S., Conte, D., Vincieri, F.F., & Galli, C. (1999). Antioxidant and other biological activities of olive mill waste waters. *Journal of Agricultural and Food Chemistry*, 47: 3397-3401.

Vlyssides A.G., Loizidou M., Gimouhopulos K & Zorpas A. (1998). Olive oil processig waste production and their characteristics in relation to olive oil method extraction. *Fresen. Environ. Bull.*7 (5-6):308-316.

Watanabe, M., Ohshita, Y., & Tsushida, T. (1997). Antioxidant compounds from buckwheat (*Fagopyrum esculentum* Möench) hulls. *Journal of Agricultural and Food Chemistry*, 45: 1039-1044.

Wasser S.P. & Weiss A.L (1999). Medicinal properties of substances occurring in higher basidiomycetes mushrooms: current perspectives (review). *Int J Med Mush* 1: 31-62.

Yadomae T., & Ohno N. (1996). Structure-activity relationship of immunomodulating (1à3)- â -D-glucan. *Rec Res Dev Chem Pharm Sci* 1: 23-33.

Yamaguchi F., Yoshimura Y., Nakazawa H., & Ariga T. (1999). Free radical scavenging activity of grape seed extract and anti-oxidants by electron spin resonance spectrometry in an $H_2O_2$/NaOH/DMSO system. *Journal of Agricultural and Food Chemistry*, 47: 2544-2548.

Xing Y., & White P. (1997). Identification and function of anti-oxidants from oat groats and hulls. *Journal of the American Oil Chemists' Society*, 74: 303-307.

Zandi P. & Gordon M.H. (1999). Antioxidant activity of extracts from old tea leaves. *Food Chemistry* 64: 285-288.

# Appendix

Adani, F., Genevini, P.L., Gasperi F., Tambone, F. (1999). Composting and humification. *Compost Science & Utilization* 7 (1): 24-33.

Ait Baddi, G., Alburquerque, J.A., Gonzalez, J., Ouatmane, A., Cegarra, J., Hafidi, M. (2002). Treatment of olive-mill wastes by composting and monitoring of end-product maturity. In: 1st *International workshop on Environmental problems in olive oil production and solutions.* Balikesir (Turkey).

Amirante P., Di Renzo G.C., Di Giovacchino L., Bianchi B., Catalano P. (1993). Technological developments in olive oil extraction plants. *Olivae,* 48: 43-53.

Amirante P., Baccioni L., Catalano P., Montel G.L. (1999). Nuove tecnologie per l'estrazione dell'olio di oliva: il decanter a cono corto a pressione dinamica variabile e controllo della velocità differenziale tamburo/coclea. *Rivista Italiana Sostanze Grasse,* 76: 129-140.

Amirante P., Catalano P. (2000). Fluid dynamic analysis of the solid-liquid separation process by centrifugation. *Journal of Agriculture Engineering Research,* 77 (2): 193-201.

Amirante P., Cini E., Montel G.L., Pasqualone A. (2001). Influence of mixing and extraction parameters on virgin olive oil quality. *Grasas y Aceites,* 52 (3-4): 198-201.

Amirante P. & Montel, G.L. (1998). Organic olives oil: extraction process optimization and by-products recovery. Technology and plants for composting. *Atti del Convegno di studio: ECOLIVA 98,* 2as Jornadas Mediterráneas de Olivar Ecológico, Puente de Génave, Jaen, Spagna, 11-15 novembre.

Andreoni N., Dellomonaco G., Spandre R., Fiorentini ER. (1996). Olive mill waste water spreading and groundwater pollution. *Agricoltura Mediterranea,* 126: 200-204.

Angelakis A.N., Marecos Do Monte M.H.F., Bontoux L., Asano T. (1999). The status of wastewater reuse practice in the mediterranean basin: need for guidelines. *Water Research.* 10 (10): 2201-2217.

Aragon J.M. (1999). Proyecto Improlive: Tratamiento del Alperujo para Mejorar su Aprovechamiento: Alimentaciòn, Equipos y Tecnologias.

Ballesteros I., Oliva J., Saez F. Ballesteros M. (2001). Ethanol production from lignocellulosic by-products of olive oil extraction. *Appl. Biochem. Biotechnol.* 91-93: 237-252.

Benavente-Garcia O., Castello J., Morente J., Ortuño A., Del Rio J.A. (2000). Antioxidant activity of phenolics extracted from *Olea europaea* L. leaves. *Food Chemistry* 68: 457-462.

Bernini R., Bovicelli P., Mincione E., Sanetti A., Villarini M. (1999). "Conversion of flavonoids present in agro-industrial wastes into useful intermediates as well bioactive compounds: an useful and simple monohydroxylation of aromatic ring of flavonoids by dimethyldioxirane (DMD)", *I° International Conference on Solid Waste,* Roma, 7-9 aprile.

Bernini R., Mincione E., Sanetti, A., Mezzetti, M., Bovicelli, P., (2000). Aromatic ring hydroxylation of flavanones by dimethyldioxirane. *Tetrahedron Letters,* 41(7): 1087-1090.

Bianco A.D., Mazzalupo I., Piperno A., Romeo G., Uccella N. (1999). Bioactive derivatives of Oleuropeine from olive fruits. *J Agric. Food Chem.* 47: 3531-3534.Bohn, J.A., & Bemiller J.H. (1995). (1à3)-b-D-glucans as biological response modifiers: a review of structure – functional activity relationships. *Carbohydrate Polymers* 28: 3-14.

Briante R., La Cara F., Tonziello M. P., Febbraio F., Nucci R. (2001). Antioxidant activity of the main bioactive derivatives from oleuropein hydrolysis by hyperthermophilic b-glycosidase. *J Agric. Food Chem.* 49: 3198-3203.

Cabrera F., Lopez R., Martinez-Bordiu A., Lome E.D., De Murillo J.M. (1996). Land treatment of olive mill wastewater. *International Biodeterioration and Biodegradation,* 38 (3-4): 215-225

Cabrera, G.L., Rodriguez, D.M.G., Maruri, A.B. (1999). Genotoxicity of extracts from the compost of the organic and the total municipal garbage using three plant bioassays. *Mutat. Res.* 426: 201-206.

Capasso R. (1997). The chemistry, biotechnology and ecotoxicology of the polyphenols naturally occurring in vegetable wastes. *Current Topics in Phytochemistry* 1: 145-156.

Capasso R., De Martino A., Arienzo M. (2002). Recovery and characterization of the metal polymeric organic fraction (polymerin) from olive oil mill waste waters. *J. Agric. Food Chem.,* 50: 4018-4025.

Capasso, R., Cristinzio, G., Evidente, A., Scognamiglio, F. (1992). Isolation, spectroscopy and selective phytotoxic effects of polyphenols from vegetable waste water. *Phytochemistry* 31: 4125-4128.

Cardelli R. & Benitez C. (1998). Changes of chemical properties in two soils amended with moist olive residues. *Agricoltura Mediterranea* 128: 171-177.

Civantos, L. (1995). Evolución de la Superficie del Olivar y de las Producciones de Aceite de Oliva en España. *Olivae* 59: 18.

FAOSTAT database (2000), in http://apps.fao.org/default.htm.

Fiorelli F., Pasetti L., Galli E. (1996). Fertility-promoting Metabolites from *Azotobacter vinelandii. Int. Biodeterioration & Biodegradation* 38(3-4):165-168 ,

Coolen W. A. (1979). Methods for the extraction of *Meloidogyne* spp. and other nematodes from roots and soil. In *Root-knot nematodes (Meloidogyne spp.). Systematics, Biology and Control (F. Lamberti and C.E. Taylor Eds.)*, pp. 317 - 329. Academic Press, London.

Cox L., Celis R., Hermosin M.C., Becker A., Cornejo J. (1997). Porosity and herbicide leaching in soils amended with olive-mill wastewater. *Agriculture, Ecosystems and Enviroment*, 65 (2): 151-161.

D'addabbo T. (1995). The nematicidal effect of organic amendments: a review of the literature,. *Nematol. medit.* 23: 299-305.

D'addabbo T., Fontanazza G., Lamberti F., Sasanelli N., Patumi M. (1997). The suppressive effect of soil amendments with olive residues on *Meloidogyne incognita. Nematol. medit.* 25: 195-198.

D'addabbo T., Sasanelli N. (1996). Effect of olive pomace soil amendments on *Meloidogyne incognita. Nematol. medit.*, 24: 91-94.

D'addabbo T., Sasanelli N. (1996). The effect of olive pomace soil amendments on *Heterodera carotae. Nematol. medit.*, 24: 205-207.

D'addabbo T., Sasanelli N., Lamberti F., Greco P., Carella A. (2000). Effect of olive and grape pomace in the control of root-knot nematodes. *Annual International Research Conference on Methyl Bromide Alternatives and Emissions Reductions. November 6 – 9, Orlando, Florida (U.S.A.) pag.55-1-2-3.*

D'addabbo T., Sasanelli N., Lamberti F., Carella A. (2000). Control of root-knot nematodes by olive and grape pomace soil amendments. *Proceedings of the fifth International Symposium on Chemical and Non-Chemical Soil and Substrate Disinfestation,* September 11-15, Turin, Italy, pp 53-57.

D'annibale A., Crestini C., Vinciguerra V. Giovannozzi Sermanni G. (1998). The biodegradation of recalcitrant effluent from an olive mill by a white-rot fungus. *Journal of Biotechnology* 61: 209-218.

D'annibale A., Stazi S.R., Vinciguerra V., Di Mattia E., Giovannozzi Sermanni G. (1999). Characterization of immobilized laccase from *Lentinula edodes* and its use in olive-mill wastewater treatment. *Process Biochem.*, 34: 697-706.

D'acqui L.P., Daniele E., Fornasier F., Radaelli L., Ristori G.G. (1999). Interaction between clay microstructure, decomposition of plant residues and humification. *European Journal of Soil Science* 49: 579-587.

De Simone C., De Marco A., Raglione M., D'ambrosio C., Pitoni P. (1999). "Utilizzazione agronomica delle acque di vegetazione. Tossicità ed influenza su alcune caratteristiche pedologiche". *Rivista di Agronomia.* 33: 8-13.

Di Vito M., Lamberti F., Carella A. (1979). La resistenza del pomodoro nei confronti dei nematodi galligeni: prospettive e possibilità. *Rivista di Agronomia,*13: 313-322.

Dias M.A., Lacerda I.C.A., Pimentel P.F., De Castro H.F., Rosa C.A. (2002). Removal of heavy metals by an *Aspergillus terreus* strain immobilized in a polyurethan matrix. *Letters in Applied Microbiology* 34: 46-50.

Durán N., Rosa M.A., D'Annibale A., Gianfreda L. (2002). Applications of laccases and tyrosinases (phenoloxidases) immobilized on different supports: a review. *Enzyme Microbial Technology,* 31(7): 907-931, Elsevier Science.

Eggen, T., Vethe O. (2001). Stability indexes for different composts. *Compost Science & Utilization,* 9(1): 19-26.

FAOSTAT database (2000), in http://apps.fao.org/default.htm.

Fenice M., D'annibale A., Giovannozzi Sermanni G., Federici F. (1999). Production of laccase, Mn-peroxidase and lipase by submerged and solid-state fermentation of *Panus tigrinus* on olive-mill wastewater International Conference "Enzyme in the Environment", 12-15 July Granada, Spain, pp. 96.

Fenice M., Giovannozzi Sermanni G., Federici F., D'Annibale A., (2002). Submerged and solid-state bioprocesses for laccase and manganese-peroxidase production on olive-mill wastewater-based media. *Journal of Biotechnology,* 100 (1): 77-85 (Elsevier Science Ireland).

Ferri D., Convertini G., Montemurro F. (2000). Agronomic benefits on crops and soil of municipal solid waste compost applied on a vertisol in southern Italy. *11$^{th}$ International Soil.*

Ferri, D., Convertini G., Montemurro F., Rinaldi M., Rana G. (2002). Olive wastes spreading in southern Italy: Effects on crops and soil. *Proceedings 12$^{th}$ International Soil Conservation Organization Conference,* Pechino – Cina.

Ferri, D., Convertini, G., Montemurro, N., Martinelli, N. (2000). Crop responses to municipal solid waste application in a 6-year field experiment in southern Italy. *Xth International Colloquium for the Optimization of Plant Nutrition. Cairo (Egitto),* 8 – 13 aprile.

Ferri D., Convertini G., Rinaldi M., De Giorgio D., La Cava P., Tartaro M.E. (1996). Effects of composted urbane waste on some characteristics of crops and soil in southern Italy. *Proc. IX International Colloquium for the Optimization of Plant Nutrition,* 526 - 530.

Filippi C., Bedini S., Levi-Minzi R., Cardelli R. & Saviozzi A. (2002). Composting ol olivemill by-products: Ghemical and microbiological evaluations. *Compost Science & utilization* 10(1), 63-71.

Fiorelli F., Pasetti L., Galli E. (1996) Fertility-promoting Metabolites from *Azotobacter vinelandii. Int. Biodeterioration & Biodegradation* 38(3-4):165-168 ,

Galli E., Di Mario F, Rapanà, Lorenzoni P., and Angelini R. (2003). Copper absorption on *Auricolaria polytrica Letters in Applied Microbiology* , 37: 133-137

Galli E., Pasetti L., Fiorelli F., Tomati U. (1997). Olive-mill wastewater composting: microbiological aspects. *Waste Management & Research* 15: 323-330

Galli E., Tomati U. (2002). A common policy to face the problem of olive oil mill wastes. In: 1$^{st}$ *International workshop on Environmental problems in olive oil production and solutions.* 7-9 June, Balikesir (Turkey).

Gao S.X., Hartnett J.P. (1993). Steady flow of non-newtonian fluids trough rectangular ducts. *International communications on Heat and Mass Transfer,* 20: 197-210.

Garcia Barrionuevo A. Morenoe. Quevedo-Sarmiento J, Gozalez Lopez J. Ramos Comenzana A (1992). Effect of wastwaters from olive oil mills on *Azotobacter* nitrogen fixation in soil. *Soil Biol. Biochem.* 24: 281-283.

Garcia Gomez, A, Roig.S., Bernal, M.P. (2003). Composting of the solid fraction of olive mill wastewater with olive leaves: organic matter degradation and biological activity. *Bioresource Technology,* 86: 59-64.

Garcia-Ortiz A., Beltran G., Uceda M., Hermoso M., Gonzalez P., Ordonez R., Giraldez J.V. (1999). Vegetation water (alpechin) application effects on soils and plants. In: *III International Symposium on Olive Growing*. ISHS *Acta Horticulturae* 474, (Metlidakis & Voyiatlis, eds), Chania, Creta (GR).

Gasparrini, R. (1999). Treatment of Olive Oil Residues. *Oils And Fats Intern.* 15 (1): 32-33.

Gordon M.H., Paiva Martins F., Almeida M. (2001). Antioxidant activity of hydroxytyrosol acetate compared with that other olive oil polyphenols. *J Agric. Food Chem.* 49: 2480-2485.

Hruschka S., Geissen K. (1997). Global Contemplation of the Olive Oil Production with Regard to the Economy, Quality and Ecology. Presentation, *Symposium "Olive Waste"*, Cultural Centre of Kalamata, Greece, 5-8 november.

Jianlong W., Xinmin•Z., Decai D., Ding Z. (2001). Bioadsorption of lead (II) from aqueous solution by fungal biomass of *Aspergillus niger*. *Journal of Biotechnology* 87(3): 273-277.

Juvonen R., Martikainen E., Schultz E., Joutti A., Ahtiainen J., Lehtokaris M. (2000). A battery of toxicity tests as indicators of decontamination in composting oily waste. *Ecotoxicology and Environmental Safety,* 47: 156-166.

Khoo K.M., Ting Y.P. (2001). Biosorption of gold by immobilized fungal biomass. *Biochemical Engineering Journal* 8(1): 51-59.

Lasaridi K., Stentiford E. (1998). A Simple respirometric technique for assessing compost stability. *Water Res.* 32 (12): 3717-3723.

Legge N. 574/1996. Nuove norme in materia di utilizzazione agronomica delle acque di vegetazione e di scarichi da frantoi oleari. *G.U.* n. 265 del 12/11/1996.

Lopez M.J., Moreno J., Ramos-Cormenzana A. (2001) *Xanthomonas campestris* strain selection for xanthan production. *J. Appl.Microbiol.* 90(5): 829-835.

Madejon E., Galli E., Tomati U. (1998). Composting of wastes produced by low consuming olive mill technology. *Agrochimica,* 42 (3): 49-60.

Marull J., Pinochet J., Rodriguez-Kabana R. (1997). Agricultural and municipal compost residues for control of root-knot nematodes in tomato and pepper. *Compost Science & Utilization,* 1: 6-15.

Mcdonald S., Prenzler P.D., Antolovich M., Robards K. (2001). Phenolic content and antioxidant activity of olive extracts. *Food Chem.* 73: 73-84.

Medina I., Satué-Gracia M.T., German J.B., Frankel E.N. (1999). Comparison of natural polyphenol antioxidants from extra virgin olive oil with synthetic antioxidants in tuna lipids during thermal oxidation. *J. Agric. Food Chem.,* 47: 4873-4879.

Monteoliva-Sanchez M., Ramos-Cormenzana A., Paredes C., Roig A., Cegarra J. (1996). The study of the aerobic bacterial microbiota and the biotoxicity in various samples of olive mill wastewaters (Alpechin) during their composting process. *International Biodeterioration & Biodegradation,* 38 (3/4): 211-214.

Mulinacci N., Romani C., Galardi C., Pinelli P., Giaccherini C., Vincieri F. (2001). Polyphenolic content in olive oil waste waters and related olive samples. *J. Agric. Food Chem.* 49 (8): 3509-3514.

Pagliai M., Vittori Antisari L. (1993). Influence of waste organic matter on soil micro and macrostructure. *Bioresouce Technology* 43: 205-213.

Paredes C., Bernal M.P., Roig A., Cegarra J. (2001). Effects of olive mill wastewater addition in composting of agroindustrial and urban wastes. *Biodegradation,* 12 (4): 225-234.

Paredes C., Bernal M.P., Roig A., Cegarra J., Sanchez-Monedero A. (1996). Influence of the bulking agent on the degradation of olive mill wastewater sludge during composting. *International Biodeterioration & Biodegradation,* 38 (3/4): 205-210.

Paredes C., Cegarra, J., Roig A., Sànchez-Monedoro M.A., Bernal M.P. (1999). Characterization of olive mill wastewater (alpechin) and its sludge for agricultural purposes. *Bioresources Technol.* 67: 111-115.

Paredes C., Roig A., Bernal M.P., Sanchez-Monedero M.A., Cegarra J. (2000). Evolution of organic matter and nitrogen during co-composting of olive mill wastewater with solid organic wastes. *Biology and Fertility of Soils.* 32 (3): 222-227.

Paredes C., Bernal M.P., Cegarra J., Roig A. (2002). Bio-degradation of olive mill wastewater sludge by its co-composting with agricultural wastes. *Bioresource Technology,* 85: 1-8.

Park J.P., Kim S.W., Hwang H.J., Yun J.W. (2001). Optimization of submerged culture conditions for the mycelial growth and exo-biopolymer production by *Cordyceps militaris. Letters in Applied Microbiology* 33: 76-81.

Ramos-Cormenzana A., Monteoliva-Sanchez M., Lopez M.J. (1992). Bioremediation of Alpechin. *Int. Biodeterioration & Biodegradation* 34:249-268.

Reverberi M., Di Mario F., Tomati U. (2004) b-glucan syntase induction in mushroom grown on olive mill wastes waters. *Applied Microbiology and Biotechnology* 253:1662-1670

Rodríguez - Kábana R. (1986). Organic and inorganic amendments to soil as nematode suppressants. *J. Nematol.* 18: 129-135.

Rodríguez-Kábana R., Estaún V., Pinochet J., Marfà O. (1995). Mixtures of olive pomace with different nitrogen sources for the control of *Meloidogyne* spp. on tomato. *J. Nematol.* 27 (4): 575-584.

Rodríguez-Kábana R., Pinochet R.J., Calvet C. (1992). Olive pomace for control of plant-parasitic nematodes. *Nematropica* 22: 149-158.

Ryan D., Robards K. (1998). Phenolic compounds in olives. *Analyst* 123: 31R-44R.

Saladino R., Bernini R., Mincione E. (2001). Innovation. Environmental oxidative catalysis for processing of agroindustrial wastewater. *La Chimica & L'Industria,* 83(1): 34-35.

Saladino, R., Bernini, R., Mincione, E. (2002). Recycling of agroindustrial wastes. *La Chimica & L'Industria,* 84(2), 47-50.

Saladino R., Neri V., Mincione E., Marini S., Coletta M., Fiorucci C., Filippone P. (2000). Environmentally benign 4-polyvinylpyridine-supported methyltrioxorhenium complexes as novel catalysts for the oxidation of phenol and methoxybenzene derivatives. *Perkin* 1(4): 581-586.

Saladino R.; Mezzetti M.; Mincione E.; Palamara A.T.; Savini P., Marini S. (1999). Synthesis, cytotoxic effect and antiviral activity of 1-(á-D-arabinofuranosyl)-5-bromo-N4-substituted cytosine and 1-(á-D-arabinofuranosyl)-5-bromo-4-methoxypyrimidin-2(1H)-one derivatives. *Nucleosides & Nucleotides,* 18: 2499-2510.

Saladino R.; Mezzetti M.; Mincione E.; Torrini I.; Paglialunga Pa.. disi M.; Mastropietro G. (1999). A New and Efficient Synthesis of Unnatural Amino Acids and Peptides by Selective 3,3-Dimethyldioxirane Side-Chain Oxidation. *Journal of Organic Chemistry,* 64(23): 8468-8474.

Sasanelli N. (1994). Tables of Nematode-Pathogenicity. *Nematol. medit.,* 22:153-157.

Sasanelli N., D'Addabbo T., Convertini G., Ferri D. (2002). Soil Phytoparasitic Nematodes Suppression and Changes of Chemical Properties Determined by Waste Residues from Olive Oil Extraction. *Proceedings of 12th ISCO Conference,* May 26-31, 2002 Beijing China. Vol. III: 588-592.

Sayadi S., Allouche N., Jaoua M., Aloui F. (2000). Detrimental effects of high molecular-mass polyphenols on olive mill wastewater biotreatment. *Process Biochem.,* 35, 725-735.

Senesi N., Brunetti G., Loffredo E., Miano T.M. (1999). Abiotic catalytic humification of organic matter in olive oil mill wastewaters. *In: Understanding Humic Substances: Advanced Methods, Properties and Applications (E.A. Ghabbour and G. Davies, eds.), Royal Society of Chemistry, London, UK, 9-17*

Sequi P., De Nobili M., Leita L., Cercignani G. (1986). A new index of humification. *Agrochimica,* 30: 175-179.

Sierra J., Marti E., Montserrat G., Cruañas R., Garau M.A. (2001). "Characterisation and evolution of a soil affected by olive oil mill wastewater disposal". *The Science of the Total Environment.* 279: 207-214.

Stazi S.R., D'Annibale, A, Giovannozzi Sermanni, G. (2002). Kinetic and degradative capability of differently-substituted chlorinated phenols by laccase. Fresenius Environ. *Microbiol.,* In corso di stampa su volume 11(9): 583-588 (PSP Publishing Monaco, Germania).

Stazi, S.R., D'Annibale, A., Fenice, M., Federici, F., Petruccioli, M., Giovannozzi Sermanni, G. (2001). Aerobic Treatment of olive-mill wastewater by the white-rot fungus *Panus tigrinus. 11th International Symposium on Environmental Pollution and its impact on the Mediterranean Region* (MESAEP), Cipro, 6-10 Ottobre, E-36.

Stazi, S.R.; D'Annibale, A.; Vinciguerra, V; Giovannozzi-Sermanni, G. (1999). Dephenolization of Olive-Mill Wastewater by Immobilized Laccase from *Lentinula edodes. International Conference "Enzyme in the Environment"*, 12-15 July Granada, Spain, pp.102.

Stirling G. R. (1991). Mode of action of organic amendments against nematodes. *Biological Control of Plant Parasitic Nematodes. Progress, Problems and Prospects.*, C.A.B. International, pp. 170–185.

Tomati U., Galli E. (1992). The fertilizing value of wastewaters from olive processing industry. In :" *Humus et Planta Proc., 107-126 (Humus, its structure and role in Agriculture and Environment (Ed.: Kubat I.), Elsevier Science, B.V. Amsterdam, The Netherlands.*

Tomati U., Galli E. (2002) An European regulation about olivemill waste industry. *Int Conf. Environ. Problems of the Mediterranean region.* Near East University TRNC.

Tomati U., Belardinelli M., Andreu M, Galli E., Capitani D, Proietti N., De Simone C. (2002). Evaluation of commercial compost quality *Waste Management & Research* 20:1-9

Tomati U., Galli E., Fiorelli F., Pasetti L. (1996). Fertilizers from composting of olive mill wastewaters. *International Biodeterioration & Biodegradation,* 38 (3/4), 155-162.

Tomati U., Galli E., Pasetti L., Volterra E. (1995). Bioremediation of olive-mill wastewaters by composting. *Waste Management & Research* 13: 509-518.

Tomati U., Galli E., Rapanà P. (2003). Trends for an industrial utilisation of olive mill waste *La Chimica e l'Industria,* 85: 1-5.

Tomati U., Madejon E., Galli E., Capitani D., Segre A.L. (2001). Structural changes of humic acids during olive mill pomace composting. *Waste Management & Research* 20:1-9

Tomati U., Madejon E., Galli E. (2000). Evolution of humic acid molecular weight as an index of compost stability. *Compost Sci. & Utilization* 8 (2):108-115

Tomati U., Rapanà P. (2002). Basidiomacromycete biopotentialities and biotechnologies. In: *Biological and Biotechnological Resources* G. Tripathi and Y.C. Tripathi, eds., Campus Book International, New Delhi, pp. 136-166.

Tomati U., Galli E., Capitani D., Mannina L., Viel C. (2003). Characterisation of a polysaccharide fraction from *Lentinula edodes* grown on olive mill waste waters. *Carbohydrate Research* 339:1129-1134

Tomati.U., Galli E. (2002). Recycling of organic matter from wastes to restore soil fertility: a critical review. In: *Bioresource & Environment* (G. Tripathi and Y.C. Tripathi, eds.),Campus Book International, New Delhi, pp. 150-181.

Vassilev N., Vassileva M., Azcon R., Fenice M., Federici F & Barea J.M. (1998). Fertilizing effect of microbiologically treated olivemill wastewaters . *Bioresource Technol.* 66(2): 133-137.

Vinciguerra V., D'annibale A., Delle Monache G., Giovannozzi Sermanni G. (1995). Correlated effects during the bioconversion of waste olive waters by *Lentinus edodes. Bioresource Technol.,* 51: 221-226.

Visioli F., Bellomo G., Galli C. (1998). Free radical-scavenging properties of olive oil polyphenols. *Biochem. Biophys. Res. Commun.* 247: 60-64.

Visioli F., Galli C. (1998). Olive oil phenols and their potential effects on human health. *J. Agric. Food Chem.*, 46: 4292-4296.

Visioli F., Romani A., Mulinacci N., Zarini S., Conte D., Vincieri F.F., Galli C. (1999). Antioxidant and other biological activities of olive mill waste waters. *J. Agric. Food Chem.*, 47: 3397-3401.

Visioli F., Vincieri F. F., Galli C. (1995). "Waste waters" from olive oil production are rich in natural antioxidants. *Experientia* 51: 32-34.

Vlyssides A.G., Bouranis D.L., Loizidou M., Karvouni G. (1996). Study of a demonstration plant for the co-composting of olive-oil-processing wastewater and solid residue. *Bioresource Technology*, 56: 187-193.

Vouyoukalou E., (1994). Use of green leaves from olive trees as soil amendment for the control of *Meloidogyne*. *EPPO Bull.*, 24: 485-488.

Zhang, P. Zhang, L., Cheng, S. (1999). Chemical structure and molecular weights of a(1®3)-D-glucan from *Lentinus edodes*. *Biosci. Biotechnol. Biochem.* 63(7), 1197-1202.

# 2

# High Rate Transpiration System—Cost Effective Model Approach for Treatment and Disposal of Pulp and Paper Mill Wastewater through Phytoextraction of Heavy Metals

*Asha Juwarkar, P.R. Thawale, S.K. Singh and D.G. Gajghate*

## Introduction

Land application is considered to be most acceptable and cost effective method of pulp and paper mill wastewater management as soil provides physical, chemical, and biological treatment to wastewater and removes 100% of the colour (1 and 41) besides helping to grow the crops (2 and 19). Presently, in India there are 305 paper mills with an installed capacity of $1.9x10^6$ tonnes per annum (TPA). Paper mill consume large quantity of water, which varies from 300-450 $m^3\ t^{-1}$-, of which 220-350 $m^3$ of water per tonne of paper produced is discharged as wastewater. The wastewater comprises; high suspended solids (SS - 290-1115 mg $L^{-1}$), Biochemical Oxygen Demand (BOD - 100-1070 mg $L^{-1}$), Chemical Oxygen Demand (COD - 600-4760 mg $L^{-1}$) and dark brown colour due to lignin (82-700 mg $L^{-1}$).

In the process of pulp and paper manufacturing, large quantity of water is required. One of the largest integrated pulp and paper mill in India, Orient Paper and Industries Limited has established on the west bank of Sone River at village Amlai in Shahdol district of Madhya Pradesh which started its production in February 1965 and an average of about 190 $m^3$ water is required to produce I MT of paper. These pollutants create problems in disposal of wastewater in water bodies. The conventional wastewater treatment method reduces SS, BOD, and to some extent COD but colour remains more or less unchanged, because of no biodegradability of lignin. Available treatment methods for colour removal i.e. use of adsorbents (activated carbon), coagulants (lime and alum) and as bleaching agent (ozone), are not so effective.

In this view extensive R&D in laboratory scale has been carried out to tackle the problem of colour pulp and paper mill wastewater, and developed an ecofriendly technology, i.e. High Rate Transpiration System (HRTS), a unique land treatment technology for utilization of pulp and paper mills wastewater, which has been demonstrated successfully for the first time at Orient Paper Mills, Amlai (M.P.) at full-scale level on an area of 80 hectares in the vicinity of the plant to treat 18,000 - 20,250 $m^3$/day of the wastewater. The present papers describes a case study for treatment and safe disposal of all the coloured wastewater for forestry development and eliminate the necessity of discharging the wastewater in river Sone during non-monsoon months with a properly designed land treatment system without any ground water contamination and absorptions of green house gases.

Many plant species heavy metals for different reasons like sequestration, draught resistance, disposal by leaf abscission, interference with other plants and pathogen/herbivore defense (7). Some of these plant species that accumulate more than 1000 ug/g dry leaf weight and, are termed hyper-accumulator plants (8). Identification of such plant species has led environmental researchers to use these plant species for remediating heavy metal contaminated soil and wastewater, which is known as phytoextraction/rhizofiltration (3). Rhizo-filtration is a type of phytoremediation whichmakes use of plant roots to sorb, concentrate and precipitate heavy metal present in the irrigated wastewater through soil plant root system into the harvestable parts of the roots and above ground shoots. Dushenkov et al., have conduced studies on use of hydropronically grown plant roots for removal of heavy metal from water (13). The same concept is used for removal of metal from wastewater by irrigating hyperaccumulating grown on land (39).

For effective planning, design and application of rhizo-filtration program mathematical models have emerged as a valuable tool. For developing a development of such mathematical model for the same will require studies on the behaviour of heavy metal movement in the rhizosphere, metal uptake process by plant species, movement of water in the unsaturated zone and its uptake rate by plant roots. Since heavy metal moves along with water phase getting sorbed/desorbed and taken up by plant roots, as a prerequisite the water movement and its uptake by the plant roots in unsaturated zone should be modeled first.

A number of studies have been carried out to understand the water uptake by plant roots which can be classified based on literature review as microscopic approach (14, 26 and 38) and macroscopic approach (11, 25). The microscopic approach helps in better understanding of water extraction process by roots but is difficult to incorporate in the model due to complex root geometry as compared to the microscopic approach (45). The macroscopic approach considers root system as a single unit and the entire root zone is assumed to extract water at a given rate, which makes it convenient to incorporate a sink term in the governing equation. The governing equation is then set to appropriate initial and boundary condition to obtain the water flow and its uptake by plant roots.

Once the water movement is simulated, governing equations for heavy metal transport in unsaturated zone taking into account sorption/desorption, metal uptake by plant species is solved to obtain temporal and spatial variation of heavy metal. Various linear and non-linear isotherms are available in literature for sorption and desorption process (9-10), which can be used for heavy metal and soil. Heavy metal ions are transported to plant roots by mass flow and diffusion. Several which simulate nutrient uptake by plant root (11, 15 and 31) follows the same principle but differ in detail and can be deployed for estimating heavy metal uptake by plant root.

## Materials and Methods

### *Study Area*

The laboratory studies were carried out at National Environmental Engineering Research Institute (NEERI), Nagpur and HRTS was installed at Orient Paper Mill, Amlai (Madhya Pradesh), India. The study area comes under hot semi-arid ecoregion and the climate of the area is characterized by hot and dry summers and mild winters. The annual precipitation ranging from 600 to 900 mm, covers 40 to 50% of annual potential evapotranspirative (PET) demand of 1600 to 2000 mm, resulting in gross annual deficit of 800 to 1200mm.

### *List of Plant Species Selected for Plantation in High Rate Transpiration System*

The details of the various species of tress and grasses planted in HRTS area is given below:

| | | |
|---|---|---|
| Acacia | - | *Acacia mangium* |
| Amaltash | - | *Cassia fistula* |
| Australian babool | - | *Acacia auriculiformis* |
| Bamboo | - | *Dendrocalamus strictus* |
| | - | *Bambusa arundinacea* |
| | - | *Bambusa vulgaris* |

| | | |
|---|---|---|
| Cassia | - | *Cassia seamea* |
| Citronella grass | - | *Cymbopogon sp.* |
| Eucalyptus | - | *Eucalyptus hybrid* |
| | - | *Eucalyptus camaldulensis* |
| Grevillea | - | *Grevillea pteridofolia* |
| Karanj | - | *Pongamia pinnata* |
| Lemon grass | - | *Cymbopogon flexuosus* |
| Neem | - | *Azadirachata indica* |
| Palamorosa grass | - | *Cymbopogon martini* |
| Siras | - | *Albizia lebbeck* |
| Vetiver grass | - | *Vetiveria zizaniodes* |

***High Rate Transpiration System***

The treatment and disposal of wastewater through High Rate Transpiration System (HRTS) provide cost effective and environmentally acceptable solution to manage the problem of coloured wastewater. The system also promotes development of environmental forests and green belt and enables creation of much required sink potential for absorption of green house gases.

The high rate transpiration system envisages the use of dynamic, multicomponent soil system as a live filtration device to renovate the wastewater through adsorption, ion exchange, precipitation and stabilization of pollutants through microbial degradation (Fig. 1). Impact network for treatment and disposal of pulp and paper mill wastewater using high rate transpiration system is presented in Fig. 2.

The high transpiration capacity of plants grown on soil matrix enables the system to serve as biopump. Various plant species like Acacia (*Acacia mangium*), Bamboo (*Bambusa arundinacea*), Neem (*Azadirachata indica*), Shishum (*Dalbergia sisso*) and Eucalyptus (*Eucalyptus hybrid)* transpire water equivalent to 5 to 13 times that of potential evapotranspiration from the soil matrix alone. The soil system works in close conjunction with plants on it that provide a bio-pump through their high transpiration capacity. The filter media provided in the furrows mainly consists of bamboo/saw dust, fly ash and gypsum at the rate of 2-5 tons per hectare. Thickness of filter media varies from 5-15 cm depending on the colour intensity of the wastewater, soil texture and permeability of the soil. In addition to this, artificial media and leaf fall from standing plants provides thick mat and forms a filter bed, which is also responsible for retention and assimilation of colour bodies.

The system works on following main principles:

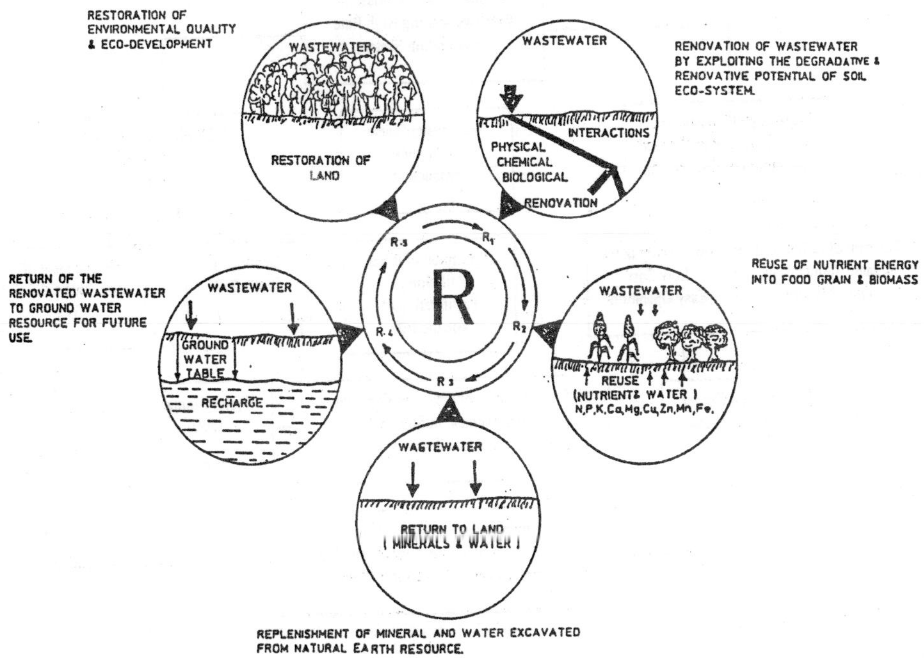


**Fig. 1. 5R Concept of Wastewater Management and Recycling, Reuse and Eco-development**

- Transpiration of large amount of water through stomatal network.
- Waste renovation using soil as a physico-bio- chemical reactor (living treatment Filter).
- Assimilation of colour in filter media. The Cross section of HRTS layout is shown in

The Fig. 3 shows the cross section of HRTS Layout developed at Orient Paper Mills, Amlai.

In this study a mathematical model is developed incorporating all the mathematical expressions for rhizo-filtration. The porous medium is assumed to be homogeneous and isotropic. In the formulation the effect of inter root competition for different metal species is not considered. This study does not account for translocation of heavy metal o the other parts of the plant since we are interested only in amount of metal uptake by plant roots. This model can be used to estimate heavy metal accumulation in plant with time thus optimizing

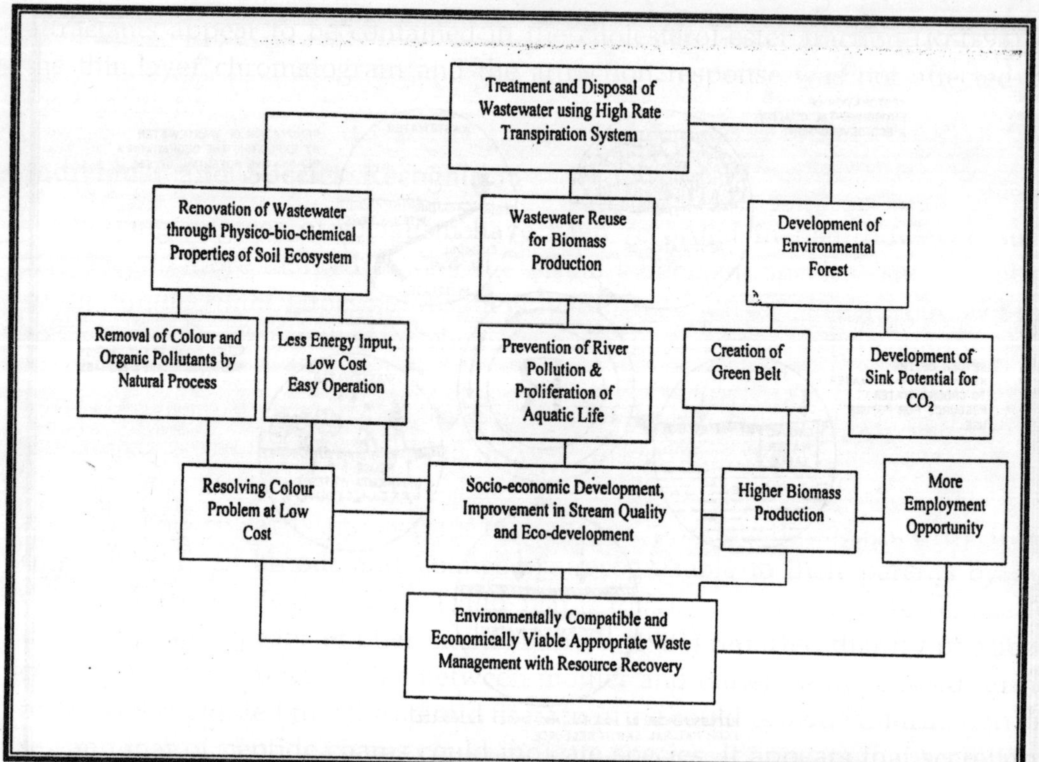


**Fig. 2. Impact Network for Treatment and Disposal of Pulp and Paper Mill Wastewater Using High Rate Transpiration System**

harvesting period and calculate heavy metal joining ground water, which can be used for optimizing irrigation schedule.

## Water Movement in Unsaturated Zone

The one dimensional mass balance equation for the flow of water in an unsaturated zone, also known as Richard's equation (37), and with a sink term (root water uptake) can be written as follows:

$$c(h)\frac{\partial h}{\partial t} = \frac{\partial}{\partial z}\left[k(h)\left(\frac{\partial h}{\partial t} - 1\right)\right] - S(z,t) \qquad (1)$$

Here h in the suction head (metric potential), is the soil moisture capacity defined as is the suction head dependent hydraulic conductivity and z is the soil depth taken to be positive upwards. S(z, t) is the sink tem for water extraction which is a spatial and temporal function and is expressed as the volume of water per unit volume of soil per unit time.

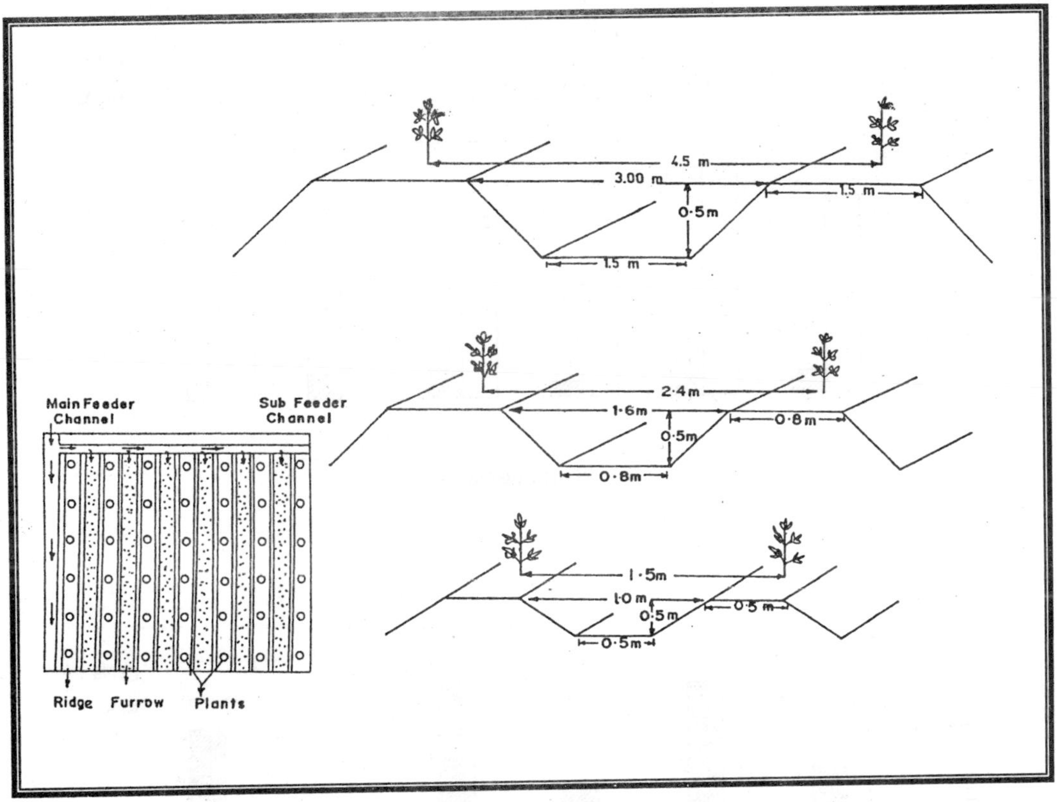


**Fig. 3. Schematic of High Rate Transpiration System for Treatment and Disposal of Pulp and Paper Mill Wastewater at Orient Paper Mills, Amlai.**

From many expressions for such functions available in literature (16, 30 and 34), the linear root water uptake model suggested by Rama Prasad (25) is adopted is adopted in this model, since it has been verified for ample soil moisture conditions (22). It is expressed as follows:

$$S(z,t) = \frac{2T_p}{Z_{max}} \alpha(h) \left[ 1 - \frac{Z_r}{Z_{max}} \right] \quad (2)$$

Her $T_p$ is the potential transpiration rate, $Z_r$ is the rooting depth, $Z_{max}$ is the maximum plant rooting depth ad (h) is the suction head dependent reduction factor. At any time the value of S(z, t) gives the uptake rate at the top of the root and varies linearly to zero at the bottom of the root as shown in Fig. 4. Empirical equations for root growth ($Z_r$) with time have been developed by many researchers (24). Some of these empirical equations require detailed information to predict

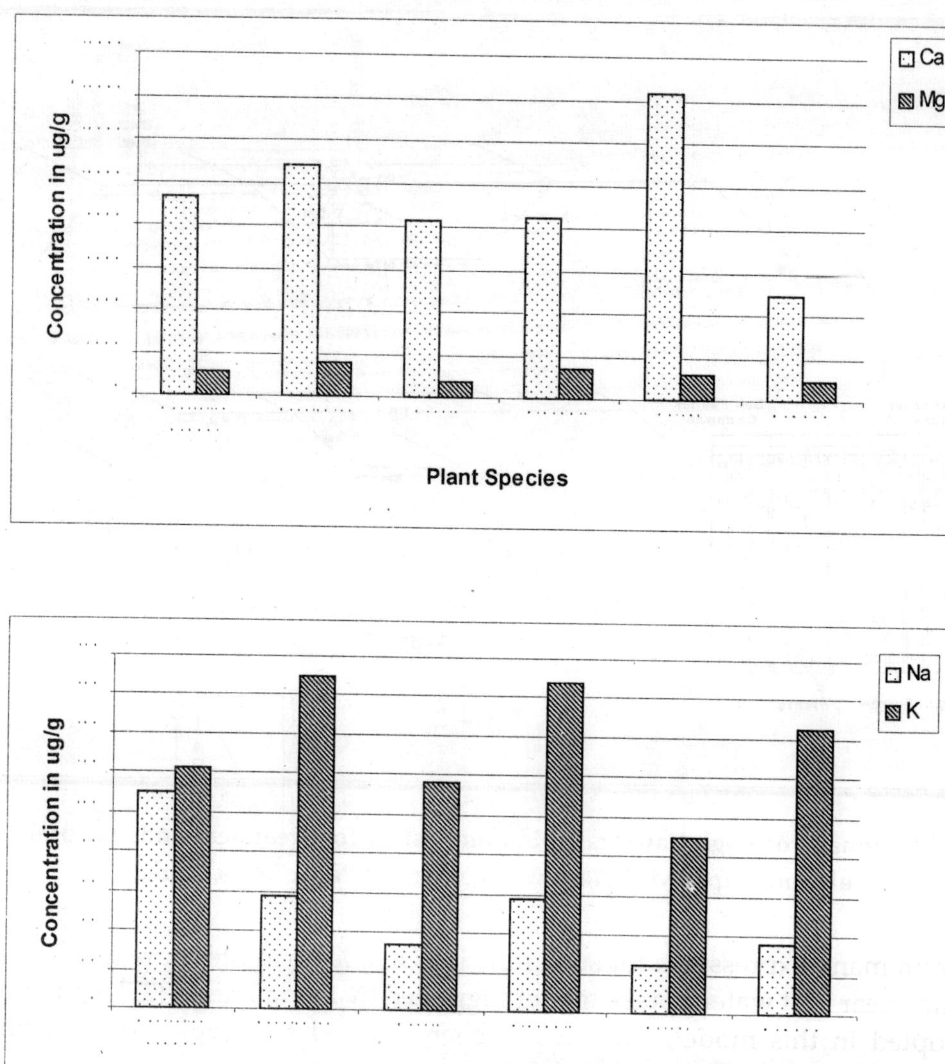


**Fig. 4. Calcium, Magnesium, Sodium and Potassium uptake by Different Plants Species Irrigated with Pulp and Paper Mill Wastewater**

the growth of root at microscopic rate and increases complexity if incorporated in the sink term and therefore a simple model suggested by Borg and Grimes (6) is adopted. It has been developed after an extensive regression analysis using 150-field observations:

$$Z_r = 0.5Z_{\max}\left[1+\sin\left(3.03\frac{DAP}{DTM}-1.47\right)\right] \quad (3)$$

Here DAP and DTM are the days after planting and days to maturity of the plant under consideration. This empirical relation does not take into account the horizontal root distribution with depth.

The solution of equation 1 requires soil moisture retention characteristic and unsaturated hydraulic conductivity. Since volumetric water content and unsaturated hydraulic conductivity are nonlinear function of suction head, explicit expressions between them are required. The closed from equation developed by Van Genuchten (42-43) from the experimental data set is adopted. These expressions can be written as:

$$\frac{\theta(h)-\theta_r}{\theta_s-\theta_r} = \frac{1}{(1+(\alpha h)^n)^m} \quad (4)$$

$$k(h) = k_s \frac{\left[1-(\alpha h)^{n-1}\left\{1+(\alpha h)^n\right\}^{-m}\right]^2}{\left[1+(\alpha h)^n\right]^{m/2}} \quad (5)$$

Here θ(h) represents the water retention curve, defining the water content as a function of the soil water pressure, $\theta_s$ and $\theta_r$ are the saturation and residual moisture content, respectively, α, n and m are the curve shape parameters. First derivative of equation 4 gives the expression for soil moisture capacity as:

$$c(h) = \frac{\partial\theta}{\partial h} = \frac{(1-n)\alpha^n h^{n-1}(\theta_s-\theta_r)}{\left[1+(\alpha h)^n\right]^{m+1}} \quad (6)$$

The parameter n, m $\theta_s$, $\theta_r$ and α of Van Genuchten expression are estimated using pedo-transfer function (PTF). Cornelis et al. (12) have evaluated the PTF's developed by several researchers and concluded that those given by Vereecken et al. (44) gives best agreement between measured and modeled values and therefore adopted here:

The water flux in the study domain can be computed using Darcy's law, which is extended for flow an unsaturated zone (4). It can be written as:

$$q(z,t) = k(h)\left(\frac{\partial h}{\partial z}-1\right) \quad (7)$$

**Heavy Metal Transport and Its Uptake by Plant Root**

Application of wastewater for irrigation and subsequent moisture uptake by plant root is modeled by creating a mass balance of metal species in the control volume. The governing one-dimensional equation for heavy metal transport through soil can be written as (36):

$$\frac{\partial}{\partial t}(\rho M_s)+\frac{\partial}{\partial t}(\theta M_w)=\frac{\partial}{\partial z}(\theta D\frac{\partial M_w}{\partial z}-qM_w)+S_p \quad (8)$$

Here $\rho$ is the bulk density of the soil, $M_s$ is the mass of the heavy metal adsorbed in the soil, $\theta$ is the volumetric moisture content, $M_w$ is the concentration of heavy metal in water phase, D is the hydrodynamic dispersion coefficient which is a function of pore velocity V, q is the Darcy velocity and Sp is the heavy metal uptake by plant roots. The pore water velocity and hyrodynamic dispersion coefficient can be written as (46):

$$V = q/n \quad (9)$$

$$D = aV + b \quad (10)$$

Here n is the porosity of the soil, a is the longitudinal dispersivity and b is the molecular diffusion coefficient.

Metals are found under several chemical forms like free ions, complexed, adsorbed, and incorporated into organic molecules and/or minerals. Their fate and availability in soil is first governed by the chemical state of the element in the contaminating material. The availability of trace elements to plant is a function of the activity of free ions in solution at root surface. Elements in solution are transferred from the soil pores to the root surface by two main processes: diffusion and mass flow (27).

Heavy metals in the unsaturated media remain in two phases, one in dissolved ionized form in the soil moisture and the other in the adsorbed phase in the soil particle. The adsorption of heavy metals in solution is assumed to be governed by linear isothemal adsorption and is proportional to the heavy metal content in soil (9).

$$M_w = kdM_s \quad (11)$$

Uptake of metals from the soil by hyperaccumulators does not appear to be a passive phenomenon. Metal contents of hyperaccumulator do not increase linearly as the metal concentration in the substrate increases. Morrision et al. (28) showed that metal hyperaccumulators will even concentrate metals in their tissues when grown on relatively low metal soils, indicating an active uptake and sequestration process. The mathematical models for uptake of metal by plants are based on transportation of the ions to the plant roots by mass flow and diffusion along the concentration gradients and are absorbed by roots at a rate that depends on the concentration of the ions at the root surface. The metal uptake is assumed to follow Michaelis-Menton type relationship as suggested by Mullins et al. for cadmium uptake (29).

$$S_p = \frac{I_{\max} M_w}{K_m + M_w} \tag{12}$$

Here $I_{max}$ is the uptake rate at infinite concentration and $K_m$ is the Michaelis constant.

## Solution of the Governing Equations

Equations 1-12 forms a complete set of governing equations which are non-linear partial differential equations, thus an analytical solution is not possible and a numerical method to obtain the solution needs to be applied. These equations are solved numerically by explicit finite difference method. Picard's iterative technique is used for linearizing the set of nonlinear simultaneous equation along with Gaussian elimination method for matrix inversion. Equations 1-6 are solved first to obtain the spatial and temporal variation of suction head (h) in soil column. The moisture content q(h) and flux (q) in the soil solved with the computed value of moisture content to obtain the heavy metal adsorbed in the soil and water phase and heavy metal uptake by plant root system. The source code for the solution of partial differential equation is written in MATLAB.

## Simulation

Plants are the essential components of the natural ecosystem and agro-system, and are the first compartment of the terrestrial food chain. When grown on polluted water and soils, they become a potential threat to human and animal health and therefore non-fodder crop should be chosen for rhizofiltration programme. After harveting the crop (plant) may be converted to bio-fuel briquette a substitute for fossil fuel (27).

A simulation is carried out to show the application of the rhizofilteration model in which waster water having arsenic concentration complying with the regulatory requirement for land disposal is used for simulating heavy metal movement in unsaturated zone. Fig. 5 shows the metal flux joining groundwater and metal and water uptake by plant roots is estimated for different scenarios.

Table 1 list the soil, plant, transport and simulation parameters used for the modeling study. Restricted irrigation for a period of 200 days is simulated and

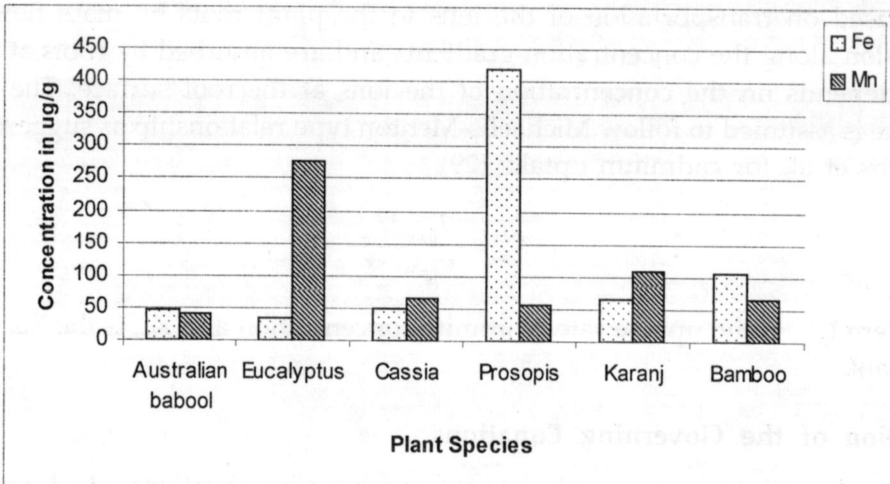


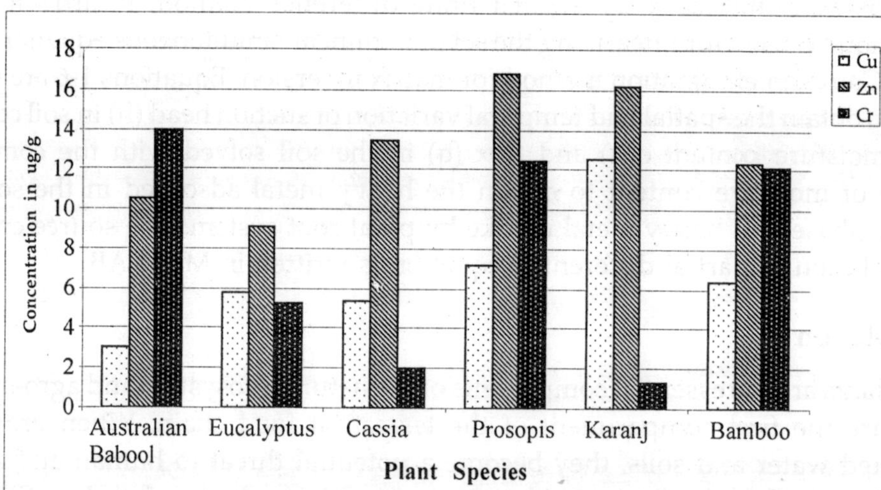


**Fig. 5. Iron, Manganese, Copper, zinc and Chromium uptake by Different Plants Species Irrigated with Pulp and Paper Mill Wastewater**

root length growth, moisture and metal uptake by plant root and metal flux to ground water is computed. The moisture content at the top boundary or ground surface is maintained at the average of saturation and residual moisture content i.e., 0.23. This moisture yields a suction head of 200 cm. The bottom boundary is kept at 0 cm suction head as it is at groundwater level. The two boundaries can be written in terms of prescribed constant suction head (Dirichlet boundary) as:

**Table 1**

**Simulation results for dirrent scenarios**

| No. | Particulars | Metal Accumu-altor Plant, $K_d$=0.3 | Non metal Accumu-altor Plant, $K_d$=0.3 | $K_d$ = 0.3 but Without Plant | $K_d$=0 and Without Plant |
|---|---|---|---|---|---|
| 1. | Total quantity of water that joins groundwater in 200 days (ml/sq.cm) | 592.34 | 592.34 | 593.163 | 593.163 |
| 2. | Maximum Concentration of metal joining groundwater (mg/l). | $0.355\times10^{-98}$ | $0.355\times10^{-98}$ | $0.355\times10^{-98}$ | $0.1992\times10^{-75}$ |
| 3. | Total quantity of metal that joins groundwater in 200 days (g/sq.cm). | $0.3471 \times 10^{-103}$ | $0.8292 \times 10^{-103}$ | $0.8673 \times 10^{-103}$ | $0.47416 \times 10^{-80}$ |
| 4. | Total metal uptake by plant in 200 days (ug). | 8.886 | Nil | Nil | Nil |

$$h(0, t) \qquad \text{at } q = q_s = 0.41 \text{ and}$$

$$h(L, t) = 200 \quad \text{gives } q = 0.23$$

Here L is the length measured form water table to ground surface. The boundary conditions for heavy metal at the ground surface and water table can be written in terms of prescribed concentration of heavy metal ($M_w$) in water phase (Dirichlet boundary) as:

$$M_w(0, t) = 0.0 \text{ ppm}$$

$$M_w(L, t) = 0.2 \text{ ppm}$$

The governing equations are then solved using the boundary condition for prescribed suction head and heavy metal in water phase to obtain the results.

### Method Used for Analysis of Wastewater, Soil and Plant Samples

Wastewater and soil samples from the identified site of M/s Orient Paper Mill, Amlai were collected and subjected for the analysis of various physico-chemical parameters. The wastewater samples used for HRTS and the well water to evaluate the effect on application of wastewater on to the soil were analyzed as per standard methods for the Examination of water and wastewater. Plant height were measured periodically and soil samples before and after application of pulp and paper mill wastewater were collected from HRTS were analysed by Standard technique (5).

**Results and Discussions**

The wastewater samples were collected from anaerobic lagoon outlet during winter and summer seasons to ascertain the wastewater for various physico-chemical parameters as depicted in Table 2. The wastewater is dark brown in colour, which indicated that the application of such highly coloured wastewater to the land may affect the soil characteristics and may also further lead to ground water pollution through leaching. The wastewater is neutral to slightly alkaline in reaction and its pH varies from 6.70 to 7.90. Soluble salt content in the wastewater, expressed as EC (mS/cm) ranges from 2.26 – 4.50. The sodium absorption ratio of wastewater is low and ranges from 2.78 – 4.56 indicating that it will not impose sodicity problem and offers its disposal on land. The wastewater also contains considerable chemical oxygen demand (404 to 675 mg/l) and appreciable quantity of chlorides, which ranges from 15.69 – 23.49 meq/l and belongs to the class 'Severe Problem'.

The results depicted in Table 3 showed that the soil from the selected field was loamy sand in texture. Bulk density of the soils ranged from 0.78 to 1.28 g/cc. Maximum water holding capacity varied from 48.60 to 16.4 %. Hydraulic conductivity of the soils was very high which ranged from $3.0 \times 10^{-2}$ to $2.0 \times 10^{-2}$ cm/sec. The soil is slightly acidic to almost neutral in reaction having pH in the range of 6.70 to 7.00. The soluble salt expressed in terms of electrical conductivity of saturation extract of soil were higher in upper layer and decreases with depth and ranged from 2.08 to 1.21 mS/cm. The cation exchange capacity of the soil in uppermost layer is good as compared to lower layer, which ranged from 30.54 to 7.60 meq/100g. The total concentration of nutrients *viz.* nitrogen, phosphorus and potassium ranged from 0.064 – 0.008%, 0.070 – 0.030% and 0.080 – 0.128% respectively. The concentration of total heavy metals like Ni, Cr, Cd, Zn, Pb and Cu are also present appreciable amount with high Fe and Mn content in the soil.

Using the above set of boundary conditions for prescribed flux and heavy metal concentration in water phase the governing equations are solved for simulating heavy metal movement, its uptake by plant roots and amount joining groundwater in the unsaturated zone. Fig. 6 shows the daily and cumulative water uptake by plant roots. Fig. 7 to 9 shows the heavy metal flux joining the groundwater for different scenarios.

**Table 2**

**Values of different parameters used for simulation**

| **Transport Parameter** | |
|---|---|
| Dispersion Coefficient (a) (cm) | 0.008 |
| Diffusion Coefficient (b) ($cm^2/hr$) | 0.002 |
| Partition Coefficient ($k_d$) | 0.30 |
| Concentration of heavy metal at upper boundary ($gm/cm^3$) | 0.0000002 |
| Concentration of heavy metal at lower boundary ($gm/cm^3$) | 0.00 |
| **Soil Parameters** | |
| Saturated moisure content, $\theta_s$ | 0.41 |
| Residual moisture content, $\theta_r$ | 0.05 |
| Empirical shape factor, n | 1.52 |
| Empirical shape factor, m | 1.0 |
| Empirical shape factor, $\alpha$(1/cm) | 0.00542 |
| Saturated hydraulic conductivity, Ks cm/h | 0.67 |
| Bulk density of soil (gm/cubic cm) | 1.67 |
| **Plant Parameters** | |
| Name of the Plant | Hypothetical |
| Maximum rooting depth ($Z_{max}$) cm | 50 |
| Days to maturity of Plant (DTM) in days | 200 |
| Transpiration rate ($T_p$) | 0.1 |
| Maximum Uptake rate ($I_{max}$) cm | 0.3 |
| Michaelis Minton Constant ($k_m$) | 0.01 |
| Reduction Factor $\alpha$(h) | 1 |
| **Simulation Parameters** | |
| Height of unsaturated zone (cm) | 500 |
| Number of nodes, N | 51.0 |
| Space step, $\Delta t$ (cm) | 10.0 |
| Time step, $\Delta t$ (hour) | 4.0 |
| Boundary condition 1 (suction head at ground water level) (cm) | 0.00 |
| Boundary condition 2 (suction head at ground water level) (cm) | 200.0 |

**Table 3**
**Physico-chemical Characteristics of Pulp and Paper Mill Wastewater (Grade- III)**

| Sr. No. | Parameters | Magnitude | Standards for Industrial Effluent on land for irrigation |
|---|---|---|---|
| 1. | pH | 6.70 – 7.90 | 5.5 – 9.0 |
| 2. | EC, mS/cm | 2.26 – 4.50 | - |
| 3 | Colour, Hz | Dark Brown | - |
| 4. | TSS, mg/l | 70.0 – 216.0 | - |
| 5. | COD, mg/l | 404 – 675 | 200 |
| 6. | Sodium, mg/l | 7.50 – 13.10 | - |
| 7. | Potassium, mg/l | 0.32 – 0.95 | - |
| 8. | Calcium, mg/l | 9.30 – 15.50 | - |
| 9. | Magnesium, mg/l | 0.40 – 8.08 | - |
| 10. | Bicarbonate, mg/l | 1.50 – 15.16 | - |
| 11. | Chloride, mg/l | 15.69 – 23.49 | 600 |
| 12. | Sodium Absorption Ratio (SAR) | 2.78 – 4.56 | - |

Taking into account sorption of heavy metal into the soil in the presence of a metal accumulating plant, metal flux joining groundwater is computed (Fig. 10). It can be inferred from the figure that in the beginning when plant roots have not grown significantly, the metal flux joining groundwater varies linearly and attains steady state for a short duration. As the root grows it takes up water and metal from the soil and thereby sharply decreases the downward metal flux joining groundwater. Thus the presence of hyperaccumulator plant and metal sorption in unsaturated zone significantly reduces the metal flux joining groundwater. Fig. 11 shows the metal flux joining groundwater in an unsaturated zone taking into consideration soil sorption in the presence of a plant that does not uptake metal. With time, the growth in the root length increases water uptake thereby changing the initially attained steady state condition of heavy metal flux joining groundwater. The sharp decrease as evident in Fig. 6 is not observed in this plot as no metal is taken up by the plant species. Fig. 8 shows

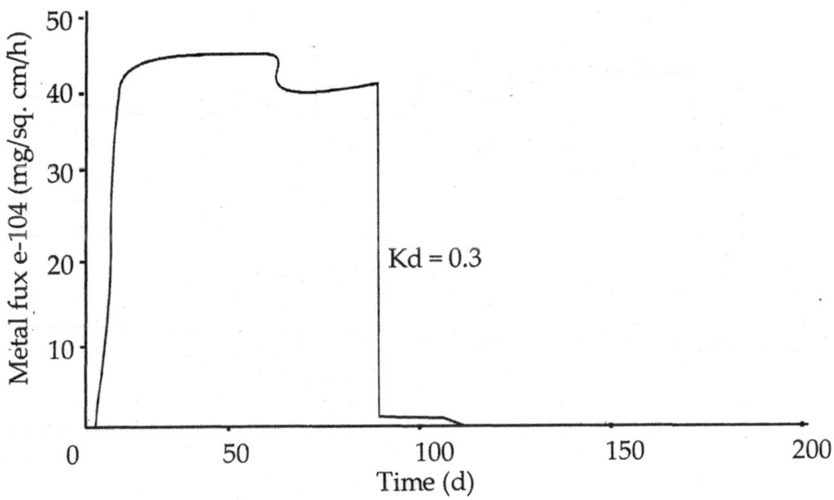


**Fig. 6. Heavy Metal (As) flux joining groundwater (with metal accumulating plant)**

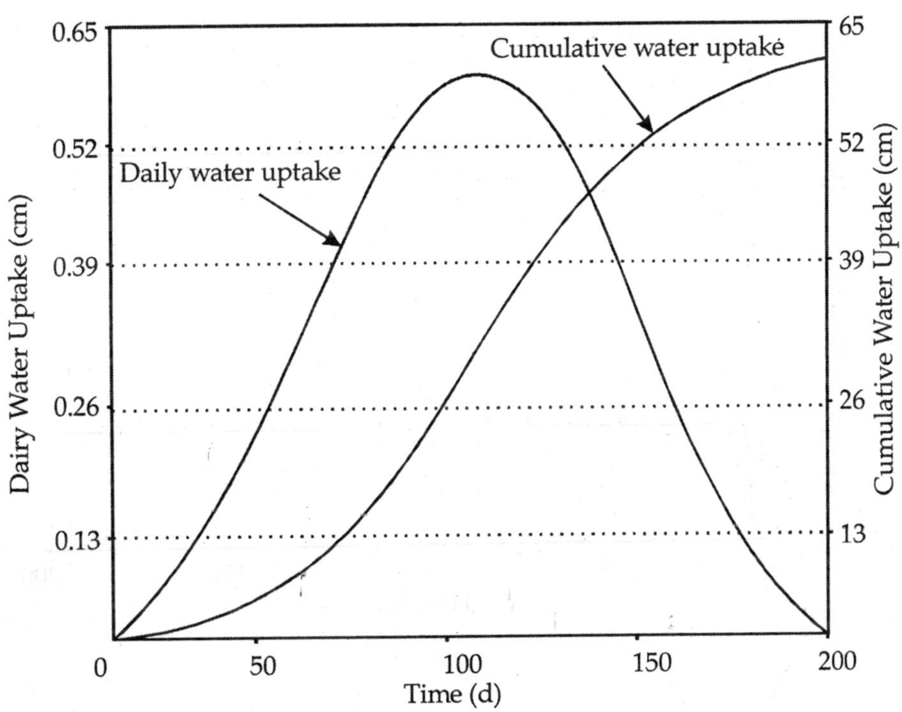


**Fig. 7. Water uptake by plant roots.**

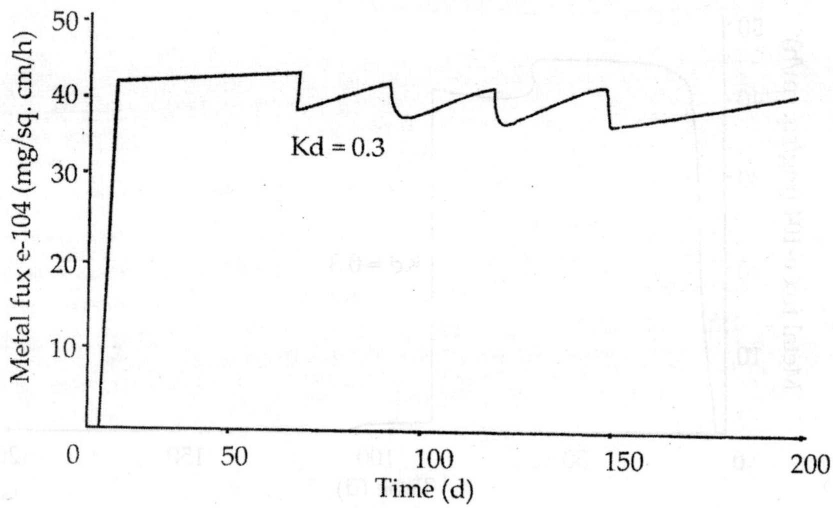


**Fig. 8. Heavy metal flux joining groundwater (without metal accumulating plant)**

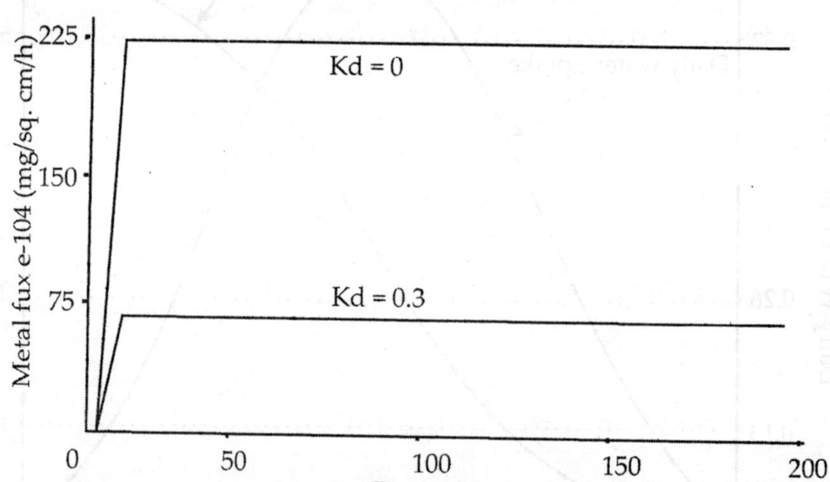


**Fig. 9. Heavy metal flux joining groundwater for different soil water partition coefficient (without any plant).**

the plot of metal flux joining groundwater for different values of soil water partition coefficient ($k_d$). It can be seen that for zero partition coefficient the metal flux joining groundwater is maximum and with the increase in partition coefficient the flux decreases. The cumulative metal taken by the plant in the simulation period is shown in Fig. 10.

## Effect on Wastewater Irrigated Soil

### *Study of Salt Movement in the HRT System*

To study the salt movement in the ridges and furrows irrigated with pulp and paper mill wastewater, soil samples were collected from center and edge of the ridges (2.5 meter wide) in the HRT system and were analyzed for their salt content. The results indicated that upper soil layer of 0-15 cm was moderately acidic, due to decomposition of leaf litter which is acidic in nature, while the soils of lower depth were neutral in reaction. The salt concentration was high in the soil from 0-15 cm depth at the edge of the ridge (EC - 3.76 I mS/cm) as compared to soil from the center portion of the ridge.

The soil from lower depth i.e. 30-90 cm of the ridge showed low in salt concentration (EC - 2.16 mS/cm). The concentration of sodium was higher in all depths (19 to 25 meq/L) followed by calcium and magnesium (Table 4). The chloride concentration was in the range of 16.92 to 33.84 meq/L. The filter media

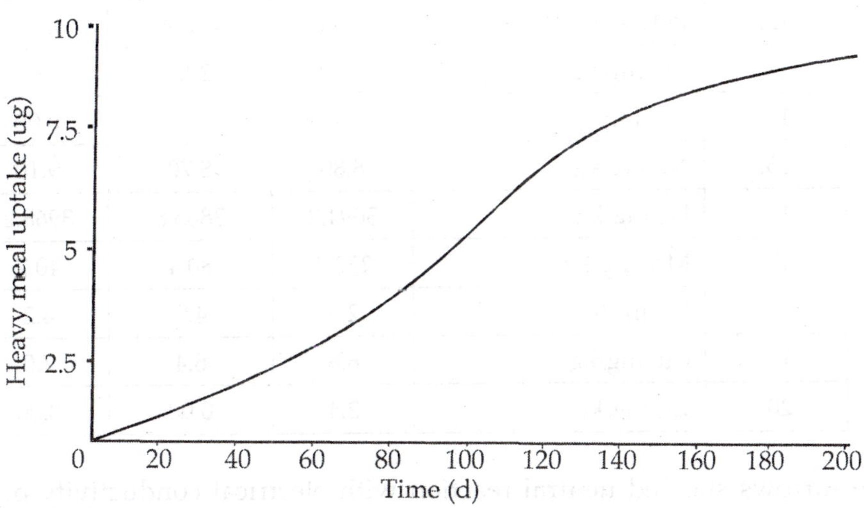


**Fig. 10. Cumulative heavy metal uptake by plant roots**

**Table 4**
**Physico-chemical Characteristics of Soil Profile Samples Collected from the Identified Site by M/s Orient Paper Mill, Amlai**

| Sr. No. | Parameters | Soil Depth, cm | | |
|---|---|---|---|---|
| | | 0-5 | 5-15 | 15-30 |
| 1. | Clay, % | - | 8.6 | 8.8 |
| 2. | Sand, % | - | 9.6 | 10.4 |
| 3. | Bulk Density, g/cc | 0.78 | 1.21 | 1.28 |
| 4. | Water Holding capacity, % | 48.6 | 24.4 | 16.4 |
| 5. | Hydraulic Conductivity, cm/sec | 0.30 | 0.218 | 0.020 |
| 6. | Cation Exchange Capacity, meq/100g | 30.54 | 8.40 | 7.60 |
| 7. | pH | 6.7 | 6.8 | 7.0 |
| 8. | ECe, mS/cm | 2.08 | 1.42 | 1.21 |
| 9. | Organic Carbon, % | 3.84 | 1.40 | 0.80 |
| 10. | Nitrogen, % | 0.064 | 0.020 | 0.008 |
| 11. | Phosphorous, % | 0.070 | 0.043 | 0.030 |
| 12. | Potassium, % | 0.080 | 0.124 | 0.128 |
| 13. | Zn, mg/kg | 40.8 | 20.8 | 16.8 |
| 14. | Cd, mg/kg | 0.40 | 0.10 | 0.00 |
| 15. | Ni, mg/kg | 8.80 | 8.70 | 9.10 |
| 16. | Fe, mg/kg | 3094.2 | 2865.8 | 3968.2 |
| 17. | Mn, mg/kg | 352.4 | 80.4 | 40.7 |
| 18. | Pb, mg/kg | 2.8 | 4.9 | 4.2 |
| 19. | Cu, mg/kg | 6.8 | 6.4 | 7.0 |
| 20. | Cr, mg/kg | 2.4 | 0.00 | 0.00 |

in the furrows showed neutral reaction with electrical conductivity of 2.1 mS/cm. This indicated that inspite of continuous application of wastewater; there was no appreciable build-up in salt concentration in the furrows because the plants on the ridges create a moisture gradient, which facilitate salt movement along with water from the furrows to the ridges.

## Physical Properties of Soil

Soil samples were collected from the field irrigated with wastewater Soil samples were collected from the field irrigated with wastewater since long period. The results as presented in Table 5, indicated that so in the irrigated areas mostly are loamy sand in texture with clay and sand content ranging from 13.8-13.9 percent and 70.8-73.8 percent respectively. The uppermost (Ao) layer of soil had low built density (0.79 g/cc) due to retention of lignin and leaf litter and it increased with the depth. Ao horizon had maximum water holding capacity of 87 percent and its hydraulic conductivity was also very high (0.430 cm/sec.). However, the soils at a depth of 5-15 c and 15-30 cm had low hydraulic conductivity and were in the order of O.122, 0.033 cm/sec, respectively.

The cation adsorption capacity of the uppermost soil horizon was high as compared to lower layer (CEC 46.32 meq/l00g), because of its organic nature. The available moisture content was also high in the upper organic rich layer (11.9 percent) and it decreased with depth.

**Table 5**
**Chemical Characteristics of Soil Samples Collected from HRTS**

| Sample Description | Depth, cm | pH | ECe, mS/cm | Water Soluble ions, meq/l | | | | | |
|---|---|---|---|---|---|---|---|---|---|
| | | | | $Na^+$ | $K^+$ | $Ca^{++}$ | $Mg^{++}$ | $HCO_3^-$ | $Cl^-$ |
| Center of the ridge | 0 – 15 | 5.5 | 2.55 | 21.0 | 0.85 | 4.4 | 1.2 | 1.2 | 25.09 |
| | 15-30 | 6.9 | 2.50 | 22.5 | 0.60 | 3.4 | 1.6 | 1.4 | 24.25 |
| | 30-60 | 7.4 | 2.68 | 22.0 | 0.40 | 1.6 | 1.4 | 2.0 | 25.38 |
| | 60-90 | 7.2 | 2.20 | 20.0 | 0.60 | 2.0 | 0.6 | 1.6 | 19.74 |
| | | | | | | | | | |
| Edge of the ridge | 0 – 15 | 5.8 | 3.76 | 25.0 | 0.60 | 4.0 | 1.80 | 1.6 | 33.84 |
| | 15-30 | 6.7 | 2.64 | 20.0 | 0.50 | 4.6 | 1.2 | 3.2 | 23.17 |
| | 30-60 | 7.2 | 2.16 | 9.0 | 0.25 | 2.0 | 0.8 | 1.6 | 16.92 |
| | | | | | | | | | |
| Filter Media | 0-10 | 7.1 | 2.10 | 14.0 | 0.45 | 5.4 | 1.8 | 4.0 | 15.22 |
| | | | | | | | | | |

## Microbial Activity and Organic Decomposition in the Soil, Irrigated with Pulp and Paper Mill Wastewater

The microbiological processes of the soil are complex, and it is neither possible nor practical to do more than pint out some principles concerning microbial decomposition in the soil in this bulletin. It should be noted that purification of organic effluents applied to the soil, or organic sludge incorporated in the soil, is accomplished by the interrelated activities of the microbial population acting in conjuction with surface adsorption phenomena in much the same manner as in activated sludge units, trickling filters and sewage and filters. Recent order to magnitude estimates of soil microbial numbers indicate the presence of about 100 million bacteria, 10 million actinomycetes, 1 million fungi, and 100 thousand algae per gram of soil, with a total live weight of 2400 pounds per acre of plow depth. Their activity is shown by the rapid decomposition of crop residues and the differences in composition of such residues and the differences in; composition of such residues and the organic matter of the soil. Of the latter, from 1/3 to ½ is estimated to be microbial in origin.

## Chemical Properties of Soil

The result presented in Table 6 showed that soil in the existing irrigated area was neutral in reaction (pH 7.0-7.4). The soluble salts expressed in terms of electrical conductivity of saturation extract of soil were higher (3.03 mS/cm) in upper layer and it decreased in the lower depth and exhibited value of 2.17 mS/cm. Increase in soluble salt and pH as observed in the present study as a result of pulp and paper mill wastewater application are also reported by Mc Carmic (1959). Juwarkar and Subramanyam (1987) also showed increase in soluble and exchangeable salts due to wastewater irrigation. The exchangeable sodium percentage of the soils was less than 15 and therefore the soils were categorized in non saline-non alkali group.

The organic carbon content in the upper (Ao) layer was high (17.7%) due to continuous accumulation of leaf litter and retention of lignin compounds of organic nature from the wastewater. The increase in organic matter content in soil was due to presence of dissolved lignin in the wastewater applied, which retained in soil. The total nitrogen and phosphorous contents were also high. Rajanan and Obliswami (1979) also reported increase in OM and available N, P, K of soil when irrigated with pulp and paper mill wastewater. Heavy metal concentration in irrigated soil at all the depths were below the toxicity level (Table 7) this might be due to the fact that toxic heavy metal are not used in any stage of process in paper manufacture.

**Table 6**
**Physico-chemical Characteristics of Wastewater Irrigated Soil Samples Collected from HRTS Site**

| Sr. No. | Parameters | Soil Depth, cm | | |
|---|---|---|---|---|
| | | 0-5 | 5-15 | 15-30 |
| 1. | Clay, % | - | 13.8 | 13.9 |
| 2. | Sand, % | - | 12.4 | 15.3 |
| 3. | Bulk Density, g/cc | 0.79 | 1.31 | 1.58 |
| 4. | Water Holding capacity, % | 86.9 | 37.3 | 27.5 |
| 5. | Hydraulic Conductivity, cm/uuu | 0.43 | 0.122 | 0.033 |
| 6. | Cation Exchange Capacity, (meq/100g) | 46.32 | 12.50 | 11.33 |
| 7. | pH | 7.0 | 7.3 | 7.4 |
| 8. | ECe, mS/cm | 3.03 | 2.65 | 2.17 |
| 9. | Exchangeable sodium percent | 7.73 | 13.66 | 11.85 |
| 10. | Organic Carbon, % | 17.73 | 2.59 | 0.41 |
| 11. | Nitrogen, % | 0.133 | 0.011 | 0.008 |
| 12. | Phosphorous, % | 0.095 | 0.060 | 0.025 |
| 13. | Potassium, % | 0.127 | 0.195 | 0.274 |
| 14. | Zn, mg/kg | 56.2 | 24.9 | 18.0 |
| 15. | Cd, mg/kg | 0.6 | 0.2 | 0.00 |
| 16. | Ni, mg/kg | 15.9 | 15.3 | 16.8 |
| 17. | Fe, mg/kg | 4592.9 | 4487.9 | 7655.0 |
| 18. | Mn, mg/kg | 452.5 | 70.6 | 53.6 |
| 19. | Pb, mg/kg | 4.1 | 13.1 | 5.2 |
| 20. | Cu, mg/kg | 8.6 | 6.8 | 7.6 |
| 21. | Cr, mg/kg | 2.6 | 0.0 | 0.0 |

**Table 7**
**Chemical Characteristics of the Well Water Located Near HRTS Site at M/s Orient Paper Mill, Amlai**

| Sr. No. | Parameters | Approximate distance from HRTS | |
|---|---|---|---|
| | | 1.0 Km | 2.0 Km |
| 1. | pH | 8.20 – 8.40 | 8.00 – 8.20 |
| 2. | EC (μS/cm) | 0.48 – 0.84 | 0.37 – 0.39 |
| 3. | Na (meq/l) | 0.67 – 3.80 | 0.70 – 0.80 |
| 4. | K, (meq/l) | 0.05 – 0.16 | 0.10 – 0.13 |
| 5. | Ca, (meq/l) | 2.76 – 3.20 | 1.52 – 2.15 |
| 6. | Mg, (meq/l) | 0.64 – 1.56 | 0.96 – 1.15 |
| 7. | $CO_3$, (meq/l) | 0.50 – 1.75 | 0.55 – 0.86 |
| 8. | $HCO_3$, (meq/l) | 1.55 – 3.55 | 1.32 – 2.35 |
| 9. | Cl, (meq/l) | 1.38 – 2.50 | 0.85 – 1.12 |
| 10. | Colour | Sparkingly clear | Sparkingly clear |
| **Heavy Metals, mg/l** | | | |
| 11. | Cd | 0.002 – 0.003 | 0.002 – 0.004 |
| 12. | Cu | 0.012 – 0.017 | 0.011 – 0.015 |
| 13. | Pb | 0.026 – 0.028 | 0.009 – 0.011 |
| 14. | Ni | BDL | 0.008 – 0.013 |
| 15. | Zn | 0.008 – 0.010 | 0.031 – 0.045 |
| 16. | Fe | 0.620 – 0.700 | 0.661 – 0.697 |

## Plant Growth

The various plant species planted in HRT system showed better growth with respect to foliage development (Plate 1). The grasses like lemon, citronella and pamarosa also developed profusely on the ridges (Plate 2). The growth of typha grown in the furrow was also good which encourages its use on large scale in the system.

Introduction of aromatic grasses like citronella and lemon in the HRTS helps to improve ambient environment in the following manner:

**Plate 1. Eucalyptus and Bamboo Plantation on the Ridges in the HRTS site at Orient Paper Mill, Amlai**

**Plate 2. Lemon and Citronella Grasses Planted in the Ridges in the HRTS site at Orient Paper Mill, Amlai**

- in controlling obnoxious odours resulting from evaporation wastewater from the furrows
- in controlling breeding of mosquitoes and other insects due to astringent and repelling aroma from the grasses
- effective check on cattle trespassing as cattle do not browse on these grasses thus preventing cattle damage to HRT system
- in controlling growth of harmful weeds like parthenium etc.

## Nutrient and Heavy Metal Uptake by the Plant Leaves

The leaves of different plants viz. Australian babul, eucalyptus, cassia, prosopis, karanj and bamboo irrigated with pulp and paper mill wastewater were analyzed for nutrient and heavy metal uptake. The leaves were analyzed for calcium, magnesium, sodium and potassium uptake. Results showed variable response with respect to nutrient and heavy metal uptake by different plant species. The uptake of calcium was higher, in the range of 2301-17591 ug/g followed by magnesium (769.4-1879.40 ug/g), potassium (87.97-65.80 ug/g) and sodium (23-115 ug/g) (Fig. 10). Among different plant species, bamboo and prosopis showed high uptake of iron while high manganese uptake was observed in case of eucalyptus. Uptake of metals like copper, zinc and chromium showed variation with the plant species (Fig. 11). The results indicated that plants differ in their capacity to accumulate nutrients and heavy metals and acts as a sink in removing different types of pollutants from soil matrix.

## Effect on Underground Water

Well water samples collected from the wells located at the distance of 1.0 and 2.0 km from the HRTS were analyzed for chemical properties. The results indicated that the well water was slightly alkaline in reaction and their pH varied from 8.2-8.4 and 8.0-8.2 respectively for the samples collected from 1.0 and 2.0 km respectively.

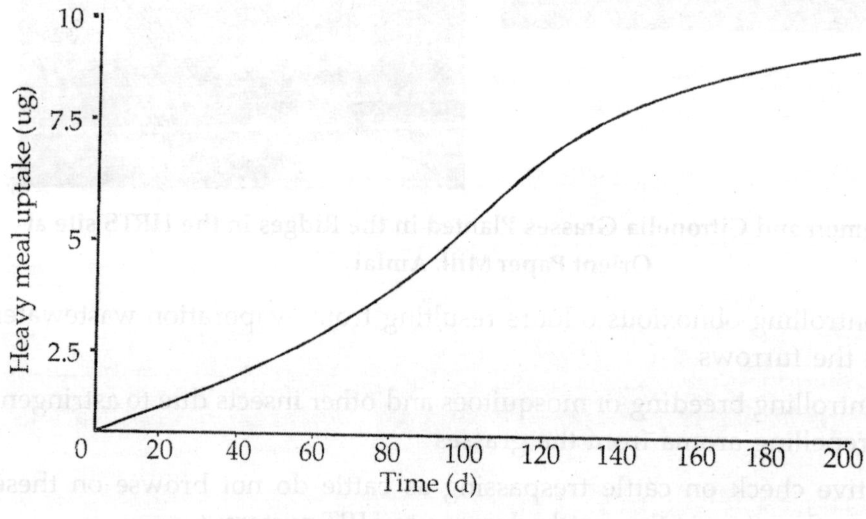


**Fig. 11. Cumulative heavy metal uptake by plant roots**

Soluble salt concentration in well water, expressed as EC (mS/cm), ranged from 0.48-0.84 and 0.37-0.39 in wells 1.0 and 2.0 km away from HRTS respectively. The significant observation was that the colour of all well waters was sparkingly clear. The concentration of heavy metals like Cd, Pb, Zn, Fe, Ni and Cu in the well waters were not in appreciable range which indicated that well waters upto a distance of 2.0 km from HRTS area were not contaminated ruling out the possibility of groundwater contamination due to wastewater.

## Conclusion

Verma et al. (21) carried out simulation for water movement and its uptake by plant roots in unsaturated zone. This paper is the extension of the earlier work and takes into account heavy metal movement in unsaturated zone and its uptake by plant roots. This model will be useful in quantifying the heavy metal uptake by plants, heavy metal flux joining the groundwater and will be helpful in design of phytoremediation programme for wastewater having very high metal concentration. Further lab and field studies are needed to estimate parameters like the maximum metal uptake rate, Michaelis – Menton constant, plant root characteristics and soil-water partition coefficient etc. for a given soil – plant system. This model can be further improved by incorporating complex root geometry, evaporation of wastewater from the ground surface and taking rainfall into consideration.

The preliminary experiment through column lysimeter conducted at NEERI, proved that the coloured Paper Mill wastewater could be successfully treated using High Rate Transpiration System. The HRTS proved that the coloured wastewater initially loaded at the rate of 150 $m^3$/ha/day and gradually increased to 250-275 $m^3$/ha/day could be successfully treated pulp and paper mill wastewater without any ground water contamination. The HRTS planted with different plant species viz. eucalyptus, bamboo, acacia mangium, cassia, gravellia, lemon grass and citronella grass showed healthy and better growth response and develop forestry which acts as a sink potential for absorptions of green house gases.

## Research Need

1. What relation exists between soil permeability, texture, and chemical characteristics and the capacity of a particular soil for treatment of various applied effluents.
2. What changes in effluent BOD, odor, and color occur during passage of different mill effluents through various soils.

3. What are the results of solids inclusion and pretreatment on the treatment capacity f various soils.
4. What effects, if any, on cover vegetation of economic value, such as tree seedlings and field crops, result from application of different mill effluents, and how may any possible adverse effects be avoided.

What effects does variation in loading and effluent composition have on physical-chemical soil characteristics.

**Acknowledgements**

The authors wish to express their gratitude to the management of M/s Orient Paper Mill, Amlai for their cooperation and assistance during the research work.

## References

Anderson, A. (1979). 'Disposal of an Integrated Pulp and Paper Mill Effluent by Irrigation', Tech. Rep. No. EPA-600/2-79-633, EPA Cincinnati, Ohio.

Anonymous (1989). 'Standard Methods for the Examination of Water and Wastewater', 17th ed. Publ. APHA, AWWA, WPCF, 1015, Washington, DC 20036.

Baker, A.J.M.; McGrath, S.P.; Reeves R.D., Smith, J.A.C. (1999). In Phytoremediation of contaminated soil and water; Terry, N., Banuelos, G.S., eds., Lewis; Boca Raton, FL, pp. 85-107.

Berar, J. (1979). Dynamics of fluids in porous media, American Elsevier Publishing Company, Inc., New York, U.S.A., pp. 482.

Black, C.A. (1965). 'Methods for Soil Analysis', Am. Soc. Agro. Inc., Publ. Madison, Wisconsin, USA.

Borg, H., Grimes D.W. Trans. (1986). ASAE, 29(1), 194-197.

Boyd, R.S. (1998). In Plants that hyperaccumulate heavy metals, Brooks, R.R., ed., CAB Internationl, Willingford, U.K., pp. 181-201.

Christensen, T.J. (1985). Water, Air Soil Pollution, 26, 255-264.

Christensen, T.J. (1987). Water, Air Soil Pullution, 34, 305-314.

Classen, N., Barber, S.A. (1976). Agronomy. J., 68, 961-964.

Cornelis, W.M., Ronsyn, J., Meirvenne, M.V., Hartmann (2001). Soil Sci. Am. J., 65, 638-648.

Dushenkov, V., Kumar, P.B.A.N., Motto, H., Raskin, I. (1995). Environ. Sci. Technol., 29, 1239-1245.

Hillel, D., Van, B., Talpaz, H. (1975). Soil. Sci., 120, 385-399.

Hoffland, E., Bolemhof, H.S., Leffelaar, P.A., Findenegg, G.R., Nelemans, J.A. (1990). Plant Soil., 124, 149-155.

Hoogland, J.C., Fedes, R.A., Belmans, C. (1981). Acta Hart., 119, 123-131.

Indian standard for Industrial and sewae effluent discharge, IS: 2490, Bereau of India Standards, New Delhi, India 1982.

Jackson, M.L. (1973). 'Soil Chemical Analysis', Publ. Prantice Hall of India Ltd, New Delhi.

Juwarkar, A.S. and Subrahamanyam, P.V.R.: (1987). 'Impact of Pulp and Paper Mill Wastewater on Crop and Soil', *Wat. Sci. Tech.* 19, 693.

Kandiah A.: 1987, 'Water Quality in Food Production', Water Quality Bulletin. F.A.O., Rome.

Marino, M.A., Tracy, J.C. (1988). Irrig, Drainage Engg., 114, 558-604.

Mathur, S. (1999). Int. J. Numer. Anal. Meth. Geomech., 23, 1349-1357.

McCormic, L.L. (1959). 'Effect of Paper Mill Wastewater on Cattle, Crop and Soil', Bull.La. *Agric. Stn.*, 529.

Molz, F.J. (1981). Water Resource Research, 17, 1245-1260.

Molz, F.J., Remson, I., Fungaroli, A.A., Drake, R.L. (1968). Water Resour. Res., 4, 1161-1169.

Morel, J.L., (1997). In Soil Ecotoxicology, eds. Tarradellas, J. Bitton, G., Rossel, D. eds. CRC Lewis Publishers, New York, U.S.A., pp. 141-176.

Morrison, R.S., Brooks, R.R., Reeves, R.D. (1980). Plant Science Letters, 17, 451-457.

Mullins, G.L., Sommers, L.E., Barber, S.A. (1986). Soil Sci. Soc. Am. J., 50, 1245-1250.

Nimha, M.N., Hanks R.J. (1973). Soil Sci. Soc. Am. Proc. Am. J., 37, 522-527.

Nye, P.H., Marriott, F.H.C. (1969). Plant Soil, 30, 459-472.

Piper, C.S. (1966). 'Soil and Plant Analysis', Univ. Adelaide, Australia.

Prasad, R.J. (1988). Hydrology, 99, 297-306.

Raats, P.A.C. (1975). J. Hydrol., 99, 297-306.

Rajanan, G. and Oblisami, G. (1979). 'Effect of Paper Factory Effluents on Soil and Crop Plants', *Indian J. Environ. Hlth.*, 21(2), 120-130.

Rao, S., Mathur, S. (1994). Journal of Irrigation and Drainage Engineering, 120, 89-96.

Richards, L.A. (1931). Physics, 1, 318-333.

Rowse, H.R., Mason, W.K., Taylor, H.M. (1983). Soil Si., 130, 218-224.

Schnoor, J.L. (1997). "Phytoremediation", Technology Evaluation Report, E Series: TE-98-01, Prepared for Groundwater Remediation Technologies Analysis Center, Pittsburgh, USA, pp. 37.

Singh, S.R., Kumar, A. (1981). Agric. Engg. ISAE, 22(2), 50-74.

Thawale, P.R., Juwarkar, A.S., Kulkarni, A.B. and Juwarkar, A.A. (1999). 'Lysimeter Studies for Evaluation of Changes in Soil Properties and Crop Yield using Wastewater', *Intern. J.Trop. Agric.* 17 (1-4), 231-244.

Van Genuchten, M.Th. (1980). Soil Sci. Soc. Am. J., 44, 892-898.

Van Genucthen, M.Th., Nielsen, D.R. (1985). Ann. Geophys. 3, 615-628.

Vereecken, H., Maes, J., Feyen, J., Darius, P. (1989). Soil Sci., 148, 389-403.

Verma, P., George, K.V., Singh, H.V., Mathew, T.P., Singh R.N. (2004). The International Journal of Environmental Studues, 61(1), 39-48.

Wang, H.F., Anderson, M.P. (1982). Introduction to groundwater modeling, W.H. Freeman and Company, San Francisco, U.S.A., 1982, pp. 180-181.

# 3

# Biological Consequences Due to Increased $CO_2$ Concentration of Nagpur City Compared with the NEERI Campus

*P. Kumar, P.R. Thawale, V.R. Sadevra and V. Sharma*

The term biological diversity refers to variability of life on earth, the ecological complexes in which species occur and the ways in which they interact with each other and their environment. Conservation of biodiversity and sustainable development of any ecosystem is a multifaceted activity, which includes various biological and ecological components. National Environmental Engineering Research Institute (NEERI) is situated at the center of Nagpur city, East longitude 79$^{O}$ 04'26.1", North latitude 21$^{O}$ 04' 26.1" and the altitude is 161 m above sea level. The total area of the NEERI campus, including the residential colony, is 32 ha. Study area is having a very rich in floral diversity, which belongs to "Southern Tropical Dry Deciduous" vegetation.

Terrestrial ecosystems exchange about 120 Gt of carbon per year with the atmosphere, through the processes of photosynthesis and respiration (Schlesinger, 1997). Roughly, half of the $CO_2$ assimilated annually through photosynthesis is released back to the atmosphere by plant respiration (Gifford, 1994; Amthor, 1995). Since terrestrial biosphere-atmosphere fluxes of $CO_2$ far outweigh anthropogenic inputs of $CO_2$ to the atmosphere, a small change in terrestrial respiration could have a significant impact on the annual increment in atmospheric $CO_2$ (Amthor, 1997). Plant collection, harvesting, thinning and other activities are banned in the campus. As a result, the species richness is excessive when compared to the surrounding areas.

Biodiversity is part of our daily lives and livelihood and constitutes the resources upon which families, communities, nations and future generations depend (Dowdeswell, 1995). Present paper comprises flora of NEERI campus and species richness.

## Materials and Methods

For the present study, International Forest Resources and Institutions (IFRI) technique has been used to analyse the vegetation (IFRI, 2002). Plots of following sizes were drawn for the study as follows:

| Radius of the plot | Vegetation study |
|---|---|
| 1m | Annual vegetation and seedlings |
| 3m | Shrubs and saplings |
| 10m | Trees and woody climbers |

A floristic survey was done to compare biodiversity of the different areas. Vegetational data were quantitatively analyzed for density, frequency, and dominance (Misra,1968). Relative density, relative frequency and relative dominance were determined following the method of Phillips (1959). The important value index (IVI) was determined as sum of relative density, relative frequency and relative dominance (Curtis, 1959). The species richness, diversity index ('H) and concentration of dominance (Cd) of the species in the forest were also analyzed. The diversity index was calculated following Shannon and Wiener (1963) method as 'H = - S (Ni/N)$\log_2$ (Ni/N). The concentration of dominance was determined by Simpson's index as (Cd) = S$(Ni/N)^2$ (Simpson,1949). Relative humidity, $CO_2$ emission and temperature was measured by $CO_2$ anlayser -Telaire 7001.

## Observations

The present study reveals the existence of 66 families of Dicotyledons and 15 families of Monocotyledons. More than 300 plant species (Table-1) were observed in wild habitation and about 50 ornamental species are growing in the gardens. Major plant families includes Acanthaceae, Amaranthaceae, Asteraceae, Ceasalpinaceae, Cucurbitaceae, Cyperaceae, Euphorbiaceae, Fabaceae, Graminae, Lamiaceae, Malvaceae and Mimosaceae.

Commonly found tree species and lianas are *Leucaena leucocephala, Azadirachta indica, Annona squamosa, Albizzia odoratisima, Butea monosperma, Butea superba, Mimosa hamata, Eugenia jambolana, Dalbergia sissoo, Santalum album Delonix regia* and *Dendrocalamus strictus*. Major shrubs and climbers of the study area are *Borassus flabellifer, Ziziphus jujuba, Lantana camara, Smilax prolifira, Tinospora cofdifolia, Bauhinia vahlii, Cryptostegia grandiflora, Securinega virosa.*

Herbs occupied 49% of the total plant species (Fig. 1), *Tridax procumbens, Oplesmus, Cassia tora, Cyperus corymbosa, Euphorbia hirta, Parthenium hysterophorus, Sida acuta, Triumfetta rhomboidea* are the major species of ground

**Table-1: List of plants in the NEERI campus**

| | Family name | Botanical Name |
|---|---|---|
| 1 | Acanthaceae | *Andrographis paniculata* |
| 2 | Acanthaceae | *Barleria cristata* |
| 3 | Acanthaceae | *Lepidagathis cristata* |
| 4 | Acanthaceae | *Peristrophe bicalyculata* |
| 5 | Acanthaceae | *Rostellularia diffusa* |
| 6 | Acanthaceae | *Rungia repens* |
| 7 | Acanthaceae | *Blepharis repens* |
| 8 | Acanthaceae | *Gentelbua urens* |
| 9 | Agavaceae | *Agave americana* |
| 10 | Agavaceae | *Furcraea foetida* |
| 11 | Agavaceae | *Sensevieria zeylanica* |
| 12 | Alismataceae | *Sagittaria tifolia* |
| 13 | Amaranthaceae | *Achyranthus aspera* |
| 14 | Amaranthaceae | *Aerva lanata* |
| 15 | Amaranthaceae | *Alternanthera sessilis* |
| 16 | Amaranthaceae | *Alternanthera pumilus* |
| 17 | Amaranthaceae | *Amaranthus blitum* |
| 18 | Amaranthaceae | *Amaranthus spinosus* |
| 19 | Anacardiaceae | *Buchanania lanzan* |
| 20 | Anacardiaceae | *Semicarpus anacardium* |
| 21 | Anacardiaceae | *Lannae coromandelina* |
| 22 | Anacardiaceae | *Mangifera indica* |
| 23 | Annonaceae | *Annona squamosa* |
| 24 | Annonaceae | *Annona reticulata* |
| 25 | Annonaceae | *Polyalthia longifolia* |
| 26 | Apocynaceae | *Alstonia scholaris* |
| 27 | Apocynaceae | *Nerium indicum* |
| 28 | Apocynaceae | *Raulwofia tetraphylla* |
| 29 | Apocynaceae | *Cataranthus rosea* |
| 30 | Aponogetonaceae | *Aponogeton natans* |

| | | |
|---|---|---|
| 31 | Araceae | *Colocasia esculenta* |
| 32 | Araceae | *Pistia stratiotes* |
| 33 | Araceae | *Amorphophallus campanulatus* |
| 34 | Araceae | *Lemna paucicostata* |
| 35 | Araceae | *Wolfia arrhiza* |
| 36 | Asclepidaceae | *Calotropis procera* |
| 37 | Asclepidaceae | *Calotropis gigantia* |
| 38 | Asclepidaceae | *Pergularia daimia* |
| 39 | Bignonaceae | *Spathodea companulata* |
| 40 | Bombacaceae | *Bombax ceiba* |
| 41 | Burseraceae | *Boswelia serrata* |
| 42 | Capparaceae | *Capparis zeylanica* |
| 43 | Capparaceae | *Capparis grandis* |
| 44 | Caricaceae | *Carica papaya* |
| 45 | Cassythaceae | *Cassitha filiformis* |
| 46 | Casuarinaceae | *Casuarina equisetifoia* |
| 47 | Ceasalpinaceae | *Bauhinia racemosa* |
| 48 | Ceasalpinaceae | *Bauhinia variegata* |
| 49 | Ceasalpinaceae | *Bauhinia vahlii* |
| 50 | Ceasalpinaceae | *Cassia absus* |
| 51 | Ceasalpinaceae | *Cassia fistula* |
| 52 | Ceasalpinaceae | *Cassia tora* |
| 53 | Ceasalpinaceae | *Caesalpina pulcherrima* |
| 54 | Ceasalpinaceae | *Cassia occidentalis* |
| 55 | Ceasalpinaceae | *Cassia siamea* |
| 56 | Ceasalpinaceae | *Cassia mimosoides* |
| 57 | Ceasalpinaceae | *Cassia pumila* |
| 58 | Ceasalpinaceae | *Delonix regia* |
| 59 | Ceasalpinaceae | *Tamarindus indica* |
| 60 | Ceasalpinaceae | *Parkinsonia aculeata* |
| 61 | Ceasalpinaceae | *Peltophorum pterocarpum* |
| 62 | Celestraceae | *Meytanus emarginata* |
| 63 | Ceratophyllaceae | *Ceratophylum demersum* |
| 64 | Chenopodiaceae | *Antigonon leptopus* |
| 65 | Cleomaceae | *Cleome viscosa* |

| | | |
|---|---|---|
| 66 | Cleomaceae | *Cleome monophylla* |
| 67 | Cleomaceae | *Cleome gynandra* |
| 68 | Cochlospermaceae | *Cochlospermum religiosum* |
| 69 | Combretaceae | *Anogiessus latifolia* |
| 70 | Combretaceae | *Combretum ovilifolium* |
| 71 | Combretaceae | *Terminalia alata* |
| 72 | Combretaceae | *Terminalia arjuna* |
| 73 | Combretaceae | *Terminalia bellirica* |
| 74 | Combretaceae | *Terminalia catappa* |
| 75 | Commelinaceae | *Commelina diffusa* |
| 76 | Commelinaceae | *Murdannia nudiflora* |
| 77 | Commelinaceae | *Rhoeo spathacea* |
| 78 | Commelinaceae | *Cyanotis cristata* |
| 79 | Commelinaceae | *Cynotis fasciculata* |
| 80 | Compositeae | *Bidens biternata* |
| 81 | Compositeae | *Tridax procumbens* |
| 82 | Compositeae | *Vernonia cinerea* |
| 83 | Compositeae | *Xanthium strumarium* |
| 84 | Compositeae | *Sonchus asper* |
| 85 | Compositeae | *Sphaeranthus indicus* |
| 86 | Compositeae | *Synedrella nodiflora* |
| 87 | Compositeae | *Blumea eiantha* |
| 88 | Compositeae | *Blumea membranacea* |
| 89 | Compositeae | *Launaea procumbens* |
| 90 | Compositeae | *Ageratum conyzoides* |
| 91 | Compositeae | *Glossogyne pinnatifida* |
| 92 | Compositeae | *Lagascea mollis* |
| 93 | Compositeae | *Parthenium hysterophorus* |
| 94 | Convolvulaceae | *Cuscuta reflex* |
| 95 | Convolvulaceae | *Evolvulus alsinoides* |
| 96 | Convolvulaceae | *Evolvulus nummularius* |
| 97 | Cucurbitaceae | *Coccinia indica* |
| 98 | Cucurbitaceae | *Lagenaria siceraaria* |
| 99 | Cucurbitaceae | *Momordica charantia* |
| 100 | Cucurbitaceae | *Tricosanthes anguina* |

| | | |
|---|---|---|
| 101 | Cucurbitaceae | *Momordica dioica* |
| 102 | Cucurbitaceae | *Cucumis melo* |
| 103 | Cucurbitaceae | *Cucumis sativus* |
| 104 | Cyperaceae | *Cynodon dactylon* |
| 105 | Cyperaceae | *Bulbostylis barbata* |
| 106 | Cyperaceae | *Cyperus corymbosa* |
| 107 | Cyperaceae | *Eleocharis acutangula* |
| 108 | Cyperaceae | *Fimbrisylis aphylla* |
| 109 | Cyperaceae | *Kyllinga brevifolia* |
| 110 | Cyperaceae | *Cyperus clarkei* |
| 111 | Cyperaceae | *Cyperus compressus* |
| 112 | Cyperaceae | *Cyperus nutans* |
| 113 | Cyperaceae | *Fimbrisylis ovata* |
| 114 | Cyperaceae | *Fimbristylis dichotoma* |
| 115 | Cyperaceae | *Fimbrisylis tetragona* |
| 116 | Dioscoreaceae | *Dioscorea bulbifera* |
| 117 | Ebenaceae | *Diospyros melanoxylon* |
| 118 | Eriocaulaceae | *Eriocaulon eleanorae* |
| 119 | Eriocaulaceae | *Eriocaulon quinquangulare* |
| 120 | Euphorbiaceae | *Kirganelia reticulata* |
| 121 | Euphorbiaceae | *Phyllanthus emblica* |
| 122 | Euphorbiaceae | *Phyllanthus virgatus* |
| 123 | Euphorbiaceae | *Securinega virosa* |
| 124 | Euphorbiaceae | *Euphorbia hirta* |
| 125 | Euphorbiaceae | *Euphorbia thymifolia* |
| 126 | Euphorbiaceae | *Euphorbia geniculata* |
| 127 | Euphorbiaceae | *Euphorbia rosea* |
| 128 | Euphorbiaceae | *Ricinus communis* |
| 129 | Euphorbiaceae | *Jatropha gossypifolia* |
| 130 | Euphorbiaceae | *Phyllanthus maderaspatensis* |
| 131 | Euphorbiaceae | *Pedilanthus tithimaloides* |
| 132 | Euphorbiaceae | *Tragia cannabina* |
| 133 | Euphorbiaceae | *Euphorbia tirucalli* |
| 134 | Euphorbiaceae | *Jatropha curcus* |
| 135 | Euphorbiaceae | *Phyllanthus amarus* |

| | | |
|---|---|---|
| 136 | Euphorbiaceae | *Phyllanthus urinaria* |
| 137 | Euphorbiaceae | *Acalypha indica* |
| 138 | Euphorbiaceae | *Chrozophora prostrata* |
| 139 | Euphorbiaceae | *Croton bonplandianum* |
| 140 | Fabaceae | *Alysicarpus monilifer* |
| 141 | Fabaceae | *Alysicarpus rugosus* |
| 142 | Fabaceae | *Alysicarpus vaginalis* |
| 143 | Fabaceae | *Abrus precatorius* |
| 144 | Fabaceae | *Butea monosperma* |
| 145 | Fabaceae | *Butea superba* |
| 146 | Fabaceae | *Aeschynomene aspera* |
| 147 | Fabaceae | *Alysicarpus longifolius* |
| 148 | Fabaceae | *Alysicarppus monilifer* |
| 149 | Fabaceae | *Alysicarpus bupleurifolius* |
| 150 | Fabaceae | *Alysicarpus longifolius* |
| 151 | Fabaceae | *Alysicarpus tetragonolobus* |
| 152 | Fabaceae | *Atylosia scarabaeoides* |
| 153 | Fabaceae | *Atylosia platycarpa* |
| 154 | Fabaceae | *Clitoria terantea* |
| 155 | Fabaceae | *Crotalaria hirsuta* |
| 156 | Fabaceae | *Crotalaria ramosissima* |
| 157 | Fabaceae | *Crotalaria medicaginea* |
| 158 | Fabaceae | *Goneogyna hirta* |
| 159 | Fabaceae | *Indigofera cordifolia* |
| 160 | Fabaceae | *Indigofera linifolia* |
| 161 | Fabaceae | *Medicago sativa* |
| 162 | Fabaceae | *Dalbergia paniculata* |
| 163 | Fabaceae | *Melilotus indica* |
| 164 | Fabaceae | *Mucuna pruriens* |
| 165 | Fabaceae | *Dalbergia sisso* |
| 166 | Fabaceae | *Desmodium dichotomum* |
| 167 | Fabaceae | *Desmodium gangaticum* |
| 168 | Fabaceae | *Desmodium triflorum* |
| 169 | Fabaceae | *Rhynchosia capitata* |
| 170 | Fabaceae | *Erythrina variegata* |

| | | |
|---|---|---|
| 171 | Fabaceae | *Indigofera linifolia* |
| 172 | Fabaceae | *Indigofera cassioides* |
| 173 | Fabaceae | *Indigofera cordifolia* |
| 174 | Fabaceae | *Phaseolus vulgaris* |
| 175 | Fabaceae | *Pongamia pinnata* |
| 176 | Fabaceae | *Pterocarpus marsupium* |
| 177 | Fabaceae | *Tephrosia purpurea* |
| 178 | Graminae | *Andropogon pumilus* |
| 179 | Graminae | *Apluda mutica* |
| 180 | Graminae | *Aristida funiculata* |
| 181 | Graminae | *Arundinella setosa* |
| 182 | Graminae | *Dendrocalmus strictus* |
| 183 | Graminae | *Chrysopogon fulvus* |
| 184 | Graminae | *Cymbopogon martinii* |
| 185 | Graminae | *Dichanthium annulatum* |
| 186 | Graminae | *Dichanthium aristatum* |
| 187 | Graminae | *Eragrostis tenella* |
| 188 | Graminae | *Heteropogon contortus* |
| 189 | Graminae | *Iseilema laxum* |
| 190 | Graminae | *Potemogeton crispus* |
| 191 | Graminae | *Aristida setacea* |
| 192 | Graminae | *Aristida depressa* |
| 193 | Graminae | *Chloris virgatus* |
| 194 | Graminae | *Coix aquatica* |
| 195 | Graminae | *Dactyloctenium aegyptium* |
| 196 | Graminae | *Desmostachya bipinnata* |
| 197 | Graminae | *Digitaria cillaris* |
| 198 | Graminae | *Eragrostis cilliaris* |
| 199 | Graminae | *Eragrostis aspera* |
| 200 | Graminae | *Oplismenus burmannii* |
| 201 | Graminae | *Setaria vericillata* |
| 202 | Graminae | *Vetiveria zizanioides* |
| 203 | Graminae | *Urochloa panicoides* |
| 204 | Hydrocharitaceae | *Hydrilla verticillata* |
| 205 | Lamiaceae | *Hyptis suaveolens* |

| | | |
|---|---|---|
| 206 | Lamiaceae | *Leucas biflora* |
| 207 | Lamiaceae | *Leonitis nepatifolia* |
| 208 | Lamiaceae | *Ocimum sanctum* |
| 209 | Lamiaceae | *Ocimum americanum* |
| 210 | Lamiaceae | *Adathoda vasica* |
| 211 | Lamiaceae | *Leucas cephalotus* |
| 212 | Lamiaceae | *Leucas stricta* |
| 213 | Lecythidaceae | *Careya arborea* |
| 214 | Leeacea | *Leea crispa* |
| 215 | Liliaceae | *Asparagus racemosus* |
| 216 | Liliaceae | *Scilla indica* |
| 217 | Lytheraceae | *Lagerstroemia parviflora* |
| 218 | Lytheraceae | *Lagerstroemia indica* |
| 219 | Malvaceae | *Malvastrum tricuspidata* |
| 220 | Malvaceae | *Sida acuta* |
| 221 | Malvaceae | *Abutilon hirtum* |
| 222 | Malvaceae | *Hibiscus cannabinus* |
| 223 | Malvaceae | *Malachra capitata* |
| 224 | Malvaceae | *Pavonia zeylanica* |
| 225 | Malvaceae | *Abelmoschus manihot* |
| 226 | Malvaceae | *Abutilon indicum* |
| 227 | Malvaceae | *Sida cordata* |
| 228 | Meliaceae | *Azadiravhta indica* |
| 229 | Meliaceae | *Melia azadirachta* |
| 230 | Meliaceae | *Soymida febrigua* |
| 231 | Menispermaceae | *Cocculus hirsutus* |
| 232 | Menispermaceae | *Tinospora cofdifolia* |
| 233 | Menispermaceae | *Cissampelos pareira* |
| 234 | Mimosaceae | *Pithecellobium dulce* |
| 235 | Mimosaceae | *Mimosa hamata* |
| 236 | Mimosaceae | *Prosopis juliflora* |
| 237 | Mimosaceae | *Acacia catechu* |
| 238 | Mimosaceae | *Acacia nilotica* |
| 239 | Mimosaceae | *Mimosa pudica* |
| 240 | Mimosaceae | *Acacia auriculiformis* |

| | | |
|---|---|---|
| 241 | Mimosaceae | *Acacia leucocephala* |
| 242 | Mimosaceae | *Albizzia lebbeck* |
| 243 | Mimosaceae | *Albizzia odoratisima* |
| 244 | Mimosaceae | *Albizzia procera* |
| 245 | Mimosaceae | *Leucaena leucocephala* |
| 246 | Mimosaceae | *Mimosa hamata* |
| 247 | Moraceae | *Ficus benghalensis* |
| 248 | Moraceae | *Ficus racemosa* |
| 249 | Moraceae | *Ficus religiosa* |
| 250 | Moraceae | *Ficus hispida* |
| 251 | Moringaceae | *Moringa critifolia* |
| 252 | Myrtaceae | *Psidium guajava* |
| 253 | Myrtceae | *Eugenia jambolana* |
| 254 | Nyctaginaceae | *Boerhavia diffusa* |
| 255 | Oxalidaceae | *Oxalis corniculata* |
| 256 | Oxalidaceae | *Bryophytum sensivatum* |
| 257 | Palmae | *Borassus flabellifer* |
| 258 | Palmae | *Phoenix sylvestris* |
| 259 | Palmae | *Roylstegia regia* |
| 260 | Papaveraceae | *Argemone mexicana* |
| 261 | Periplocaceae | *Cryptolepis buchanani* |
| 262 | Periplocaceae | *Cryptostegia grandiflora* |
| 263 | Periplocaceae | *Hemidesmus indicus* |
| 264 | Polygalaceae | *Polygala arvensis* |
| 265 | Polygalaceae | *Polygala elongata* |
| 266 | Pontederiaceae | *Eichhornia crassipes* |
| 267 | Portulaceae | *Portulaca oleraceae* |
| 268 | Portulaceae | *Portulaca quadrifida* |
| 269 | Potamogetonaceae | *Potamogeton crispus* |
| 270 | Rananculaceae | *Cleome viscosa* |
| 271 | Rhamnaceae | *Ventilago denticulata* |
| 272 | Rhamnaceae | *Ziziphus jujuba* |
| 273 | Rhamnaceae | *Zizphus glaberrima* |
| 274 | Rhamnaceae | *Ziziphus oneoplia* |
| 275 | Rubiaceae | *Borreria articularis* |

| | | |
|---|---|---|
| 276 | Rubiaceae | *Oldenlandia corymbosa* |
| 277 | Rubiaceae | *Borreria diffusa* |
| 278 | Rubiaceae | *Mitragyna parvifolia* |
| 279 | Rubiaceae | *Anthocephalus chinensis* |
| 280 | Rutaceae | *Chloroxylon swietenia* |
| 281 | Rutaceae | *Aegle marmelos* |
| 282 | Rutaceae | *Murraya koenigii* |
| 283 | Santalaceae | *Santalum album* |
| 284 | Sapindaceae | *Cardiospermum helicacabum* |
| 285 | Sapindaceae | *Sapindus emarginatus* |
| 286 | Sapindaceae | *Schleichera oleosa* |
| 287 | Sapotaceae | *Madhuca longifolia* |
| 288 | Simarubiaceae | *Ailanthus excelsa* |
| 289 | Smilacaceae | *Smilax prolifira* |
| 290 | Solanaceae | *Datura metal* |
| 291 | Solanaceae | *Solanum indicum* |
| 292 | Sterculiaceae | *Helicteres isora* |
| 293 | Tiliaceae | *Grewia hirsuta* |
| 294 | Tiliaceae | *Corchorus aestuans* |
| 295 | Tiliaceae | *Grewia tilifolia* |
| 296 | Tiliaceae | *Triumfetta rhomboidea* |
| 297 | Typhaceae | *Typha angustata* |
| 298 | Verbenaceae | *Lantana camara* |
| 299 | Verbenaceae | *Tectona grandis* |
| 300 | Verbenaceae | *Vitex nigunda* |
| 301 | Violaceae | *Hybanthus enneaspermus* |
| 302 | Vitaceae | *Cissus quadrangularis* |
| 303 | Zygophyllaceae | *Tribulus terrestris* |
| 304 | Zygophyllaceae | *Guaicacum officinale* |

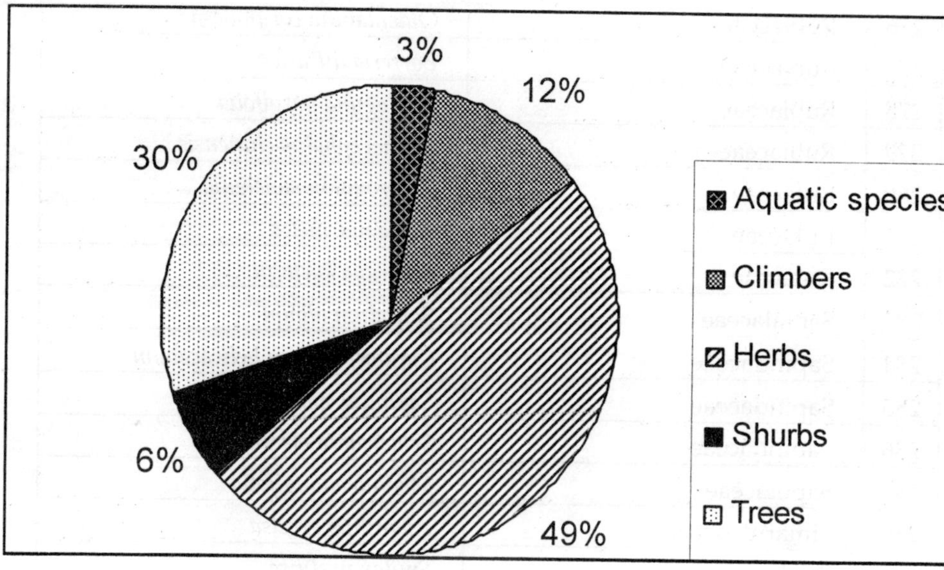


**Fig. 1. Floral biodiversity of different habits**

flora. *Ceratophylum demersum* and *Ipoemea aquatica* are the dominant aquatic species growing in small ponds.

*Leucena leucocephala* is the dominant species in all forms viz., trees, saplings and seedlings. This species showed 100% frequency and 0.41 Importance Value Index. Density analysis also shows that *Leucena leucocephala* is dense. Within the planted species, *Roystegia regia* and *Eugenia jambolina* distributed along the internal roads. Saplings of Sandal wood (*Santalum album)* were observed with 20% frequent and 0.014 IVI, meanwhile the trees (DBH<10) were found with less in number. High biomass plants (more than 50 DBH) were observed all around the area, these plants are *Leucena leucocephala, Eugenia jambolina, Azadirachta indica, Tamerindus indicus* etc.

The vast difference in $CO_2$ concentration was observed in different regions of NEERI campus. The range of ground $CO_2$ concentration observed during day time was 410 to 475 ppm, whereas aerial $CO_2$ concentration range was 325 to 395 ppm (Fig. 2). Higher concentrations of $CO_2$ at ground level may be because of root respiration and microbial respiration. Atmospheric $CO_2$ concentration of forest zone when compared with road side/main building zone was in the range 400 to 600 ppm. It showed a difference of around 200 ppm $CO_2$ concentration. This indicates that the forest zone removed about 200 ppm of $CO_2$ through photosynthetic activity. Soil humus layer has occupied 3cm to 10cm of soil profile. Relative humidity is varied from 60% to 80% (Rainy season) and temperature

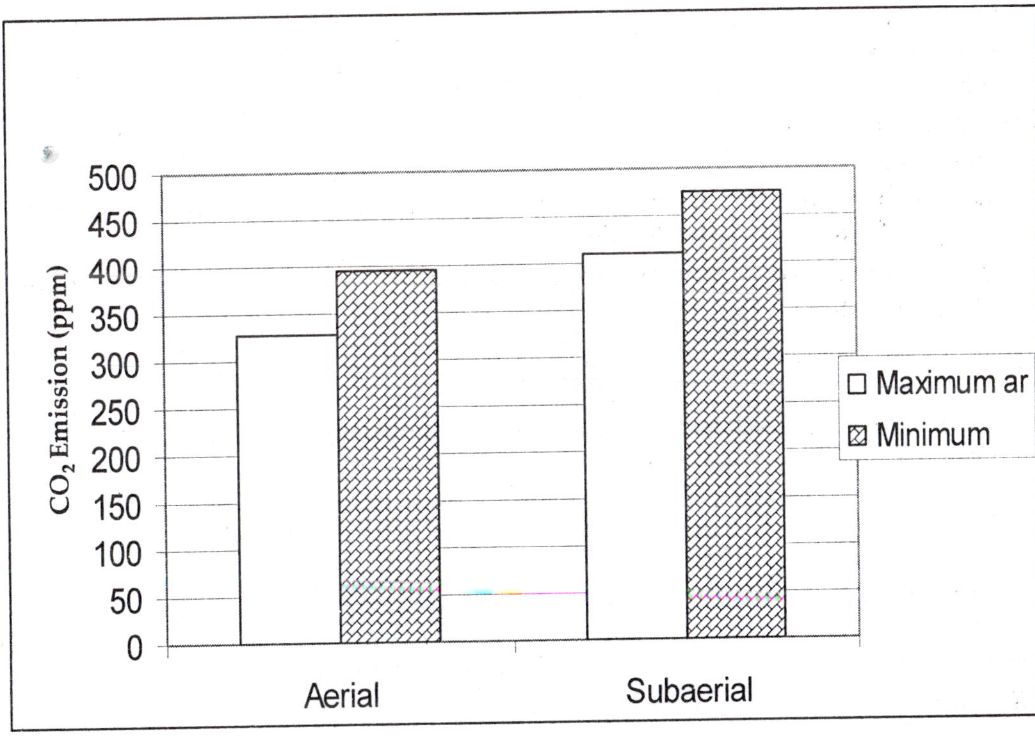


**Fig. 2. Emission of $CO_2$ in different regions**

ranged from 26°C to 33°C (Day time). Some Mushrooms (Fig. 3), Pteridophytes (*Selagenella, Pteris* etc.) and Bryophytes (*Porella, Anthoceros* and other mosses) growing abundanty, are the best indicators of high relative humidity.

**Fig. 3. Mushroom growth at high humid area**

GIS image of campus two pattern of bands, a very dark region and a medium dark region. Field results showed near about same concentration of dominance in both regions (Simpson index- 0.02), while species richness differs around very dark and medium dark regions (Shannon-wiener index: 3.62 and 3.55 respectively). Very dark bands in GIS image is due to 100% crown cover by tree canopy (higher chlorophyll concentration), while medium dark band region is occupied by 70-80% canopy. The above study showed that NEERI campus is rich in flora with high frequency, density and species richness which is comparable to natural forest.

## References

Amthor J.S. (1995). Terrestrial higher-plant responses to increasing atmospheric $CO_2$ in relation to the global carbon cycle. *Global Change Biology* 1: 243-274.

Amthor J.S. (1997). Plant respiratory responses to elevated $CO_2$ partial pressure. In: Allen LHJ, Olszyck DM, eds. *Adäances in Carbon Dioxide Eåects Research*. Madison, WI, USA: American Society of Agronomy Special Publication, 61, 35-77.

Curtis, J.T. (1959). *The vegetation of Wisconsin. An ordination of Plant Communities,* Madison: Wisconsin Press.

Dowdeswell, E. (1995). Global Biodiversity Assessment, UNEP, CUP, UK.

Gifford R.M. (1994). The global carbon cycle: a view point on the missing sink. Australian Journal of Plant Physiology 21: 1–15.

Schlesinger, W.H. (1997). Biogeochemistry: an analysis of global change, 2nd edn. San Diego, CA: Academic Press.

IFRI, (2002). *International Forest Research and Institution (IFRI) Field Manual (version* 10.5). Bloomington: Indiana University, Workshop in Political Theory and Policy Analysis.

Misra, R. (1968). *Ecology Work Book*: Oxford and IBH Publ.Co. New Delhi.

Phillips, E.A. (1959). *Methods of Vegetation study.* A Holt-dryden Book, Henry Holt and Co.Inc. New York.

Shannon, C.E. and W. Wiener (1963). *The Mathematical Theory of Communication*. Univ. Illinois Press, Urbana.

Simpson, E.H. (1949). Measurement of Diversity. *Nature,* 163-188.

# 4

# Monitoring Colour and COD Removal Capacity of Soil and Assessment of Growth Performance of Crop Grown with Pulp and Paper Mill Wastewater

*P. R. Thawale, Asha Juwarkar,*
*S.K. Singh and A.V. Kulkarni*

Land application is considered to be most acceptable and cost effective method of pulp and paper mill wastewater management as soil provides physical, chemical, and biological treatment to wastewater and removes 100% of the colour (Anderson, 1979) besides helping to grow the crops (Yakunshenko *et al.* 1971; Lumbely, 1977; Hansen *et al.* 1980; Juwarkar and Subrahamanyam, 1987 and NEERI-UNEP Report, 1994). Presently, in India there are 305 paper mills with an installed capacity of $1.9 \times 10^6$ tonnes per annum (TPA). Paper mill consumes large quantity of water which varies from 300-450 $m^3 t^{-1}$, of which 220-350 $m^3$ of water per tonne of paper produced is discharged as wastewater. The wastewater comprises high suspended solids (SS - 290-1115 mg $L^{-1}$), Biochemical Oxygen Demand (BOD - 100-1070 mg $L^{-1}$), Chemical Oxygen Demand (COD - 600-4760 mg $L^{-1}$) and dark brown colour due to lignin (82-700 mg $L^{-1}$). These pollutants create problems in disposal of wastewater in water bodies. Conventional wastewater treatment methods reduces SS, BOD, and to some extent COD but colour remains more or less unchanged, because of no biodegradibility of lignin. Available treatment methods for colour removal, i.e., use of adsorbents (activated carbon), coagulants (lime and alum) and as bleaching agent (ozone), are not so effective. Therefore, the land application and disposal of pulp and paper mill wastewater through crop irrigation for treatment of colour and COD removal were studied in detail. This paper highlights the characterization of pulp and paper mill wastewater, COD and colour removal capacity of soil, changes in percolate quality, soil property and crop growth.

Different approaches may be used to study soil, wastewater and plant interaction and retentive batch experiments are frequently used to establish retention equilibrium under lab condition (Kogelknabner *et al.*1991, Reed and Cline *et al.*1994), Nevertheless, because of the mixing and the high solutions to soil ratios used, these experiments typically fail to reproduce well field conditions. These can be approximated more closely by lysimeter experiments that additionally allow the evaluation of some relevant transport parameters. Also the partition coefficient of isotherm parameters may be obtained from corresponding break-through curve ( Qualls and Haines *et al.* 1992, Burgisser *et al.* 1993). In this sense this work deals with the study of soil wastewater and retention by soils with lysimeter . In subsequent paper, a model is applied on result to study the retention phenomenon as well as the solid-liquid contact efficiency in system.

## Materials and Methods

### *Study Area*

The laboratory studies were carried out at National Environmental Engineering Research Institute (NEERI), Nagpur and the field lysimeter unit was established at an integrated large pulp and paper mill in Madhya Pradesh (India) which lies on 21-22° North latitude and 75-76° East longitude. The study area comes under hot semi-arid ecoregion and the climate of the area is characterized by hot and dry summers and mild winters. The annual precipitation, ranging from 600 to 900 mm, covers 40 to 50% of annual potential evapotranspirative (PET) demand of 1600 to 2000 mm, resulting in gross annual deficit of 800 to 1200mm.

### *Wastewater*

The mill produces 150 tonnes of paper per day employing kraft process and generates about 220-320 $m^3$ of wastewater per tonne of paper manufactured. The wastewater generated from various sections in the mill is collected and treated in aerated lagoons before its discharge into the river. The combined wastewater after secondary treatment was used in the study and plain water (river water) was employed for control treatment. The characteristics of wastewater and plain water are presented in Table I.

### Crops

Jowar, wheat and moong crops were selected and tested for germination study and response to wastewater application. Crops namely jowar (*Sorghum vulgare*), wheat (*Triticum aestium*) and moong (*Vigna radiata*) were grown in each lysimeter respectively for three years. The details given in Table II:

**Table 1**
**Characteristics of wastewater and plain water used in lysimeter study**

| Parameters | Wastewater | | | Plain water | | |
|---|---|---|---|---|---|---|
| | Minimum | Maximum | Average | Minimum | Maximum | Average |
| pH | 7.2 | 8.2 | 7.7 | 7.5 | 8.4 | 7.9 |
| ECw (mS $cm^{-1}$) | 1.6 | 2.6 | 2.1 | 0.2 | 0.4 | 0.3 |
| $Ca^{++}$ (meq $L^{-1}$) | 7.2 | 9.9 | 8.5 | 0.65 | 1.56 | 1.10 |
| $Mg^{++}$ (meq $L^{-1}$) | 3.0 | 4.8 | 3.9 | 0.24 | 0.53 | 0.38 |
| $Na^{+}$ (meq $L^{-1}$) | 5.7 | 11.0 | 7.5 | 0.32 | 1.97 | 1.14 |
| $K^{+}$ (meq $L^{-1}$) | 0.4 | 1.0 | 0.7 | 0.02 | 0.35 | 0.18 |
| $HCO_3^{-}$ (meq $L^{-1}$) | 6.4 | 7.2 | 6.8 | 0.26 | 2.01 | 1.13 |
| $Cl^{-}$ (meq $L^{-1}$) | 13.0 | 16.0 | 14.5 | 0.58 | 1.74 | 1.16 |
| $SO_4^{--}$ (meq $L^{-1}$) | 1.1 | 2.9 | 2.0 | 0.03 | 0.65 | 0.34 |
| SAR | 3.6 | 4.6 | 4.1 | 0.56 | 2.19 | 1.37 |
| COD (mg $L^{-1}$) | 295.0 | 485.0 | 380.0 | 5.4 | 7.3 | 6.3 |
| Lignin (mg$L^{-1}$) | 40.0 | 52.0 | 46.0 | - | - | - |
| Colour (Pt-Co units) | 560.0 | 680.0 | 620.0 | - | - | - |

SAR : Sodium Adsorption Ratio.
COD : Chemical Oxygen Demand.

**Table II**
**Details of crops grown in lysimeters**

| Name of the Crops | Variety | Duration of Crop | Method of Sowing | Fertilizer Dose (kg $ha^{-1}$) | | |
|---|---|---|---|---|---|---|
| | | | | N | $P_2O_5$ | $K_2O$ |
| Jowar | CSH-9 | June-September | Dibbling | 120 | 60 | 60 |
| Wheat | HDM-1553 | November-February | Dibbling | 100 | 60 | 40 |
| Moong | S-8 | March-May | Dibbling | 25 | 50 | 25 |

### *Installation of Field Lysimeter*

Field lysimeters were prepared by excavating pit of size 5mx2mx2m lined with polyethylene sheet to avoid the seepage of water. Soils of varying physico-chemical characteristics were used in the field lysimeter study (Table III). Soils were collected from three different locations around the mill and bulk samples were collected separately from different layers. The soil collected from each location was filled into duplicate lysimeters. During filling care was taken to make soil layers in lysimeters resemble those in the field conditions. Bulk density of soil in each layer was adjusted to field values by compaction. Lysimeters I, II and III were filled with sandy loam, clay and loamy sand respectively. Channels were provided length wise by placing half cut pipe to collect percolate from lysimeters in central percolate collection chamber. In all, six lysimeters were installed. One set of 3 lysimeters was used for wastewater application while other for plain water application, which served as control. Schematic of a field lysimeter is depicted in Figure 1.

### *Operation of Field Lysimeter*

Crops were irrigated with wastewater and plain water @ 5 cm/hectare i.e. 500 litre/lysimeter/irrigation and allowed to grow till maturity. Irrigation frequency was adjusted to crop water demand. Post harvest soil samples were collected in the different cropping seasons and tested for changes in physico-chemical properties. The amount of percolates, collected daily, was measured and analysed for chemical characteristics.

### *Methods of Analysis*

Wastewater and percolate samples were analysed as per Standard Methods for examination of water and wastewater, 17th ed. (APHA, 1989). The effect of effluent on germination of crop seeds of jowar, wheat and moong was assessed following standard procedure (IRST, 1966). Methods described by Piper (1996), Black (1965) and Jackson (1973) were employed for physico-chemical analysis of soil. Analysis of variance was carried out as described by Panse and Sukhatme (1967).

## Results and Discussion

### *Evaluation of Wastewater Quality for Irrigation*

In management of irrigated crop production "Scientific Management Modification Approach" plays important role in evaluation of suitability of water/

**Table III**
**Physico-chemical characteristics of soils used in field lysimeters**

| Parameters | Soil Depth, cm | Lysimeter I | Lysimeter II | Lysimeter III |
|---|---|---|---|---|
| Texture | 0-30 | Sandy loam | Clay | loamy Sand |
| Sand (%) | | 60.8 | 21.9 70.8 | |
| Silt (%) | | 23.2 | 30.0 17.2 | |
| Clay (%) | | 16.0 | 48.0 12.0 | |
| M.W.H.C. (%) | | 31.0 | 46.9 25.0 | |
| B.D. (g $cc^{-1}$) | | 1.29 | 1.41 1.30 | |
| pH | | 6.8 | 7.3 6.5 | |
| ECe (mS $cm^{-1}$) | | 0.2 | 0.6 | 0.2 |
| CEC (meq 100 $g^{-1}$) | | 12.8 | 40.8 | 9.8 |
| Texture | 30-90 | Sandy loam | Clay | Loamy sand |
| Sand (%) | | 63.5 | 28.0 70.8 | |
| Silt (%) | | 20.4 | 26.8 13.0 | |
| Clay (%) | | 16.0 | 45.2 16.2 | |
| M.W.H.C. (%) | | 30.0 | 43.8 23.8 | |
| B.D. (g $cc^{-1}$) | | 1.30 | 1.46 1.31 | |
| pH | | 6.7 | 7.2 | 6.4 |
| ECe (mS $cm^{-1}$) | | 0.3 | 0.5 | 0.2 |
| CEC (meq 100 $g^{-1}$) | | 9.8 | 38.4 11.1 | |
| Texture | 90-160 | Sandy loam | Clay | Loamy sand |
| Sand (%) | | 64.0 | 34.0 64.6 | |
| Silt (%) | | 21.0 | 26.0 20.2 | |
| Clay (%) | | 15.0 | 40.0 15.1 | |
| M.W.H.C. (%) | | 32.1 | 38.0 28.0 | |
| B.D. (g $cc^{-1}$) | | 1.31 | 1.48 1.33 | |
| pH | | 6.97.4 | 6.3 | |
| ECe (mS $cm^{-1}$) | | 0.5 | 0.3 | 0.2 |
| CEC (meq 100 $g^{-1}$) | | 10.8 | 30.1 12.8 | |

M.W.H.C. - Maximum water holding capacity ($M_{water}/M_{total}$ x 100)
B.D. - Bulk density
ECe - Electrical conductivity of saturation extract
CEC - Cation exchange capacity

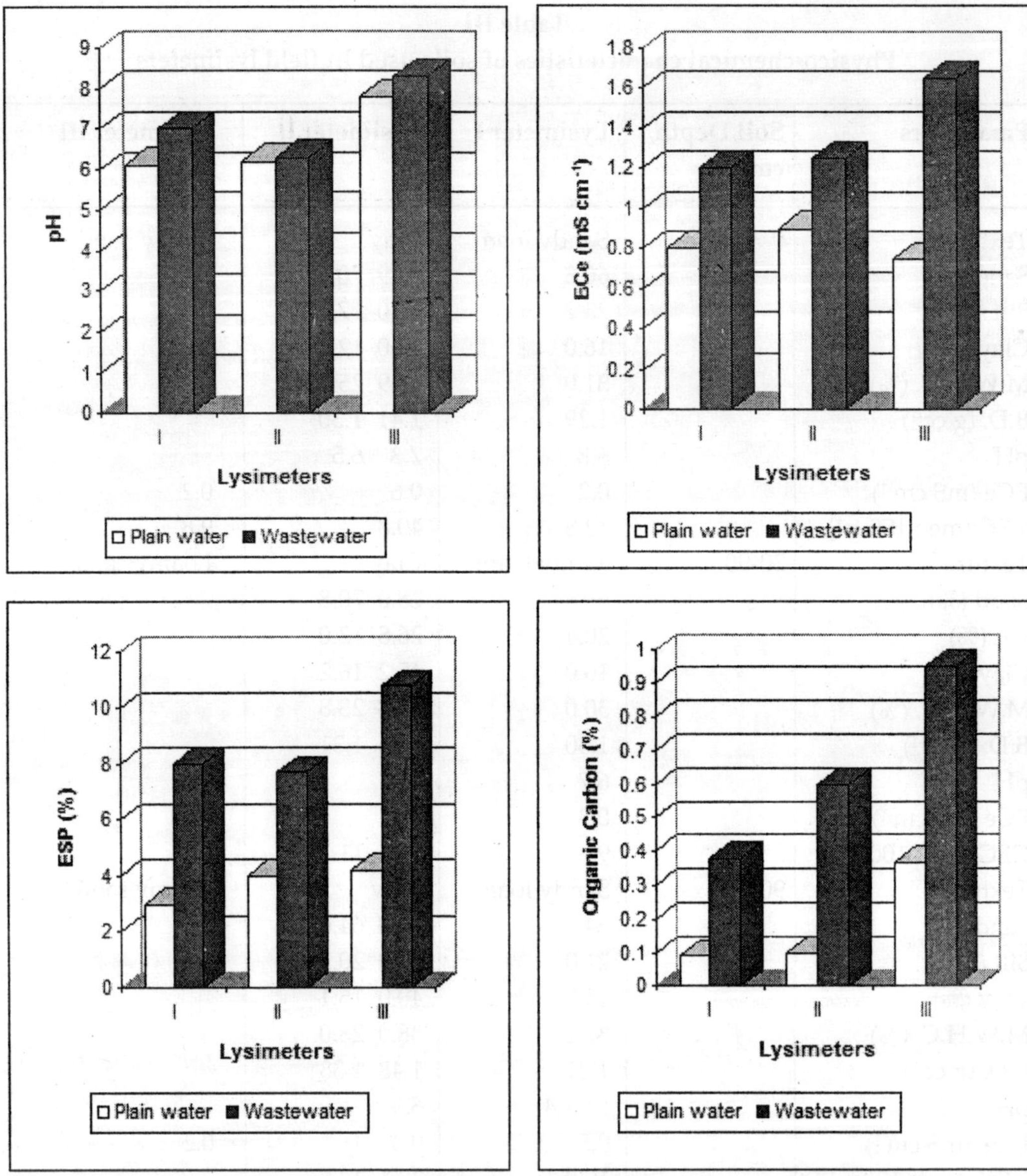


**Fig. 1. Chemical characteristics of soils irrigated with plain water and wastewater**

wastewater for crop irrigation. In general, irrigation quality of wastewater has been evaluated with respect to salinity, SAR and specific ion toxicity.

The pulp and paper mill wastewater categorised as "Moderate Restriction to Severe Restriction" on their use with respect to salinity and specific ion toxicity

(ECw, 1.6 - 2.6 mS $cm^{-1}$ and Chloride, 13 - 16 meq $L^{-1}$). Considering the combined effect of EC and sodium adsorption ratio (SAR) ascertain the impact on infiltration rate, the wastewater was observed to belong to category `No Restriction' (SAR, 3.6 - 4.6) as per Kandiah (1987). The plain water was used under control treatment grouped in $C_1S_1$ and $C_2S_2$ class as per USSL classification (USSL staff 1954) and in `No Restriction' class as per Ayers and Westcot (1976).

### *Removal of COD and Colour*

Percolate quality and quantity from field lysimeter was monitored for 3 years to evaluate COD and colour assimilation through soil-wastewater interaction and data are reported in Table IV. Results indicated that sandy loam and loamy sand soil in the lysimeter I and III percolate was 17 & 20% of irrigation respectively, while clay soil in lysimeter II it was only 15% because of higher retention of wastewater in clay soil which depends on maximum water holding capacity of the soil. The quantity of percolates collected from different lysimeters which received wastewater and plain water applications in Table V.

Quantitative assessment of percolate from soils received plain water did not show much variation in quantity compared to soil which received wastewater indicates that hydraulic conductivity of soil remained unaffected due to wastewater application. The proportions of water existing as percolate were calculated to include only the volume of applied irrigation water. Loamy sand and sandy loam soils respectively showed 96 & 97% removal of COD, while clay soil removed 99% of COD because of high CEC and presence of calcium carbonate in clay loam soil. The colour removal in both the cases was 100%. Similar findings were reported by Malo (1967), Narum *et al.* (1979) and Thawale *et al.* (1999).

### *Changes in Percolate Quality*

Percolate initially showed leaching of calcium and magnesium and retention of sodium, which gets stabilized subsequently. The results reported by Juwarkar and Subhramanayam (1987) and Narum *et al.* (1979) also support the phenomena of retention of sodium and leaching of calcium and magnesium in the soil as a result of soil-wastewater interaction. The magnitude of removal varies from soil to soil. Sandy loam and loamy sand soil reach equilibrium after 50 applications, while clay soil required 80 application to reach the equilibrium state because clay soil having higher percent of clay and CEC than sandy loam and loamy sand (Figures 2, 3 & 4). Results further indicates that chlorides were not retained in the soils as the chlorides in wastewater and percolate did not show any variation.

**Table IV**

**Volume of wastewater applied and percolate collected and COD removals from lysimeters**

| Parameters | Lysimeters | | |
|---|---|---|---|
| | I | II | III |
| Number of applications given* | 105 | 105 | 105 |
| Volume of wastewater applied per application, litres | 500 | 500 | 500 |
| Total volume of wastewater applied, litres | 52,500 | 52,500 | 52,500 |
| Total volume of percolate, litres | 9,082.6 | 7,876.2 | 10,712.0 |
| Total COD applied, gm/lysimeter (tonnes/hectare) | 18,007.5 (18.0) | 18,007.5 (18.0) | 18,007.5 (18.0) |
| Percent COD removed | 97 | 99 | 96 |
| Wastewater colour (Pt-Co units) | Dark brown (585) | Dark brown (585) | Dark brown (585) |
| Percolate colour | Colourless | Colourless | Colourless |

* Wastewater application was given during the period of three years for jowar, wheat and moong crop adjusted to their water requirement.

**Table V**
**Quantity of percolates collected from different lysimeters**

| Lysimeters | Quantity of Percolate Collected (%) | |
|---|---|---|
| | Wastewater Irrigated | Plain water Irrigated |
| I | 17.0 | 17.8 |
| II | 15.0 | 15.6 |
| III | 20.0 | 20.9 |

### *Changes in Physico-chemical Properties of Soil*

Application of wastewater resulted in some changes in soil quality. There was slight increase in maximum water holding capacity (MWHC) of soils irrigated with wastewater, which was due to the deposition of lignin (46.0 mg $L^{-1}$) contributed through wastewater irrigation. Bulk density of soils in all the lysimeters did not show much variation when irrigated with wastewater. Changes in bulk density and water holding capacity of soils irrigated with wastewater and plain water are reported in Table VI.

Such type of variation in soil physical characteristics might be attributed to the addition of organic materials in the soil through the effluent and interaction between the various constituents as suggested by Baver *et al.* (1972).

The pH of soil was raised by 0.6-1.3 units when irrigated with alkaline effluent. There was accumulation of soluble salts and increase in ECe of soils. Salt accumulation was more in clay soil as compared to sandy loam and loamy sand. Although there was increase in ESP (8.1 to 11.6) of soil, none of soils exhibited alkalinity problem. Rajanan and Oblisami (1979) also reported increase in pH, ECe, ESP of soil when irrigated with pulp and paper mill wastewater. Increase in ECe and ESP of soils which received wastewater was because of presence of high amount of soluble salts specially sodium as compared to calcium and magnesium. Monsoon rains contributes to reduce the impact of soil salinity, which is build up due to wastewater application (Figure 5). With the onset of monsoon rains, the accumulated salts are leached down beyond the rhizosphere, reducing soil ECe from 0.72 to 1.20 mS $cm^{-1}$ which is quite low as compared to ECe of soil in summer indicating effective leaching of soluble salts. Singh *et al.* (1994) also reported salt leaching in soil irrigated with saline water during monsoons. The organic carbon content increased appreciably in wastewater irrigated soil, due to presence of dissolved lignin in wastewater applied, which

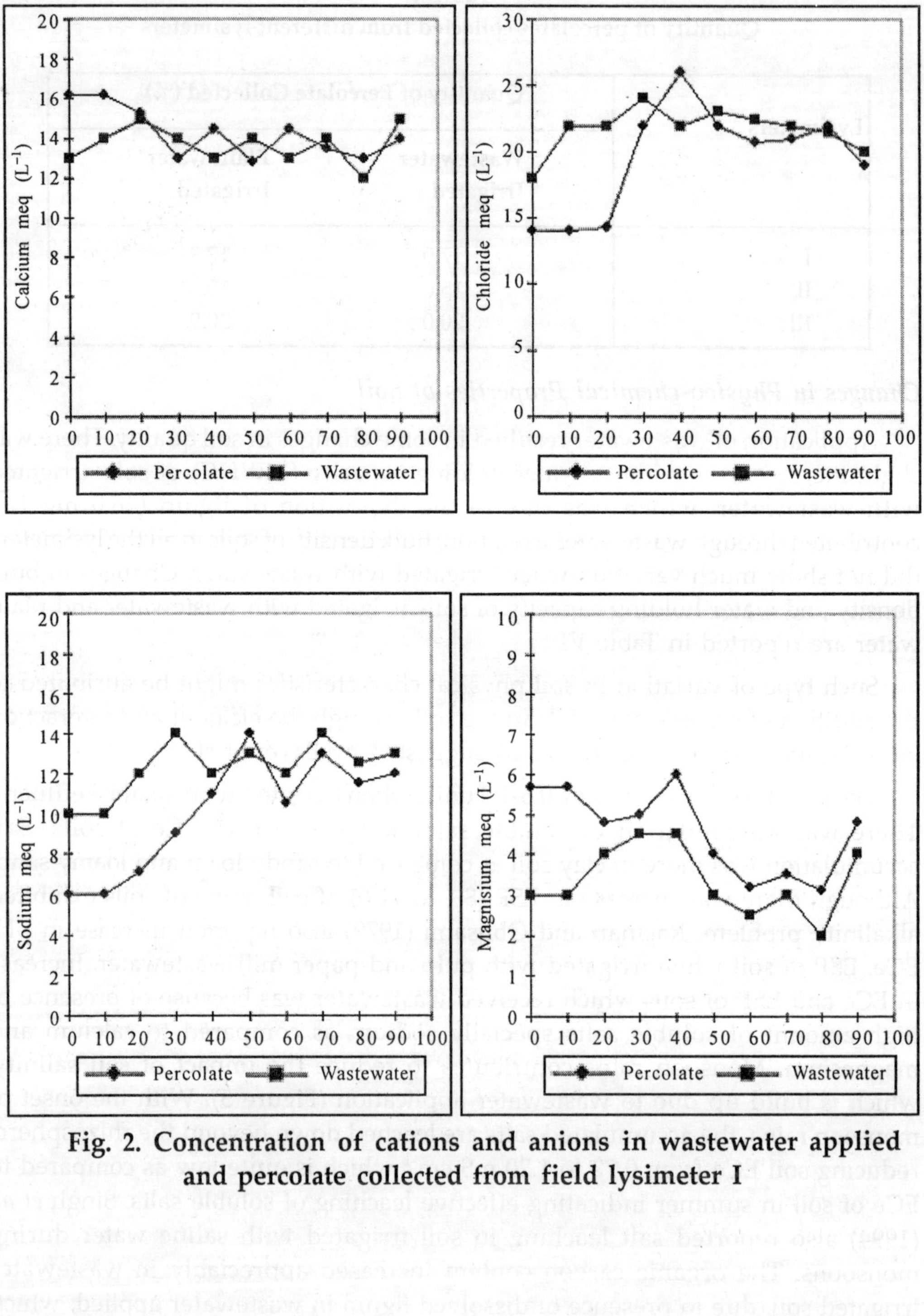


**Fig. 2. Concentration of cations and anions in wastewater applied and percolate collected from field lysimeter I**

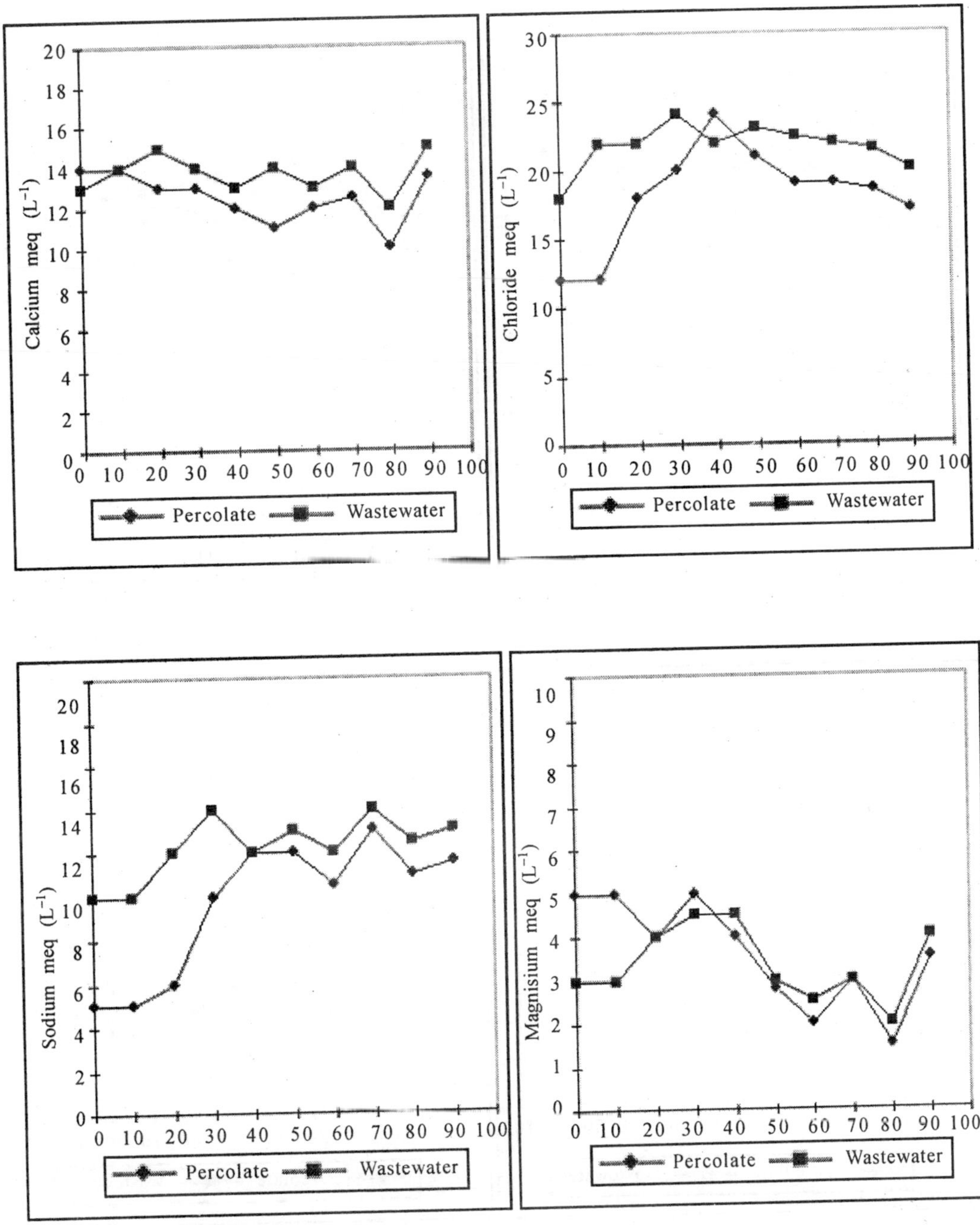


**Fig. 3. Concentration of cations and anions in wastewater applied and percolate collected from field lysimeter II**

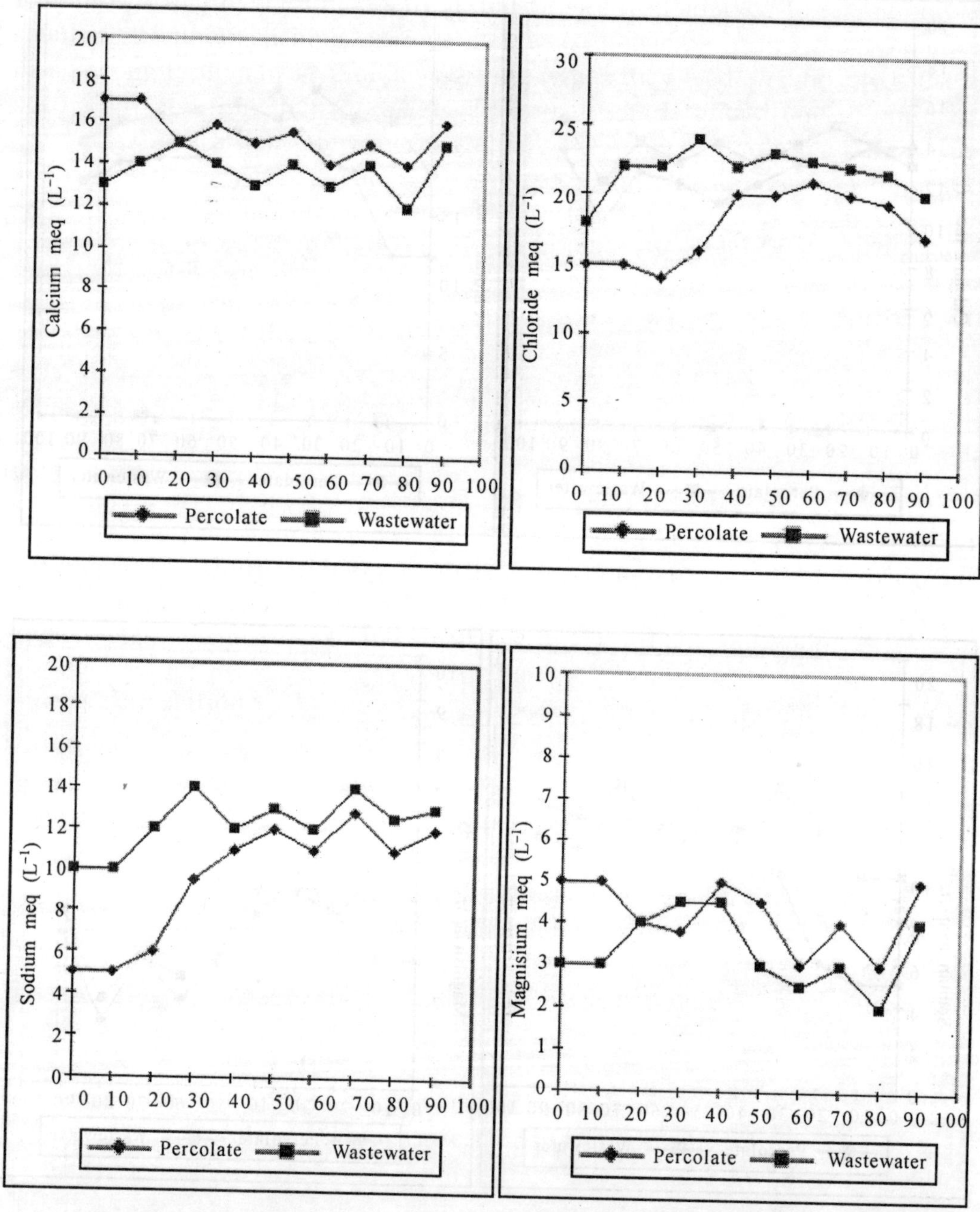


**Fig. 4. Concentration of cations and anions in wastewater applied and percolate collected from field lysimeter III**

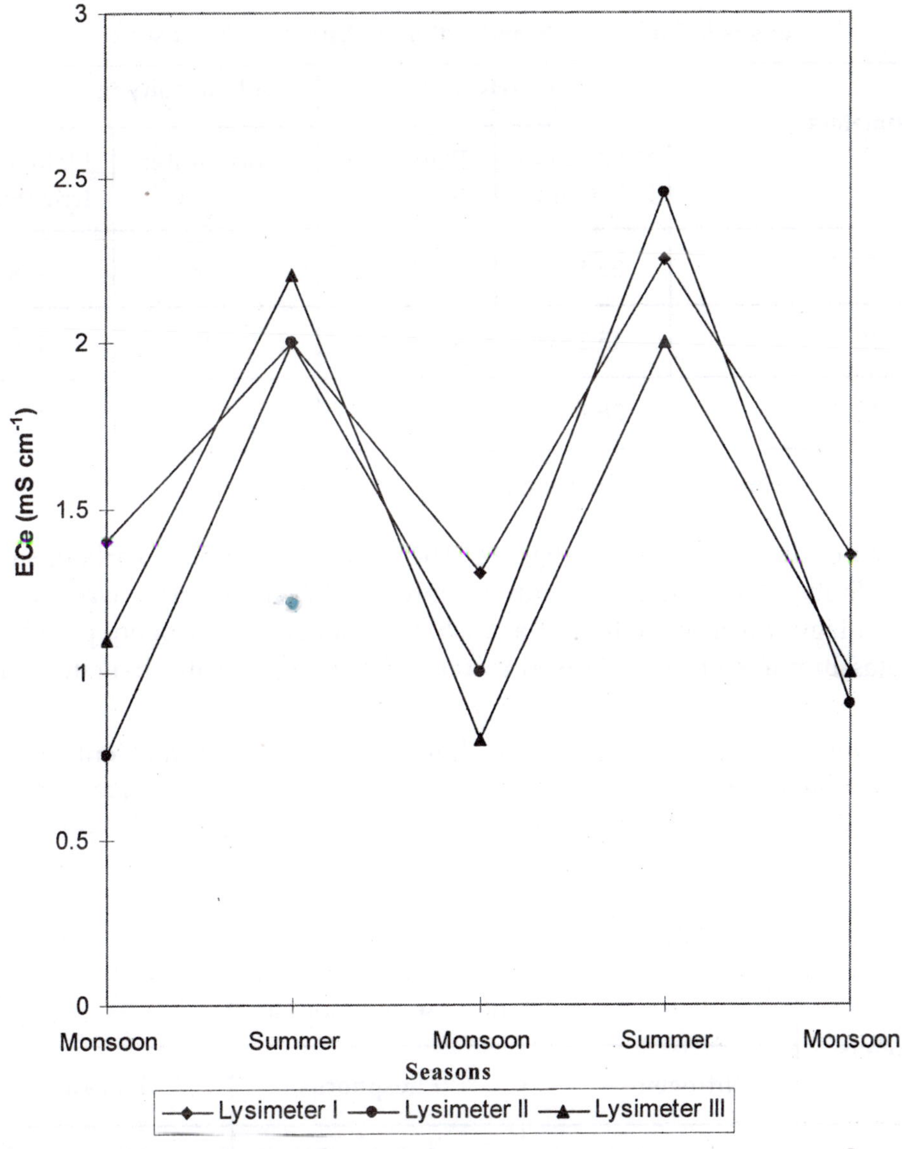


*Fig. 5* Changes in electrical conducvitiy of saturation extract of soils in field lysimeters

**Table VI**

**Changes in bulk density and water holding capacity of soils**

| Lysimeters | MWHC (%) | | Bulk density (g $cc^{-1}$) | |
|---|---|---|---|---|
| | Wastewater Irrigated | Plain Water Irrigated | Wastewater Irrigated | Plain Water Irrigated |
| I | 33.1 | 31.0 | 1.28 | 1.30 |
| II | 45.2 | 42.9 | 1.42 | 1.45 |
| III | 28.1 | 25.7 | 1.30 | 1.31 |

MWHC – Maximum Water Holding Capacity ($M_{water}/M_{total}$ x 100).

was retained in soil as calcium lignate. The changes in organic carbon, salinity and alkalinity of soils irrigated with wastewater (WW) and plain water (PW) is shown in Figure 6. The content of available nutrients viz. nitrogen, phosphorus and potassium in soils also showed increase due to wastewater irrigation (Table VII).

The mineralization of organic material in the soils as well as nutrients available in the wastewater might be responsible for such increase in the available plant nutrients. Such type of increased levels of nutrients in soils was also reported by

**Table VII**

**The content of available nutrients in soils**

| Lysimeters | Available Nutrient (mg $100g^{-1}$) | | | | | |
|---|---|---|---|---|---|---|
| | Nitrogen | | Phosphorus | | Potassium | |
| | PW | WW | PW | WW | PW | WW |
| I | 2.45 | 3.80 | 1.52 | 2.68 | 5.01 | 22.62 |
| II | 3.79 | 7.21 | 4.01 | 6.00 | 6.00 | 26.30 |
| III | 2.38 | 3.21 | 1.32 | 2.02 | 4.55 | 22.11 |

PW : Plain Water; WW : Wastewater

other workers (Igbounamba, 1972 and Patra and Behera, 1974). This findings also accord with the findings of Qualls *et al.* (1991), Hongve (1999) and Hongve *et al.* (2000)

### *Effect of Wastewater on Germination, Yield of Crops and Nutritional Quality of Food Grain*

#### *Seed Germination*

The germination test of jowar, wheat and moong seeds irrigated with wastewater and plain water did not show significant differences. Germination percentage of jowar seeds was 72.43 and 71.73% respectively when irrigated with wastewater and plain water, while wheat showed 94.53 and 95.19%, and moong showed 70.42 and 72.19% germination. These results are comparable with the data reported by Spulnik (1940).

#### *Grain and Straw Yield*

The wastewater application resulted in an increase of crop yield (Table VIII). Jowar crop showed significant increase in grain yield grown in different types of soils and irrigated with wastewater and plain water for three years. Eight samples from the same lysimeter i.e. eight replications were used to generate mean and SEM values. A significant difference was observed for the first year of soil-wastewater interaction. Jowar showed 17.5% increase in grain yield with wastewater in Lysimeter I having sandy loam soil over control followed by Lysimeter III (loamy sand) and Lysimeter II (clay) with increase of 16.3 and 12.8% respectively compared to control. Straw yield of jowar also showed a significant difference for consecutive 3 years, irrigated with wastewater and plain water in three different types of soils. However, there was no significant impact of soil and wastewater interactions on jowar straw yield. Jowar straw showed maximum increase of 11.2% in sandy loam and increase of 8.8 and 8.3% was recorded for clay (Lysimeter II) and loamy sand soil (Lysimeter III) respectively over control.

Wheat crop grown in sandy loam soil showed maximum increase in grain yield of 18.3% with wastewater over control and followed by clay and loamy sand soils which showed increase of 15.1 and 6% respectively over control. For the first and second years, no significant impact of soil type on the grain yield was observed. However, wastewater and its interaction with soil types showed significant difference in the grain yield in the third year. In the case of straw yield, except for third year for soil; soil, wastewater and their interaction had significant impact on the straw yield. Wheat showed an increase of 16.5% in straw in case

**Table VIII**

**Effect of pulp and paper mill wastewater on grain and straw yield of crops grown in field lysimeters**

| Lysimeters and crops | Grain Yield (q ha$^{-1}$) | | | | | | Straw Yield (q ha$^{-1}$) | | | | | |
|---|---|---|---|---|---|---|---|---|---|---|---|---|
| | Ist year | | IInd year | | IIIrd year | | Ist year | | IInd year | | IIIrd year | |
| | PW | WW | PW | WW | PW | WW | PW | WW | PW | WW | PW | WW |
| **Jowar** | | | | | | | | | | | | |
| Lysimeter I | 33.20 | 40.00 | 29.62 | 33.52 | 31.31 | 39.19 | 52.10 | 58.30 | 48.13 | 51.19 | 50.96 | 58.19 |
| Lysimeter II | 37.80 | 42.21 | 34.25 | 37.31 | 35.42 | 42.32 | 56.12 | 61.12 | 53.62 | 57.16 | 55.82 | 61.15 |
| Lysimeter III | 42.41 | 50.43 | 37.73 | 42.42 | 41.31 | 50.72 | 62.01 | 66.32 | 56.25 | 61.13 | 60.15 | 67.32 |
| Soil | | | | | | | | | | | | |
| SEm + | 0.215 | | 0.410 | | 0.468 | | 0.412 | | 0.562 | | 0.391 | |
| CD 5% | 0.813 | | 1.222 | | 1.315 | | 1.241 | | 1.679 | | 1.221 | |
| Wastewater | | | | | | | | | | | | |
| SEm + | 0.328 | | 0.272 | | 0.391 | | 0.419 | | 0.415 | | 0.432 | |
| CD 5% | 0.956 | | 0.745 | | 1.143 | | 1.331 | | 1.213 | | 1.392 | |
| Interaction | | | | | | | | | | | | |
| SEm + | 0.571 | | 0.450 | | 0.698 | | 0.730 | | 0.712 | | 0.779 | |
| CD 5% | 1.700 | | NS | | NS | | NS | | NS | | NS | |
| **Wheat** | | | | | | | | | | | | |
| Lysimeter I | 23.46 | 24.95 | 20.61 | 21.31 | 25.18 | 27.18 | 26.30 | 27.31 | 22.52 | 23.52 | 27.31 | 29.75 |
| Lysimeter II | 21.97 | 27.46 | 19.46 | 23.12 | 23.32 | 27.32 | 24.25 | 29.30 | 21.31 | 25.31 | 25.17 | 29.72 |
| Lysimeter III | 23.32 | 27.33 | 20.31 | 23.71 | 24.14 | 27.19 | 25.13 | 29.11 | 22.11 | 25.69 | 26.32 | 29.21 |
| Soil | | | | | | | | | | | | |
| SEm + | 0.241 | | 0.213 | | 0.248 | | 0.250 | | 0.218 | | 0.362 | |
| CD 5% | NS | | 0.639 | | NS | | 0.891 | | NS | | NS | |
| Wastewater | | | | | | | | | | | | |
| SEm + | 0.198 | | 0.213 | | 0.179 | | 0.191 | | 0.216 | | 0.152 | |
| CD 5% | 0.588 | | 0.616 | | 0.531 | | 0.539 | | 0.651 | | 0.451 | |
| Interaction | | | | | | | | | | | | |
| SEm + | 0.361 | | 0.312 | | 0.317 | | 0.321 | | 0.381 | | 0.266 | |
| CD 5% | 1.111 | | 1.132 | | 0.913 | | 0.932 | | 1.099 | | 0.719 | |
| **Moong** | | | | | | | | | | | | |
| Lysimeter I | 6.94 | 8.51 | 5.31 | 6.12 | 7.23 | 8.93 | 8.16 | 9.62 | 6.28 | 7.15 | 8.16 | 9.22 |
| Lysimeter II | 7.85 | 8.93 | 5.52 | 6.47 | 7.624 | 8.31 | 8.31 | 9.32 | 6.27 | 7.32 | 8.39 | 9.82 |
| Lysimeter III | 8.32 | 9.84 | 5.63 | 8.61 | 9.12 | 9.11 | 9.44 | 10.01 | 8.11 | 9.12 | 10.12 | 10.03 |
| Soil | | | | | | | | | | | | |
| SEm + | 0.132 | | 0.121 | | 0.169 | | 0.121 | | 0.160 | | 0.212 | |
| CD 5% | 0.371 | | 0.362 | | 0.513 | | 0.373 | | 0.331 | | 0.637 | |
| Wastewater | | | | | | | | | | | | |
| SEm + | 0.114 | | 0.131 | | 0.113 | | 0.127 | | 0.112 | | 0.120 | |
| CD 5% | 0.314 | | 0.253 | | 0.283 | | 0.399 | | 0.312 | | 0.253 | |
| Interaction | | | | | | | | | | | | |
| SEm + | 0.182 | | 0.146 | | 0.163 | | 0.234 | | 0.176 | | 0.145 | |
| CD 5% | NS | | NS | | NS | | NS | | NS | | NS | |

PW - plain water, WW - wastewater, NS - non significant (1 q = 100 kg).

SEm – Standard Error mean, CD 5% – Critical difference at 5% level.

of sandy loam soil (Lysimeter I), 12.2% for clay soil (Lysimeter II) and 5.3% increase with respect to loamy sand soil (Lysimeter III) respectively.

The significant difference in the grain and straw yield of moong was observed for three years grown in different types of soils and irrigated with plain water and wastewater. Except for third year grain yield, no significant effect of soil and wastewater irrigation was observed on grain and straw yield. The moong crop grown in sandy loam soils (Lysimeter I) showed increase of 20.9% while 20.6 and 12.4% was recorded in clay (Lysimeter II) and loamy sand (Lysimeter III) soils respectively over control. Increase of 14.0% in straw yield was observed as a result of wastewater application in sandy loam soil while 12.8 and 10.0% increase was recorded for clay and loamy sand soils. Similar results were reported by Juwarkar and Subrahmanyam (1987), NEERI (1994) and Thawale *et al.* (1999). The presence of root promoting phenolic compounds in the wastewater might have played a role in influencing the beneficial effect on plant growth as suggested by Bose *et al.* (1971).

*Nutritional Quality of Food Grain*

Food grain of crop from wastewater irrigated and control lysimeters were collected and analysed for macronutrient (N, P, K Ca, Mg & Na) and micronutrient (Cu, Zn, Fe & Mn) to ascertain the changes in nutrient quality (Table IX).

The presences of micronutrient in grains were slightly higher in control as compared to wastewater application. A marginal decrease in micronutrient in foodgrains, which received wastewater, was due to addition of organic matter, which forms insoluble chelates. The lignin incorporated in soil through wastewater forms insoluble complexes with these micronutrients and reduced their bioavailability. However, at no stage of crop growth was nutrient deficiency observed. This substantiates the fact that a slight variation in mineral nutrition will not affect normal physiological functions in crops. Further there were no accumulations of toxic heavy metals like Cd, Ni, Pb, Cr etc. in food grains because it was found below detectable level. Pulp and paper mill wastewater used for crop irrigation did not affect the nutritional quality of grains. This was earlier confirmed through an experiment conducted at East Texas Pulp and paper company (McCormic, 1959). Similar results were reported by Juwarkar *et al.* (1987) on nutritional quality of pulses, cereals, vegetables and cash crops which received paper mill wastewater for 16 years.

## Conclusions

The treated pulp and paper mill effluent used in the study belongs to irrigation water class 'Moderate' to 'Severe Restriction on use' with respect to salinity and

**Table IX**
**Effect of pulp and paper mill wastewater on nutritional quality of food grain of crops grown in field lysimeters**

| Lysimeters and Treatments | Macronutrient, % | | | | | | Micronutrient, mg $Kg^{-1}$ (grain dry weight) | | | |
|---|---|---|---|---|---|---|---|---|---|---|
| | N | P | K | Na | Ca | Mg | Zn | Cu | Fe | Mn |
| JOWAR | | | | | | | | | | |
| Lysimeter I (Sandy Loam) | | | | | | | | | | |
| With Wastewater | 1.75 | 0.31 | 0.39 | 0.033 | 0.17 | 0.26 | 25.9 | 8.4 | 82.5 | 19.4 |
| With Plainwater | 1.64 | 0.33 | 0.34 | 0.029 | 0.20 | 0.31 | 26.3 | 10.3 | 99.8 | 21.3 |
| Lysimeter II (Clay) | | | | | | | | | | |
| With Wastewater | 1.71 | 0.33 | 0.36 | 0.037 | 0.17 | 0.32 | 25.1 | 8.0 | 74.3 | 23.3 |
| With Plainwater | 1.65 | 0.35 | 0.36 | 0.038 | 0.19 | 0.35 | 26.9 | 8.9 | 81.5 | 24.2 |
| Lysimeter III (Loamy Sand) | | | | | | | | | | |
| With Wastewater | 1.70 | 0.29 | 0.36 | 0.034 | 0.13 | 0.29 | 24.1 | 7.0 | 83.2 | 22.6 |
| With Plainwater | 1.60 | 0.33 | 0.40 | 0.029 | 0.12 | 0.36 | 26.3 | 9.0 | 96.2 | 24.4 |
| WHEAT | | | | | | | | | | |
| Lysimeter I (Sandy Loam) | | | | | | | | | | |
| With Wastewater | 2.28 | 0.35 | 0.23 | 0.034 | 0.12 | 0.28 | 37.4 | 9.1 | 73.1 | 49.2 |
| With Plainwater | 2.29 | 0.34 | 0.29 | 0.033 | 0.16 | 0.30 | 43.3 | 11.5 | 79.5 | 52.3 |
| Lysimeter II (Clay) | | | | | | | | | | |
| With Wastewater | 2.23 | 0.33 | 0.24 | 0.037 | 0.22 | 0.29 | 39.3 | 12.7 | 95.6 | 51.1 |
| With Plainwater | 2.29 | 0.24 | 0.23 | 0.024 | 0.18 | 0.31 | 44.9 | 16.2 | 97.2 | 59.4 |
| Lysimeter III (Loamy Sand) | | | | | | | | | | |
| With Wastewater | 2.37 | 0.33 | 0.32 | 0.036 | 0.30 | 0.34 | 29.1 | 9.5 | 92.2 | 53.1 |
| With Plainwater | 2.25 | 0.25 | 0.23 | 0.025 | 0.17 | 0.40 | 35.9 | 12.7 | 97.1 | 53.5 |

| MOONG | | | | | | | | | | |
|---|---|---|---|---|---|---|---|---|---|---|
| Lysimeter I (Sandy Loam) With Wastewater | 3.22 | 0.43 | 1.40 | 0.050 | 0.24 | 0.20 | 23.3 | 12.3 | 58.1 | 23.3 |
| With Plainwater | 3.37 | 0.49 | 1.61 | 0.051 | 0.23 | 0.26 | 26.2 | 13.7 | 60.3 | 26.7 |
| Lysimeter II (Clay) With Wastewater | 3.23 | 0.40 | 1.61 | 0.061 | 0.23 | 0.20 | 26.2 | 11.8 | 60.3 | 24.7 |
| With Plainwater | 3.03 | 0.40 | 1.58 | 0.063 | 0.24 | 0.24 | 29.2 | 13.3 | 62.1 | 26.2 |
| Lysimeter III (Loamy Sand) With Wastewater | 3.20 | 0.35 | 1.43 | 0.050 | 0.23 | 0.15 | 21.4 | 10.1 | 56.1 | 22.1 |
| With Plainwater | 3.01 | 0.35 | 1.43 | 0.053 | 0.18 | 0.20 | 23.2 | 13.1 | 57.8 | 23.2 |

specific ion toxicity (Kandiah, 1981). When SAR is considered it was categorized under the group 'No Restriction on use'. Soils used in the study had capacity to retain colour and COD of wastewater which may otherwise contaminate the ground water. Presence of calcium and magnesium in the percolate from lysimeters indicated their nonretention in the soil profile. On the other hand sodium was retained in the soil profile. Wastewater irrigated soils showed increase in the water holding capacity, pH, ECe, ESP, organic carbon and available nutrient content of soils but decrease in the bulk density. Monsoon rains leached the accumulated salts beyond the rhizosphere, thereby reducing the soil ECe and indicating effective leaching of soluble salts. Wastewater irrigation did not affect the germination and nutritional quality of food grain of crops. The application of wastewater infact increased the yield of the crop. The study clearly demonstrated that crops like jowar, wheat and moong can be successfully grown under irrigation with pulp and paper mill wastewater without affecting the soil quality. The surface layers may not show build-up of salts, due to the flushing action of monsoon rains, but salinisation of the soil profile will occur under long-term application of the wastewater.

The evaluation of the hazards associated to this kind of contamination would require a good understanding of the different phenomena involved, not only the chemical equilibria, bi\ut also the hydrodyanamics and the contact between the soil and the contaminant solution. This will be further studied in a subsequent paper, where the mathematical model to explore the relative importance of the parameters involved will be presented.

**Acknowledgements**

Authors wish to express their gratitude to management of Paper Mill for their cooperation during the research work. The research work has been supported by Network for Industrial Environmental Management, United Nations Environment Programme (NIEM, UNEP).

## References

American Public Health Association: 1989, 17th ed, 'Standard Methods for the examination of water and wastewater', APHA, Washington, DC.

Anderson, A.: 1979, 'Disposal of an Integrated Pulp and Paper Mill Effluent by Irrigation', Tech. Rep. No. EPA-600/2-79-633, EPA Cincinnati, Ohio.

Ayers, R.S. and Westcot, D.W.: 1976, 'Water Quality for Agriculture', Irrig. Drainage Paper No. 29 FAO, Rome.

Baver, L.D., Gardner, W.H. and Gardner, W.R.: 1972, 'Soil Physics', John Willey and Inc. New York.

Black, C.A.: 1965, 'Methods for Soil Analysis', Am. Soc. Agro. Inc., Publ. Madison, Wisconsin, USA.

Bose, T.K., Basu, S. and Basu, R.N.: 1973, 'Changes in Rooting on Cutting of Bangainvielea and Hibiscus', Indian J. Pl. Physol. 16, 127-139.

Burgisser, C.S.; Cernik, M,; Borkovec, M,: Sticher, H., 1993, Determination of non linear adsorption isotherms from column experiments. An alternative to batch experiments. Environ. Sci. Technology. 27,943-948

Hansen, E.A., Dawson, H.D. and Tolsted, N.D.: 1980, 'Irrigation of Intensively Cultured Plantation with Paper Mill Effluent', Tappi, 63(11), 139.

Hongve, D.: 1999, 'Production of Dissolved Organic Carbon in Forested Catchments, Journal of Hydrology, 224, 91-99.

Hongve, D., Van Hees P.A.W. and Lundstrom U.S.: 2000, 'Dissolved Components in Precipitation Water Percolated through Forest Litter', European Journal of Soil Science, 51, 667-677.

Igbounamba, O.: 1972, 'Exchangeable Reserve Potassium and Other Cation Reserves in Agodi and Ado Soils', Soil Sci. 113, 394-409.

International Rules for Seed Testing: 1966, ' IRST ', Proc. I. t. Seed Test. Assoc., 31, 1-152.

Jackson, M.L.: 1973, 'Soil Chemical Analysis', Publ. Prantice Hall of India Ltd, New Delhi.

Juwarkar, A.S. and Subrahamanyam, P.V.R.: 1987, 'Impact of Pulp and Paper Mill Wastewater on Crop and Soil', Wat. Sci. Tech. 19, 693.

Kandiah A.: 1987, 'Water Quality in Food Production', Water Quality Bulletin. F.A.O., Rome.

Kogelknabner, I : Hatcher, P.G. ; Zech, W. Chemical structural, studies of forest soil humic acids- Aromatic carbon fraction .Soil Sci. Soc. Am. J. 1991, 55, 241- 247.

Lumbely, W.G.: 1977, 'The Simplest Effluent Permit', Pulp and Pap. Int. 12, 165.

Malo, B.A.: 1967, 'Semi-chemical Hard Wood Pulping and Effluent Treatment', J. Wat. Pollut. Control Fed. 39, 1875.

McCormic, L.L.: 1959, 'Effect of Paper Mill Wastewater on Cattle, Crop and Soil', Bull.La. Agric. Stn., 529.

Narum, Q.A., Nickelson, D.P. and Roehne Nils.: 1979, 'Disposal of an Integrated Pulp and Paper Mill Effluent by Irrigation', EPA. Cincinnati Ohio. Tech. Rep. No. EPA-600/2-79-633.

NEERI-UNEP Report: 1994, 'Studies on Pulp and Paper Mill Wastewater and Soil Interaction using Lysimeter', Bangkok, Thailand.

Panse, S.V. and Sukhatme, P.V.: 1967, 'Statistical Methods for Agricultural Workers', ICAR, New Delhi.

Patra, S.K. and Behera, S.: 1974, 'Effect of Paper Mill Sludge on Mineralization of Soil Nitrogen', Agra Univ. J. Res. 4, 111-113.

Piper, C.S.: 1966, 'Soil and Plant Analysis', Univ. Adelaide, Australia.

Qualls, R.G., Haines, B.L. and Swank, W.T.: 1991, 'Fluxes of Dissolved Organic Nutrients and Humic Substances in a Deciduous', Forest Ecology, 72, 254-266.

Rajanan, G. and Oblisami, G.: 1979, 'Effect of Paper Factory Effluents on Soil and Crop Plants', Indian J. Environ. Hlth., 21(2), 120-130.

Reed, B.E.; Cline, S.R., 1994, Retention and release of lead by a very fine sandy loam. I. Isotherm modeling, Sep., Sci. technol.

Singh, R.B., Minhas, P.S., Chauhan, C.P.S. and Gupta, R.K.: 1994, 'Salt Leaching with Monsoons and Yield of Indian Mustard as Affected by Saline Irrigation Waters of Varying Cl : $SO_4$ Ratios', J. Indian Soc. Soil. Sci. 42(3), 436-441.

Spulnik, J.B.: 1940, 'Effect of Waste Sulphite Liquor on Soil Properties and Plant Growth', Soil Sci. 49, 37.

Thawale, P.R., Juwarkar, A.S., Kulkarni, A.B. and Juwarkar, A.A.: 1999, 'Lysimeter Studies for Evaluation of Changes in Soil Properties and Crop Yield using Wastewater', Intern. J.Trop. Agric. 17 (1-4), 231-244.

USSL Staff: 1954, 'Diagnosis and Improvement of Saline and Alkali Soils', U.S. Dept. Agric. Handbook No. 60.

Yakunshenko, I.K., Kazantav, I.Y. and Ovsyannikava, V.G.: 1971, 'Wastewater from Sulphate Pulp Production and their Use for Irrigation', Chem. Abstr. 74, 196. 79305n.

# 5

# Agro-Economic Evaluation of Water Resource

*P.R. Thawale, T.K. Ghosh,*
*S.K. Singh and A.V. Kulkarni*

## Introduction

Today, in developing countries, the domestic demand for agricultural products is largely met through differnt means. This success could not have been achieved without the last half-century's investment in irrigation. India - one of the principal inputs to food production in South Asia, irrigated agriculture continues to play a critical role in achieving food security and poverty alleviation and improving the quality of life (Campbell, 1995).

Irrigation is the largest public investment in many countries in the developing world (Campbell, 1995). Apart from economic considerations for site identification, dam structures, hydrology and hydraulics of irrigation, it is important for the policy makers and planners to evaluate how productively land and water resources are being used for agriculture, and to make more informed strategic decisions regarding irrigation and food production. The planning of crop production in command area and its economic feasibility is a complex problem that requires thorough analysis. Although the benefits from irrigation can be dramatic, consideration of various technical, economic and financial factors must be made before the capital investment can be justified.

The paper describes the approach in economic evaluation and cost benefits analysis with respect to agricultural produces of Jigaon Dam (JD), which is proposed to be build on the Purna River, mainly for irrigation purpose. The proposed site is located near Adol village (20°56′46″ N and 76°22′3″ E) in the Jigaon block of Buldana district in Maharashtra (India). The project envisages to utilise 516.77 million cubic meters ($MM^3$) yield from Purna river for irrigation.

The catchment area of the proposed Jigaon Dam Project (JDP) is 17690 Sq. kms and the total command area stretches to irrigate an area of 63015 ha of land. The total submergence area in the proposed JDP covers an area of 18937 ha, which will include 43 villages for full submergence and 9 villages for partial submergence in Buldana district, while one village will be in Akola district. The existing irrigation benefits in the study area are meager (6.57%).

## Methodology

### *Agricultural Returns in the Existing Scenario (Unirrigated)*

In order to review the agricultural returns due to the executions of the new dam project, it is required to compare the existing value of agricultural produces with respect to the value of agricultural produces in the area, after the execution of the project. The crop use, the existing prices of agricultural commodities and current average agricultural yield of each crop cultivated in the region were collected from the concern authorities. The values of unirrigated produces per ha in the area was estimated as:

$$AP = Y..V \tag{1}$$

where

AP s value of agricultural produces in for the crop i per hectare (Rs./ha.)

Y is the average yield of crop i in the present scenario (unirrigated) in quintals/ ha.

V is the current value of agricultural produces of crop i in Rs./quintals.

The gross benefits due to agricultural production for a crop per ha was estimated considering the deduction in farmhouse practices and the revenues generated by fodder and farmyard manure. The deduction in farmhouse practices was considered as 10% of the value of agricultural produces in for the crop. The fodder yielding crops cultivated in the area were jowar, groundnut and wheat and the fodder receipts were estimated as 15, 10 and 10% of the value of agricultural produces after deduction of the farmhouse practices. In the region, the sowing, ploughing and all intercultural operations are done using animal power i.e. bull, which yields to farmyard manure. Therefore, farmyard manure receipts were considered in the gross benefits as 3% of the value of agricultural produces after deduction of the farmhouse practices. Thus, the gross benefits were estimated as:

$$G = (AP - D) + FR + FY \tag{2}$$

where

G is the gross benefits due to agricultural practices for the crop i per ha in Rs.

D is the cost of farmhouse practices for the crop i per ha in Rs.

FR is the value of fodder receipts for the crop i per ha in Rs.

FY is the value of farmyard manure/ dung receipts for the crop i per ha in Rs.

In order to estimate the expenditure in agricultural production, the parameters considered were cost of seeds, pesticides, fodder, fertilizer and manure, labour charges, implement charges, irrigation and interest in working capital. The cost of the seeds per hectare was estimated as:

$$S = Sc.Sr \tag{3}$$

where

S is the cost of the seeds per ha. for crop i in Rs.

Sc is the cost of the seeds per kg. for crop i in Rs.

Sr is the seed rate for crop i in kg/ha.

The estimated cost of pesticides for each crop collected from Divisional Joint Director of Agriculture (Extn.), Amravati for the year 1993-94. A 50% increase in prices was considered for the year 2000-01 on the base price (1993-94) taking into account the inflation.

Estimation the cost of fertilizers applied was based on the nutrient required for the each crop in area in terms of NPK i.e. Nitrogen (N), Phosphorus (P) and Pottasium (K). The data of nutrient requirement of various crops were obtained from packages, practices and recommendations of agricultural crops by Punjabrao Agricultural University (PAU,1999). The amount different fertilizers required by each crop were estimated based on NPK contents of the fertilizers and the NPK requirement of the crops. For example, the nutrient requirement of Hybrid Jowar is N = 80 kg/ha and P = 40 kg/ha. Applying 200 kg of Sufala (20 -20 - 0), which contains 20% N, 20% P and 0% K will substitute the nutrient requirements by 40 kg of N and 40 kg of P. The balance 40 kg of N requirements is accomplished by applying 87 kg of Urea (46 – 0 – 0). The economic considerations and fertilizers available in the region were taken into account to estimate the fertilizer requirement. The value of the fertilizer and manure application per ha for a crop was estimated as:

$$F = \sum_{j=1}^{t} W_j.C_j \tag{4}$$

where

F is the total cost of fertilizer and manure application per ha for crop i.

t is the number of types of fertilizers/manures applied for crop i.

$W_j$ is the weight of fertilizer/manure j (in quintals for fertilizers/ cartload for manures/ kg for groundnut cakes) to be applied to crop i

$C_j$ is the cost of fertilizer/manure j (in quintals for fertilizers/ cartload for manures/ kg for groundnut cakes) to be applied to crop i

The fodder expenditure of the farm animals, implement charges and labour charges were considered as 10%, 3% and 20% of the gross benefits respectively. The interest on capital was considered as 13% per annum on the sum of all the expenditures for the duration of crops. The expenditure due to agricultural practices for each crop were estimated as:

$$E = S + F + Fe + \text{Im} + L + P + Ir + I \tag{4}$$

where

E is the expenditure in due to cultivation of the crop i per ha in Rs.

Fe is the fodder expenditure per ha. for crop i in Rs.

Im is the implement charge per ha. for crop i in Rs.

L is the labour charges per ha. for crop i in Rs.

P is the cost of pesticides per ha. for crop i in Rs.

Ir is the irrigation charges per ha. for crop i in Rs.

I is the interest on working captial per ha. for crop i in Rs.

The net return per ha for each crops were computed as

$$\mathrm{N} = (\mathrm{G} - \mathrm{E}) \tag{5}$$

where N is the net returns due to agricultural practices for crop i in Rs.

In order to standardize the computations involved to derive the cost benefits, initially the costs were calculated considering 1000 ha. The net returns due to agricultural practices for 1000 Ha in the unirrigated area was estimated as:

$$NR = \sum_{i=1}^{n} \mathrm{N.P_c} x10 \tag{6}$$

where

NR is the net returns due to agricultural practices for 1000 Ha in the unirrigated area in Rs.

Pc is the percentage of area under crop i

The net returns due to agricultural practices in the unirrigated area was estimated as:

$$NR_{ui} = \frac{NR.(SA + QA + CA + DA + IA)}{1000} \quad (7)$$

where

SA is the proposed submergence area in ha.

QA is the proposed quarry area in ha.

CA proposed colony area in ha.

DA is the proposed dam site a rea in ha.

IA is the proposed irrigable area in ha.

## Agricultural Returns after Execution of Project (Irrigated)

### *Cropping Pattern Design and Project Formulation*

Within a few years of project completion, actual cropping patterns usually differ considerably from the originally conceived pattern (Campbell, 1995). Nevertheless, economic feasibility and the design of an irrigation system requires assumptions at least as to the class of crops to be grown in each season, and economic and financial analyses (farm budgets) require assumptions that are more specific. A projected cropping pattern needs to be designed considering the various parameters involved. These include the advent of a new type of crop in the area, balancing between seasonal water supply and demand, change in price structure of agricultural commodities and inputs, and the expected agricultural returns. Considering these aspects, a proposed cropping pattern was designed based on the climatic and the soil characteristics of the area, taking into consideration the local cultural and eating habits.

### *Total Water Requirement for Crops*

In order to estimate the water demand of each proposed crop, the most important parameter, is the potential evapotranspiration (PET) of the crops in the area. The PET is often predicted based on climatological data. To compute the PET, the reference evapotranspiration ($ET_o^*$) i.e. the PET of grass/alfalfa needs to be estimated. The modified Penman equation (Doorenbos *et al.* 1975) was used to estimate $ET_o^*$ and is expressed as

$$ETo^* = \underbrace{W.R_n}_{\text{radiation term (mm/day)}} + \underbrace{(1 - W).f(u).(e_a - e_d)}_{\text{aerodynamic term (mm/day)}} \quad (8)$$

where

$ET_o^*$ is the reference evapotranspiration [mm $day^{-1}$] need to be adjusted/ corrected for the day/night time weather conditions.

W is the values of weighing factor for the effect of radiation on reference crop evapotranspiration at different temperatures and altitudes

$R_n$ is the net radiation [mm $day^{-1}$]

(1 – W) is the values of weighing factor for the effect of wind and humidity on reference crop evapotranspiration at different temperatures and altitudes

f(u) is a wind related function

$e_a$ is the saturation vapour pressure [m bar]

$e_d$ is the actual vapour pressure [m bar]

The formula can be classified on two terms i.e. radiation terms (Solar Radiation) and aerodynamic terms (Wind related parameters). The parameters in formula needs to be derived based on the available meteorological data (IMD) (i.e. temperature, Wind speed, relative humidity and sunshine hours) and geographical parameters (i.e. latitude and altitude) of the place. The data from the nearest meteorological station at Buldana were considered for the purpose.

*Radiation Terms*

The values of W and $R_n$ were derived using the meteorological data and geographical parameters. The weighing factor W is estimated as:

$$W = \Delta/(\Delta + \gamma) \tag{9}$$

where

$\Delta$ is the slope vapour pressure curve [kPa°$C^{-1}$] i.e. the rate of change of the saturation vapour pressure with temperature.

$\gamma$ psychrometric constant [kPa °$C^{-1}$]

The Slope Vapour Pressure Curve was derived using the formula (Tetens,1930 & Murray,1967) taking in to account the temperature of the place. The psychrometric constant ($\gamma$) is derived by considering altitude and temperature of the place. (Bunt, 1952, Burman et. al., 1987, Harrison, 1963).

$R_n$ is derived by

$$R_n = R_{ns} \downarrow - R_{nl} \uparrow \tag{10}$$

where $R_{nl}$ is the net longwave radiation [mm $day^{-1}$] and $R_{ns}$ is the net incoming shortwave solar radiation [mm $day^{-1}$]. The net incoming short wave solar radiation $R_{ns}$ is estimated considering reflection coefficient as 0.25 and using the available parameters like latitude, month of the year, sunlight hours

(Doorenbos et. al. 1997, ASCE , Duffie et al. 1991, Gommes 1983, Craig 1984, Jensen et al. 1990, London 1982). $R_{nl}$ is computed as:

$$R_{nl} = f(t).f(ed).f(\tfrac{n}{N}) \tag{12}$$

where

f(t) is the correction for temperature on long wave radiation

f(ed) is the correction for vapour pressure on long wave radiation

$f(\frac{n}{N})$ is the correction for the ratio of actual and maximum bright sunshine hours on long wave radiation [W $m^{-2}$]

f(t), f(ed) and $f(\frac{n}{N})$ were estimated as:

$$f(t) = \sigma K^4 \tag{13}$$

$$f(ed) = 0.34 - .444\sqrt{e_d \times 0.01} \tag{14}$$

$$f\left(\frac{n}{N}\right) = 0.1 + .9\frac{n}{N} \tag{15}$$

where

$\sigma$ is the Stefan-Boltzman constant = 5.6698 $x10^{-8}$ [W $m^{-2}K^{-4}$]

K is the air temperature [$^oK$]

$e_d$ is the actual vapour pressure [m bar]

n is the actual hours of bright sunshine [hrs]

N is the maximum possible duration of bright sunshine [hrs]

*Radiation Terms*

The values of f(u) $e_a$ and $e_d$ were derived using the meteorological data. The saturation vapour pressure was estimated by

$$e_d = e_a \times \frac{RH}{100} \tag{16}$$

where RH is the relative humidity [%] and $e_a$, the saturation vapour pressure [m bar] is estimated from air temperature (Tetens, 1930).

f(u) was estimated as:

$$f(u) = 0.27(1 + \frac{u_2}{100}) \tag{17}$$

where $u_2$ is the wind speed at hieght of 2 meter [km $hr^{-1}$]. Since, wind speed were not recorded at 2 meter, correction were made on wind speed based on factors given by Doorenbos *et al.* (1977). The monthly ET* were adjusted/ corrected for the day/night time weather conditions to obtain corrected reference monthly evapotranspiration ET (Doorenbos *et al.* 1977) considering monthly daytime average wind speed (m/s), monthly day and night ratio of wind speed and monthly relative humidity (%). The monthly crop factors of each crop were estimated based on FAO methods (Doorenbos *et al.* 1977). The crop factors (CF) of each crop were multiplied with monthly ET to obtain the crop evapotranspiration $ET_C$ of each crop for the corresponding month. The effective rainfall (ERF) was estimated based on USDA methodology (USDA, 1969) based on monthly rainfall and consumptive use. The monthly water requirement (WR) was estimated as

$$WR = ET_c - ERF \qquad \{\text{If WR} <0 \text{ then WR}=0\} \qquad (18)$$

where

WR is monthly water requirement of crop i for month k

$ET_c$ is the evapotranspiration of crop i for month k

ERF is the effective rainfall for month k

The irrigation scheduling was done on fortnightly basis. The irrigation requirement for fortnightly basis was estimated, if the period of crop cultivation begins at second fortnight or extends upto first fortnight. The estimation of fortnightly WR is similar to monthly WR (eqn. 18) except for fortnightly values of $ET_c$ and ERF. The first and second fortnightly values of $ET_c$ and ERF were obtained from sliding scale method. The monthly net water requirement NWR of the crops is the same as WR except the last fortnightly irrigation has been curtailed.

The total water requirement TWR in mm was estimated as 1000 Ha was estimated as:

$$TWR = IE.\sum_{i=1}^{n}\sum_{k=1}^{d} NWR.P_c x10 \qquad (19)$$

where

IE is the irrigation efficiency

NWR is the net water requirement for crop i for the month k

d is the number of calendar months of cultivation for crop i

The irrigation efficiency was considered as 50% considering evaporation from canals and dam, conveyance and seepage losses. Based on NWR obtained above the water requirement of total area in million cum was estimated.

### *Economic Considerations*

The change in price structures is a crucial and sensitive parameter that is often unpredictable. Several external parameters govern the prices of agricultural commodities, which includes:

1. Production of the crops or some alternate crops in the area, in the state, or in the country.
2. The government policies regarding import and export of agricultural commodities.
3. The purchasing power of the consumers.
4. The cost of the agricultural inputs and the government policies regarding agricultural subsidies.
5. The shift or changes in consumer habits.

On the other hand, the cost of agricultural inputs is influenced by various factors:

1. The cost of fertilizers, in India is dependent on government policies like subsidy and import regulations and cost of raw materials.
2. Estimating requirement of pesticides for crop protection is more complex and unpredictable as new pesticides are being launched and the use of existing pesticides being restricted or banned due to environment and health concerns.
3. The labour charges are depended on the migration of labours and the industrilisation.

Considering the complexity and the unpredictability, the prices of the agricultural commodities and the prices of agricultural inputs after execution of the project were considered as the current prices. The estimation of the value of irrigated produces were estimated in the similar fashion than that of the current trend except in equation 7, the values of SA, QA, CA and DA were considered as zero as these areas would not be cultivable and the irrigation cost which was not considered as the area was not irrigated. The irrigation requirement of various crops estimated above were reviewed to identify the calendar months in which irrigation is required. The irrigation charges of a crop are the sum of the monthly charges of the month in which irrigation is required.

## Results and Discussions

### *Land Use Pattern*

The agricultural practices is dominant in the area with 78.33% of the area under crops followed by forest (12.85%). The land use pattern of the area is depicted in Fig 1. Rainfed agriculture is traditional practice and mainstay of the population in the area. The natural vegetation in this region comprises tropical, dry deciduous and tropical thorny forests.

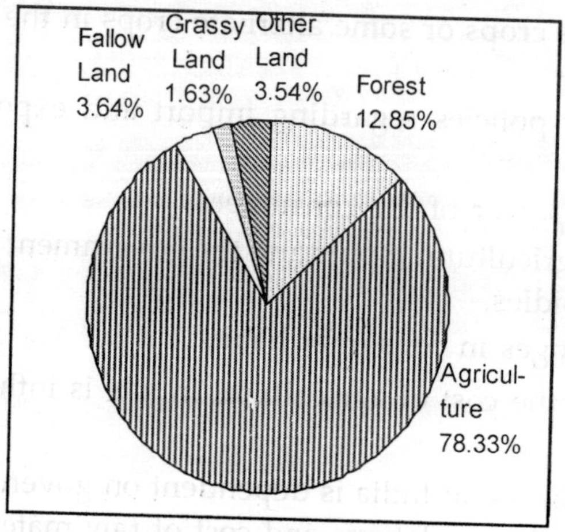


**Fig. 1. Land Use Pattern in the Study Area at JDP**

### *Existing Crop Use Pattern*

The existing cropping pattern based on the average of the four years (1994-95 to 1997-98) is depicted in Fig. 2. Cotton (cash crop) is the major crop grown in this region (41.9%) followed by jowar (20.1%), which is the main cereal crop. The other cereal crops include bajra and wheat. Tur, udid and gram are the

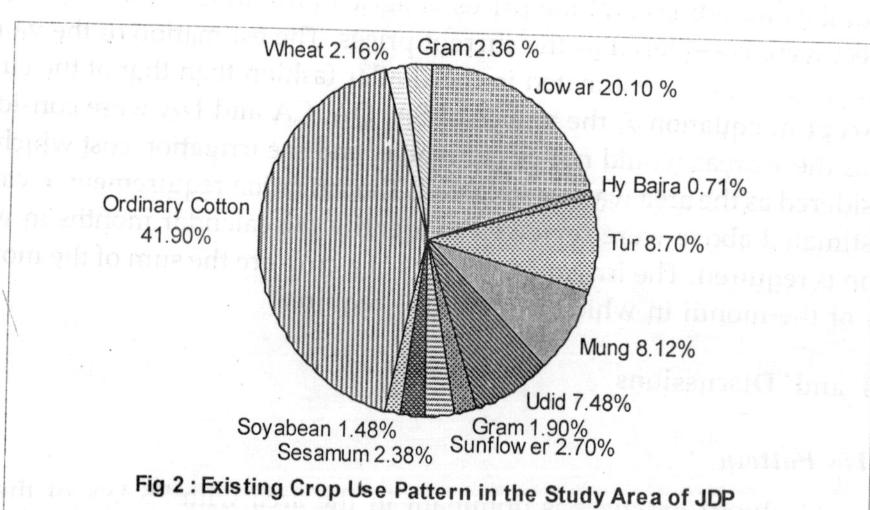


**Fig 2 : Existing Crop Use Pattern in the Study Area of JDP**

pulses cultivated in this area. The oilseeds grown in this area are groundnut sesamum (til) and sunflower.

### *Fertilizer Requirements and Costs*

The cost of fertilizers and their NPK contents is shown in Table 1. The nutrient requirements, cost of fertilizers and manures per hectare for existing and proposed crops is depicted in Table 2. The unit fertilizer and manure cost for turmeric was the highest (Rs.17,100.00) and was lowest for groundnut (Rs.548.00).

**Table 1**

**NPK contents and Prices of Fertilizers for the year 2000–01 in the study area.**

| Sr. No | Fertilizers | NPK Contents | | | Prices in Rs. /100 kg |
|---|---|---|---|---|---|
| | | N | P | K | |
| 1 | Urea | 46 | 0 | 0 | 476.00 |
| 2 | Single Super Phosphate (SSP) | 0 | 16 | 0 | 300.00 |
| 3 | Sufala[1] | 20 | 20 | 0 | 660.00 |
| 4 | Sufala[2] | 19 | 19 | 19 | 660.00 |
| 5 | Farmyard Manure/Compost | | | | 150.00* |
| 6 | Groundnut Cakes | | | | 800.00 |

* per cartload

### *Gross Benefits, Expenditure and Net Benefits in the Current Scenario*

In the existing scenario, the cultivation pulse crop Tur, which is cultivated in 8.7% of the area was found to be the most profitable crop and capital-intensive crop. The gross returns, cultivation cost and net profits from agricultural produces per ha were the highest for Tur i.e. Rs. 17341.65, 11538.30 and 5803.35 (Table 3 through 5). Bajra – a cereal crop, cultivated in meager 0.71% of the area was found to be economically unfeasible for cultivation. This crop is generally grown by farmers for their personal consumption and since the most farm labours belongs to the farmers family, the labour charges for cultivation are not accounted by the farmers. Cotton cultivated in around 41.9% area generates 52.7% of the net revenues in the region.

### Benefits from Unirrigated Produces

The breakup of total area under cultivation, total revenue generated per 1000 ha of cultivation and total area is given below:

**Table 2**

**Nutrient requirements, cost of fertilizers and manures per hectare for existing and proposed crops (Rs./ha)**

| Crop | Organic Manure Requirements | | Fertilizer Requirements per ha. in Kgs. | | | Manure cost Rs. | Name of the Fertilizer | Rate of Application kg/ha | Fertilizer Cost (Rs./ha) | Fertilizer & Manure Cost (Rs./ha) |
|---|---|---|---|---|---|---|---|---|---|---|
| | FarmYard Manure (cartload) | Groundnut Cakes (Kg) | N | P | K | | | | | |
| Hy. Jowar | 10 | | 80 | 40 | 0 | 1500.00 | Sufala[1] | 200 | 1320.00 | |
| | | | | | | | Urea | 87 | 414.12 | 3234.00 |
| Bajra | 10 | | 60 | 20 | 0 | 1500.00 | Sufala[1] | 100 | 660.00 | |
| | | | | | | | Urea | 87 | 414.12 | 2574.00 |
| Tur | 10 | | 25 | 50 | 0 | 1500.00 | Sufala[1] | 125 | 825.00 | |
| | | | | | | | SSP | 157 | 471.00 | 2796.00 |
| Mung & Udit | 10 | | 20 | 40 | 0 | 1500.00 | Sufala[1] | 100 | 660.00 | |
| | | | | | | | SSP | 125 | 375.00 | 2535.00 |
| Groundnut | - | | 16.5 | 16.5 | 0 | - | Sufala[1] | 83 | 547.80 | 548.00 |
| Sunflower | - | | 40 | 60 | 0 | - | Sufala[1] | 200 | 1320.00 | |
| | | | | | | | SSP | 125 | 375.00 | 1695.00 |
| Sesamum | | | 25 | 25 | 0 | - | Sufala[1] | 125 | 825.00 | 825.00 |
| Soyabean | - | | 30 | 75 | 30 | - | Sufala[2] | 158 | 1042.80 | |
| | | | | | | | SSP | 282 | 846.00 | 1889.00 |
| Cotton | 12 | | 30 | 15 | 0 | 1800.00 | Sufala[1] | 75 | 495.00 | |
| | | | | | | | Urea | 33 | 157.08 | 2452.00 |
| Wheat | 5 | | 40 | 20 | 0 | 750.00 | Sufala[1] | 100 | 660.00 | |
| | | | | | | | Urea | 44 | 209.44 | 1619.00 |
| Gram (unirrigated) | - | | 25 | 40 | 0 | - | Sufala[1] | 125 | 825.00 | |
| | | | | | | | SSP | 94 | 282.00 | 1107.00 |
| LS Cotton | 20 | | 100 | 50 | 50 | 3000.00 | Sufala[2] | 263 | 1735.80 | |
| | | | | | | | Urea | 110 | 523.60 | 5259.00 |
| Gram (irrigated) | - | | 25 | 50 | 0 | - | Sufala[1] | 125 | 825.00 | |
| | | | | | | | SSP | 157 | 471.00 | 1296.00 |
| Chillies | - | | 150 | 50 | 50 | - | Sufala[2] | 263 | 1735.80 | |
| | | | | | | | Urea | 218 | 1037.68 | 2773.00 |
| Tomatoes | | | 125 | 0 | 0 | | Urea | 272 | 1294.72 | 1295.00 |
| Turmeric | 50 | 1200.00 | | | | 17100.00 | | | | 17100.00 |
| Sunflower (Hybrid) | - | | 80 | 60 | 0 | - | Sufala[1] | 300 | 1980.00 | |
| | | | | | | | Urea | 44 | 209.44 | 2189.00 |
| Cauliflower | - | | 100 | 50 | 0 | - | Sufala[1] | 250 | 1650.00 | |
| | | | | | | | Urea | 109 | 518.84 | 2169.00 |

**Table 3**

**Cropwise Gross Benefit per Hectare for Existing Unirrigated Crops in the Command Area of JDP.**

| Sl. No. | Crop | Yield in Quintals/ ha | Rate in Rs. per Qtls | Total Value Agricultural Produce in Rs./ha. | Cost of Farm House Practices in Rs./ha. | Value after Deduction in Farm House Practices in Rs./ha. | Value of Fodder Receipts in Rs./ha | Value of Farm Yard Manure/ Dung Receipts in Rs./ha. | Gross Benefits in Rs./ha. |
|---|---|---|---|---|---|---|---|---|---|
| 1 | Hy.Jowar | 14.00 | 442.36 | 6193.08 | Nil | 6193.08 | 928.96 | 185.79 | 7307.84 |
| 2 | Hy. Bajra | 10.00 | 369.78 | 3697.83 | Nil | 3697.83 | | 110.93 | 3808.76 |
| 3 | Tur | 10.00 | 1,683.65 | 16836.55 | Nil | 16836.55 | | 505.10 | 17341.65 |
| 4 | Mung | 6.00 | 1,721.29 | 10327.74 | Nil | 10327.74 | | 309.83 | 10637.57 |
| 5 | Udid | 6.00 | 902.25 | 5413.47 | Nil | 5413.47 | | 162.40 | 5575.88 |
| 6 | Groundnut | 7.00 | 1,165.56 | 8158.92 | Nil | 8158.92 | 815.89 | 244.77 | 9219.58 |
| 7 | Sunflower | 10.00 | 1,155.43 | 11554.28 | Nil | 11554.28 | | 346.63 | 11900.91 |
| 8 | Sesamum (Til) | 4.00 | 2,371.60 | 9486.41 | Nil | 19486.41 | | 284.59 | 9771.00 |
| 9 | Soyabean | 10.00 | 780.11 | 7801.06 | Nil | 7801.06 | | 234.03 | 8035.09 |
| 10 | Ordinary Cotton | 8.00 | 1,445.47 | 11563.78 | Nil | 11563.78 | | 346.91 | 11910.70 |
| 11 | Wheat | 10.00 | 706.80 | 7068.02 | Nil | 7068.02 | 706.80 | 212.04 | 7986.86 |
| 12 | Gram | 10.00 | 1,061.41 | 10614.09 | Nil | 10614.09 | | 318.42 | 10932.51 |

Table 4

Cropwise Expenditure per Hectare for Existing Unirrigated Crops in the Command Area of JDP

| Sl. No. | Crop | Duration of Crops in Days | Seed Cost per ha in Rs | Fertilizer and Manure Cost in Rs. | Fodder Expenses | Implement Charges in Rs. | Labour Charges in Rs. | Cost of Pesticides in Rs. | Irrigation Charges in Rs. | Interest in Working Capital in Rs. | Total Expenses in Rs. |
|---|---|---|---|---|---|---|---|---|---|---|---|
| 1 | Hy.Jowar | 120 | 320.00 | 3234.00 | 619.31 | 185.79 | 1238.62 | 466.50 | 0.00 | 259.18 | 6323.40 |
| 2 | Hy. Bajra | 105 | 120.00 | 2574.00 | 369.78 | 110.93 | 739.57 | 337.50 | 0.00 | 159.01 | 4410.79 |
| 3 | Tur | 170 | 2250.00 | 2796.00 | 1683.65 | 505.10 | 3367.31 | 277.50 | 0.00 | 658.74 | 11538.30 |
| 4 | Mung | 70 | 600.00 | 2535.00 | 1032.77 | 309.83 | 2065.55 | 277.50 | 0.00 | 170.05 | 6990.70 |
| 5 | Udid | 70 | 600.00 | 2535.00 | 541.35 | 162.40 | 1082.69 | 277.50 | 0.00 | 129.62 | 5328.56 |
| 6 | Groundnut | 120 | 1800.00 | 548.00 | 815.89 | 244.77 | 1631.78 | 435.00 | 0.00 | 234.02 | 5709.46 |
| 7 | Sunflower | 100 | 1750.00 | 1695.00 | 1155.43 | 346.63 | 2310.86 | 396.00 | 0.00 | 272.61 | 7926.52 |
| 8 | Sesamum (Til) | 90 | 200.00 | 825.00 | 948.64 | 284.59 | 1897.28 | 105.00 | 0.00 | 136.57 | 4397.08 |
| 9 | Soyabean | 110 | 1125.00 | 1889.00 | 780.11 | 234.03 | 1560.21 | 366.00 | 0.00 | 233.28 | 6187.63 |
| 10 | Ordinary Cotton | 180 | 225.00 | 2452.00 | 1156.38 | 346.91 | 2312.76 | 996.00 | 0.00 | 480.12 | 7969.17 |
| 11 | Wheat | 120 | 1375.00 | 1619.00 | 706.80 | 212.04 | 1413.60 | 892.50 | 0.00 | 265.80 | 6484.74 |
| 12 | Gram | 105 | 1250.00 | 1107.00 | 1061.41 | 318.42 | 2122.82 | 390.00 | 0.00 | 233.72 | 6483.37 |

Table 5

**The Net Benefits from the Unirrigated Agricultural Produces per 1000 ha in the Present Scenario of the Study Area of JDP**

| Sl. No. | Crop | Net Benefit per ha in Rs. | Percentage Area of Cultivation | Area of the Crop in ha/ 1000 ha of Cultivated Area | Benefit in Rs. for 1000 ha |
|---|---|---|---|---|---|
| 1 | Hy.Jowar | 984.44 | 20.10 | 201.01 | 197883.81 |
| 2 | Hy. Bajra | -602.03 | 0.71 | 7.12 | -4284.77 |
| 3 | Tur | 5803.35 | 8.70 | 86.98 | 504781.76 |
| 4 | Mung | 3646.87 | 8.12 | 81.21 | 296144.35 |
| 5 | Udid | 247.31 | 7.48 | 74.77 | 18491.57 |
| 6 | Groundnut | 3510.12 | 1.90 | 18.98 | 66630.94 |
| 7 | Sunflower | 3974.39 | 2.70 | 26.98 | 107244.26 |
| 8 | Sesamum (Til) | 5373.92 | 2.38 | 23.81 | 127966.01 |
| 9 | Soyabean | 1847.46 | 1.48 | 14.83 | 27400.10 |
| 10 | Ordinary Cotton | 3941.53 | 41.90 | 419.03 | 1651616.77 |
| 11 | Wheat | 1502.12 | 2.16 | 21.65 | 32517.34 |
| 12 | Gram | 4449.14 | 2.36 | 23.63 | 105119.23 |
| | Total | | 100.00 | 1000.00 | 3131511.38 |

1. Proposed Submergence Area in ha. = 18937.00
2. Proposed Quarry Area in ha. = 125.00
3. Proposed Colony Area in ha. = 20.00
4. Proposed Dam Site Area in ha. = 76.50
5. Proposed Irrigable Area in ha. = 76500.00
6. Total Area Under Cultivation in ha (Sum 1-5) = 95658.50
7. Benefits from Unirrigated Produces for1000 ha. in Rs. = 3131511.38
8. Benefits from Unirrigated Produces for total area in Rs. = 299555681.65
9. Benefits from Unirrigated Produces for total area in Rs. (lakhs) = 2995.56

## Proposed Cropping Pattern

The proposed cropping pattern was designed based on the climatic and soil characteristics.

### *Climate*

The area comes under Deccan plateau and it is subjected to tropical climate. Broadly, the year can be divided into three seasons, viz. winter, summer and monsoon. The climate is dry with large variation in temperatures ranging from 12°C in winter to 42°C during summer seasons. The winter season occurs from November to February, while December and January are the coldest months. The summer season extends from March to June. During this season, the maximum temperatures arc recorded during the month of May. The precipitation in this area is mainly due to southwest monsoon with a medium rainfall zone of 600 to 1000 mm and the weighted average rainfall being about 750 mm, which is more or less, assured but its distribution has shown erratic trend. The monsoon precipitation, which commences generally in the last week of June, continues till end of September, with occasional spell of post monsoon showers during October. Out of the total rainfall, about 94% of the precipitations occur during June to October. During June to September the percentage of relative humidity is high which is of the order of 88% in month of August.

### *Soil Type and Soil Quality of Command Area*

The physical properties of soil denote that soils of the area are clayey in texture with very high amount of clay followed by silt and sand contents. The clay content of soils varies from 48.3 to 74.5 percent. The soils have medium to good water holding capacity and the maximum water holding capacity ranges from 53.9 to 86.4%. The available water for plants ranges between 31.44 and 44.65%. The hydraulic conductivity of the soils varies from $7.5 \times 10^{-5}$ to $8.9 \times 10^{-4}$ cm/sec, in command area. The porosity of the soils varies from 56.9 to 75.9% in command area. The bulk density of the soils varies from 1.19 to 1.49 g/cc.

Chemical analysis of the soils in the study area reveals that the soils are slightly to moderately alkaline with pH varying between 7.6 and 8.5. The electrical conductivity of the soils of command area varied from 0.40 to 1.80 mS/cm. No salinity was observed and all kinds of crops can be grown without any toxicity. The availability of nitrogen in the soil was medium, however, the soils were very rich in phosphorus and potassium. The metals present in the soils were within the limits and with no toxic potential. The soil has a very high adsorption capacity. The exchangeable sodium percentage in the soil indicates that the soil has a good permeability. The organic carbon in soil was found to be varying between low and medium.

## *Proposed Cropping Pattern*

The proposed cropping pattern is designed for two seasons kharif (rainy season) and rabi (winter season). It is proposed to irrigate 60% and 80% of the area during kharif and rabi season respectively. 40% of the kharif crops are proposed to rainfed with draught varieties or rainfall will take care of the irrigation requirements, which includes groundnut (20% of area), pulses (20%) and sheshum (10%). And 20% of land will remain uncultivated during rabi season. The irrigated kharif season crops include hybrid jowar (20% of area), sunflower (10%) and tomatoes (10%). The irrigated rabi crops includes oilseeds (20% of area), wheat (15%), gram (15%) and tomatoes (10%). Two seasons crops like cotton (10% of area), chillies (8%) and turmeric (2%) are proposed to be cultivated and irrigated during kharif and rabi season, which constitutes 20% of the area (Fig 3 & 4).

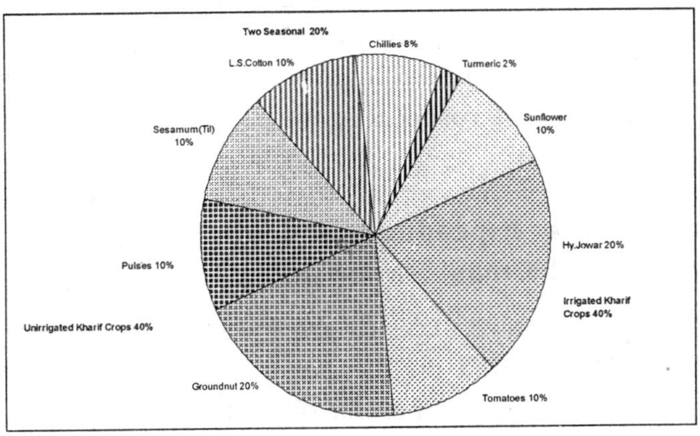


**Fig. 3 Proposed Crop Use Pattern for Kharif Season in the study area.**

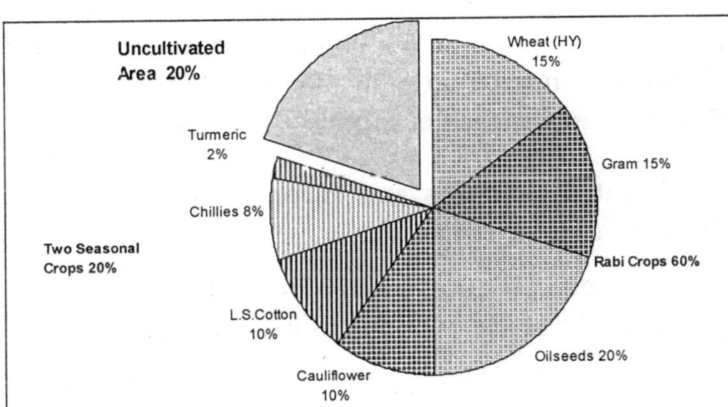


**Fig. 4 Proposed Crop Use Patern for Rabi Season in the study area.**

The monthly reference evapotranspiration of the study area is depicted in Table 6. The ET was the highest during May (291.67 mm) followed by April (250.78 mm) and the lowest during December and November (125.58 and 128.63 mm respectively).

## Irrigation Requirement and Costs

The total water requirement for the irrigated area is depicted in Table 7. The total water requirement for 1000 ha and command area are 4.32 and 330.59 million cum respectively. The evapotranspiration for rabi crop gram was the lowest (256.56 mm) and the two seasonal crop Turmeric was the highest (784.56 mm). In the proposed cropping pattern the total water requirement (considering irrigation efficiency) of wheat was 55.31 million cum, the highest and the lowest was sunflower (3.52 million cum).

The estimated irrigation charges of various crops are shown in Table 8. The charges for irrigating cotton were the highest (Rs. 735.00) followed by turmeric and rabi crops (Rs. 660.00). The cost of irrigating kharif crops was the lowest (Rs. 240.00).

## Predicted Gross Benefits, Expenditure and Net Benefits of Irrigated Crops

The gross benefit per hectare, expenditure and net benefits per ha for various crops proposed to be cultivated on the execution of the irrigation project is shown in Table 9. Gross benefits per ha on Turmeric were the highest (Rs. 246,582.00) but is the most capital-intensive crop with expenditure per ha of Rs. 187,840.14 (Table 10). The gross benefit and net benefits per hectare (Table 11) of unirrigated groundnut was the lowest (Rs. 9,219.58 and Rs. 3,510.12 respectively). The cultivation cost per ha of unirrigated Seshum (Til) was the lowest 4397.08. The net returns from cultivation of 1000 ha irrigation land amounts to Rs. 2,32.57 lakhs - an increase of around 642.68%.

## Benefits from Irrigated Produces and Increase in income in the area due to Irrigation

Benefits from irrigated produces per 1000 ha and total cultivable area and the increase in income in the study area due to irrigation is shown below:

| | | |
|---|---|---|
| 1 | Proposed Irrigable Area in ha. | = 76,500.00 |
| 2 | Benefits from Irrigated Produces/1000 ha. in Rs. | = 2,32,57,389.27 |
| 3 | Benefits from Irrigated Produces for 76500 ha in Rs. | = 1,77,91,90,278.85 |

**Table 6**

**Evapotranspiration of Reference Crop at the Study Area ( Location - 23° N Altitude of 652 m)**

| Sr. No | Month | Mean Monthly Maximum Possible Duration of Sun Shine Hours | Mean Monthly Actual Duration of Sun Shine Hours | Mean Monthly Average Temperature in °C | Mean Monthly Maximum Relative Humidity in % | Mean Monthly Minimum Relative Humidity in % | Mean Monthly Wind Speed kms/hr | Wind Run Measurement Height | Average Daily Evapo-transpiration mm | Monthly Evapo-transpiration mm |
|---|---|---|---|---|---|---|---|---|---|---|
| 1 | January | 10 | 7 | 21.4 | 54 | 35 | 8.3 | 10 | 4.61 | 142.96 |
| 2 | February | 11.5 | 8.05 | 24.3 | 42 | 26 | 8.5 | 10 | 5.8 | 162.45 |
| 3 | March | 12 | 8.4 | 27.9 | 42 | 27 | 9.1 | 10 | 6.96 | 215.81 |
| 4 | April | 12.6 | 8.82 | 30.95 | 39 | 24 | 10.7 | 10 | 8.36 | 250.78 |
| 5 | May | 13.1 | 9.17 | 30.25 | 51 | 25 | 15 | 10 | 9.41 | 291.67 |
| 6 | June | 13.3 | 9.31 | 28.7 | 75 | 50 | 14.8 | 10 | 7.65 | 229.57 |
| 7 | July | 13.2 | 9.24 | 25.3 | 86 | 71 | 14 | 10 | 6.09 | 188.93 |
| 8 | August | 12.8 | 8.56 | 24.05 | 88 | 78 | 11.5 | 10 | 5.24 | 162.32 |
| 9 | September | 12.3 | 8.61 | 24.65 | 86 | 69 | 10.9 | 10 | 5.26 | 157.66 |
| 10 | October | 11.7 | 8.19 | 23.4 | 67 | 49 | 6.4 | 10 | 4.77 | 147.8 |
| 11 | November | 11.2 | 7.84 | 22.65 | 54 | 41 | 5.8 | 10 | 4.29 | 128.63 |
| 12 | December | 10.9 | 7.63 | 21.35 | 53 | 37 | 6.4 | 10 | 4.05 | 125.58 |
| Total | | | | | | | | | | 2204.16 |

**Table 7**
**Water requirement of various crops in the study area.**

**1. Reference Monthly Evapotranspiration (mm) and Rainfall (mm)**

| Months | Jun | Jul | Aug | Sep | Oct | Nov | Dec | Jan | Feb | Total |
|---|---|---|---|---|---|---|---|---|---|---|
| Reference Monthly Evapotranspiration (mm) | 229.57 | 188.93 | 162.32 | 157.66 | 147.80 | 128.63 | 125.58 | 142.96 | 162.45 | 1445.9 |
| Average Rainfall (mm) | 134.60 | 184.60 | 211.80 | 119.20 | 45.00 | 18.60 | 9.60 | 8.10 | 6.10 | 737.60 |
| Average First Fortnight Rainfall (mm) | 45.98 | 85.44 | 113.50 | 72.50 | 27.98 | 11.31 | 5.43 | 4.35 | 3.15 | |
| Average Second Fortnight Rainfall (mm) | 88.62 | 99.16 | 98.30 | 46.70 | 17.02 | 7.29 | 4.17 | 3.75 | 2.95 | |

**2. Water Requirement of different crops.**

| Sr. No. | Crops | Para-meters | Period of Cultivation | | Duration of Cultiva-tion | Water Requirement for various months in mm | | | | | | | | | | Net Water Requirement | |
|---|---|---|---|---|---|---|---|---|---|---|---|---|---|---|---|---|---|
| | | | From | To | | Jun | Jul | Aug | Sep | Oct | Nov | Dec | Jan | Feb | Total | 1000 ha | Total Area |
| **Two Seasonal Crops** | | | | | | | | | | | | | | | | | |
| 1 | L.S. Cotton | CF | 16-Jun | 27-Dec | 195 | 0.220 | 0.410 | 0.990 | 0.930 | 1.125 | 0.850 | 0.650 | | | | | |
| | | ETc | | | | 23.22 | 77.46 | 160.70 | 146.62 | 166.28 | 109.34 | 81.63 | | | 765.24 | | |
| | | ERF | | | | 23.22 | 77.46 | 141.12 | 86.53 | 36.00 | 13.52 | 6.74 | | | 384.59 | | |
| | | WR | | | | 0.00 | 0.00 | 19.58 | 60.09 | 130.28 | 95.82 | 74.89 | | | 380.65 | | |
| | | NWR | | | | 0.00 | 0.00 | 19.58 | 60.09 | 130.28 | 95.82 | 37.44 | | | 343.20 | 0.3432 | 26.2552 |
| 2 | Chillies | CF | 16-Jun | 12-Dec | 180 | 0.220 | 0.410 | 0.480 | 0.975 | 1.125 | 0.658 | 0.625 | | | | | |
| | | ETc | | | | 23.22 | 77.46 | 77.91 | 153.72 | 166.28 | 84.64 | 38.17 | | | 621.39 | | |
| | | ERF | | | | 23.22 | 77.46 | 77.91 | 87.53 | 36.00 | 10.98 | 3.37 | | | 316.47 | | |
| | | WR | | | | 0.00 | 0.00 | 0.00 | 66.19 | 130.28 | 73.66 | 34.80 | | | 304.92 | | |
| | | NWR | | | | 0.00 | 0.00 | 0.00 | 66.19 | 130.28 | 73.66 | | | | 270.12 | 0.2161 | 16.5315 |
| 3 | Turmeric | CF | 16-Jun | 26-Jan | 225 | 0.220 | 0.410 | 0.480 | 0.478 | 1.150 | 1.150 | 0.840 | 0.750 | | | | |
| | | ETc | | | | 23.22 | 77.46 | 77.91 | 75.36 | 169.97 | 147.92 | 105.49 | 107.22 | | 784.56 | | |
| | | ERF | | | | 23.22 | 77.46 | 77.91 | 75.19 | 36.21 | 14.65 | 6.95 | 5.96 | | 317.55 | | |
| | | WR | | | | 0.00 | 0.00 | 0.00 | 0.17 | 133.76 | 133.27 | 98.54 | 101.26 | | 467.00 | | |
| | | NWR | | | | 0.00 | 0.00 | 0.00 | 0.17 | 133.76 | 133.27 | 98.54 | 50.63 | | 416.37 | 0.0833 | 6.3705 |
| **Kharif Crops** | | | | | | | | | | | | | | | | | |
| 1 | Sunflower | CF | 16-Jun | 23-Sep | 100 | 0.220 | 0.410 | 1.000 | 0.500 | | | | | | | | |
| | | ETc | | | | 23.22 | 77.46 | 162.32 | 78.83 | | | | | | 341.83 | | |
| | | ERF | | | | 23.22 | 77.46 | 141.64 | 74.13 | | | | | | 316.45 | | |
| | | WR | | | | 0.00 | 0.00 | 20.68 | 4.70 | | | | | | 25.38 | | |
| | | NWR | | | | 0.00 | 0.00 | 20.68 | 2.35 | | | | | | 23.03 | 0.0230 | 1.7618 |
| 2 | Hv.Jowar | CF | 16-Jun | 12-Oct | 120 | 0.220 | 0.410 | 0.885 | 0.830 | 0.600 | | | | | | | |
| | | ETc | | | | 23.22 | 77.46 | 143.65 | 130.86 | 45.38 | | | | | 420.57 | | |
| | | ERF | | | | 23.22 | 77.46 | 137.75 | 85.26 | 17.70 | | | | | 341.39 | | |
| | | WR | | | | 0.00 | 0.00 | 5.90 | 45.60 | 27.68 | | | | | 79.18 | | |
| | | NWR | | | | 0.00 | 0.00 | 5.90 | 45.60 | 0.00 | | | | | 51.50 | 0.1030 | 7.8797 |

| | | | | | | | | | | | | | | | |
|---|---|---|---|---|---|---|---|---|---|---|---|---|---|---|---|
| **3 Tomatoes** | CF | 16-Jun | 23-Sep | 100 | 0.220 | 0.410 | 1.050 | 0.600 | | | | | | | |
| | ETc | | | | 23.22 | 77.46 | 170.44 | 94.60 | | | | | 365.71 | | |
| | ERF | | | | 23.22 | 77.46 | 143.92 | 77.37 | | | | | 321.97 | | |
| | WR | | | | 0.00 | 0.00 | 26.52 | 17.23 | | | | | 43.74 | | |
| | NWR | | | | 0.00 | 0.00 | 26.52 | 17.23 | | | | | 43.74 | 0.0437 | 3.3463 |
| Rabi Crops | | | | | | | | | | | | | | | |
| **1 Wheat** | CF | 16-Oct | 12-Feb | | | | | 0.280 | 0.300 | 0.920 | 0.700 | 0.225 | | | |
| | ETc | | | | | | | 19.79 | 38.59 | 115.53 | 100.17 | 16.87 | 290.86 | | |
| | ERF | | | | | | | 8.08 | 12.06 | 7.00 | 5.33 | 1.42 | 34.39 | | |
| | WR | | | | | | | 11.71 | 26.53 | 108.53 | 94.24 | 15.45 | 256.47 | | |
| | NWR | | | | | | | 11.71 | 26.53 | 108.53 | 94.24 | | 241.02 | 0.3615 | 27.6567 |
| **2 Gram** | CF | 16-Oct | 28-Jan | | | | | 0.280 | 0.300 | 1.100 | 0.420 | | | | |
| | ETc | | | | | | | 19.79 | 38.59 | 138.14 | 60.04 | | 256.56 | | |
| | ERF | | | | | | | 8.08 | 12.06 | 7.39 | 5.37 | | 32.90 | | |
| | WR | | | | | | | 11.71 | 26.53 | 130.75 | 54.67 | | 223.66 | | |
| | NWR | | | | | | | 11.71 | 26.53 | 130.75 | 27.34 | | 196.33 | 0.2945 | 22.5284 |
| **3 Groundnut** | CF | 16-Oct | 12-Feb | | | | | 0.280 | 0.300 | 0.844 | 0.800 | 0.575 | | | |
| | ETc | | | | | | | 19.79 | 38.59 | 105.99 | 75.05 | 43.11 | 282.53 | | |
| | ERF | | | | | | | 8.08 | 12.06 | 6.95 | 5.64 | 1.98 | 34.71 | | |
| | WR | | | | | | | 11.71 | 26.53 | 99.04 | 69.41 | 41.13 | 247.82 | | |
| | NWR | | | | | | | 11.71 | 26.53 | 99.04 | 69.41 | | 206.70 | 0.3514 | 26.8807 |
| **4 Sunflower** | CF | 16-Oct | 23-Jan | | | | | 0.280 | 0.300 | 1.050 | 0.525 | | | | |
| | ETc | | | | | | | 19.79 | 38.59 | 131.86 | 75.05 | | 265.29 | | |
| | ERF | | | | | | | 8.08 | 12.06 | 7.23 | 5.64 | | 33.01 | | |
| | WR | | | | | | | 11.71 | 26.53 | 124.63 | 69.41 | | 232.28 | | |
| | NWR | | | | | | | 11.71 | 26.53 | 124.63 | 34.71 | | 197.58 | 0.0593 | 4.5344 |
| **5 Cauliflower** | CF | 15-Oct | 28-Jan | | | | | 0.280 | 0.300 | 1.250 | 0.700 | | | | |
| | ETc | | | | | | | 19.79 | 38.59 | 156.98 | 100.07 | | 315.43 | | |
| | ERF | | | | | | | 8.08 | 12.06 | 7.78 | 5.83 | | 33.75 | | |
| | WR | | | | | | | 11.71 | 26.53 | 149.20 | 94.24 | | 281.68 | | |
| | NWR | | | | | | | 11.71 | 26.53 | 149.20 | 94.24 | | 281.68 | 0.2817 | 21.5484 |
| Total Net Water Requirement | | | | | | | | | | | | | | 2.1607 | 165.2936 |
| Total Water Requirement Considering an Irrigation Efficiency of 50% | | | | | | | | | | | | | | 4.3214 | 330.5871 |

**Table 8**

**Irrigation Charges for irrigating various Crops in the study area.**

| Sl. No. | Crop | Duration of Crops in Days | Irrigation Charges for Kharif Season @ Rs.120 per month | | Irrigation Charges for Rabi Season @ Rs.165 per month | | Total Irrigation Charges in Rs. |
|---|---|---|---|---|---|---|---|
| | | | Months to be irrigated | Amount in Rs. | Months be irrigated | Amount in Rs. | |
| **I) Two Seasonal Crops** | | | | | | | |
| 1 | LS Cotton | 190 | 2 | 240.00 | 3 | 495.00 | 735.00 |
| 2 | Chillies | 180 | 1 | 120.00 | 2 | 330.00 | 450.00 |
| 3 | Turmeric | 225 | | 0.00 | 4 | 660.00 | 660.00 |
| **II) Kharif Crops** | | | | | | | |
| 4 | Sunflower | 130 | 2 | 240.00 | | | 240.00 |
| 5 | Hy.Jowar | 120 | 2 | 240.00 | | | 240.00 |
| 6 | Tomatoes | 100 | 2 | 240.00 | | | 240.00 |
| **III) Rabi Crops** | | | | | | | |
| 7 | Wheat | 120 | | | 4 | 660.00 | 660.00 |
| 8 | Gram | 75 | | | 4 | 660.00 | 660.00 |
| 9 | Groundnut | 120 | | | 4 | 660.00 | 660.00 |
| 10 | Sunflower | 130 | | | 4 | 660.00 | 660.00 |
| 11 | Cauliflower | 90 | | | 4 | 660.00 | 660.00 |

4 Benefits from Irrigated Produces for 76500 ha in Rs. (in lakhs) = 1,77,91.90

5 Increase in income in the area due to irrigation in Rs. = 1,47,96,34,597.20

6 Increase in income in the area due to irrigation in Rs. (in lakhs) = 1,47,96.35

## Conclusions

The paper describes an approach to estimate the agricultural benefits/ economics due to irrigation projects. This methodology, the environmental and ecological economics due to dam construction of dam needs to integrated with economics of construction of dam to project the feasibility of the dam. As the execution of irrigation project takes years and the prices of agricultural inputs and outputs fluctuate, this approach needs to be reviewed from time to time.

Table 9

Crop wise Gross Benefit per Hectare for the proposed Irrigated Crops in the Command Area of JDP

| Sl. No. | Crop | Yield in Quintals/ ha | Rate in Rs. per Qtls | Total Value Agricultural Produce in Rs./ha | Cost of Farm House Practices in Rs./ha | Value after Deduction in Farm House Practices in Rs./ha | Value of Fodder Receipts in Rs./ha | Value of Farm Yard Manure/ Dung Receipts in Rs./ha | Gross Benefits in Rs./ha |
|---|---|---|---|---|---|---|---|---|---|
| I) | Two Seasonal Crops | | | | | | | | |
| 1 | LS Cotton | 20 | 1,445.47 | 28909.46 | 2890.95 | 26018.51 | | 780.56 | 26799.07 |
| 2 | Chillies | 10 | 2,598.71 | 25987.15 | 2598.71 | 23388.43 | | 701.65 | 24090.08 |
| 3 | Turmeric | 200 | 1,330.00 | 266000.00 | 26600.00 | 239400.00 | | 7182.00 | 246582.00 |
| II) | Irrigated Kharif Crops | | | | | | | | |
| 4 | Sunflower | 15 | 1,155.43 | 17331.42 | | 17331.42 | | 519.94 | 17851.37 |
| 5 | Hy.Jowar | 25 | 442.36 | 11059.07 | | 11059.07 | 1658.86 | 331.77 | 13049.71 |
| 6 | Tomatoes | 160 | 282.00 | 45120.00 | 4512.00 | 40608.00 | | 1218.24 | 41826.24 |
| III) | Unirrigated Kharif Crops | | | | | | | | |
| 7 | Groundnut | 7 | 1,165.56 | 8158.92 | | 8158.92 | 815.89 | 244.77 | 9219.58 |
| 8 | Tur | 10 | 1,683.65 | 16836.55 | | 16836.55 | | 505.10 | 17341.65 |
| 9 | Sesamum (Til) | 4 | 2,371.60 | 9486.41 | | 9486.41 | | 284.59 | 9771.00 |
| IV) | Rabi Crops | | | | | | | | |
| 10 | Wheat | 25 | 706.80 | 17670.05 | | 17670.05 | 1767.01 | 530.10 | 19967.16 |
| 11 | Gram | 15 | 1,061.41 | 15921.14 | | 15921.14 | | 477.63 | 16398.77 |
| 12 | Groundnut | 15 | 1,165.56 | 17483.40 | | 17483.40 | 1748.34 | 524.50 | 19756.24 |
| 13 | Sunflower | 15 | 1,155.43 | 17331.42 | | 17331.42 | | 519.94 | 17851.37 |
| 14 | Cauliflower | 200 | 869.00 | 173800.00 | 17380.00 | 156420.00 | | 4692.60 | 161112.60 |

**Table 10**

**Crop wise Expenditure per Hectare for Proposed Irrigated Crops in the Command Area of JDP**

| Sl. No. | Crop | Duration of Crops in Days | Seed Cost per Hec. in Rs | Fertilizer and manure cost in Rs. | Fodder Expenses | Implement Charges in Rs. | Labour Charges in Rs. | Cost of pesticides In Rs. | Irrigation Charges in Rs. | Interest in Working Capital in Rs. | Total Expenses in Rs. |
|---|---|---|---|---|---|---|---|---|---|---|---|
| **I) Two Seasonal Crops** | | | | | | | | | | | |
| 1 | LS Cotton | 20 | 250.00 | 5259.00 | 2601.85 | 780.56 | 5203.70 | 2080.50 | 735.00 | 1174.48 | 18085.09 |
| 2 | Chillies | 10 | 400.00 | 2773.00 | 2338.84 | 701.65 | 4677.69 | 2647.50 | 450.00 | 896.81 | 14885.49 |
| 3 | Turmeric | 200 | 75000.00 | 17100.00 | 23940.00 | 7182.00 | 47880.00 | 2142.00 | 660 | 13936.14 | 187840.14 |
| **II) Irrigated Kharif Crops** | | | | | | | | | | | |
| 4 | Sunflower | 15 | 1750.00 | 2189.00 | 1733.14 | 519.94 | 3466.28 | | 240.00 | 352.54 | 10250.91 |
| 5 | Hy.Jowar | 25 | 320.00 | 3234.00 | 1105.91 | 331.77 | 2211.81 | 936.00 | 240.00 | 358.14 | 8737.63 |
| 6 | Tomatoes | 160 | 250.00 | 1295.00 | 4060.80 | 1218.24 | 8121.60 | 1005.00 | 240.00 | 576.65 | 16767.29 |
| **III) Unirrigated Kharif Crops** | | | | | | | | | | | |
| 7 | Groundnut | 7 | 1800.00 | 548.00 | 815.89 | 244.77 | 1631.78 | 435.00 | 0.00 | 234.02 | 5709.46 |
| 8 | Tur | 10 | 2250.00 | 2796.00 | 1683.65 | 505.10 | 3367.31 | 277.50 | 0.00 | 658.74 | 11538.30 |
| 9 | Sesamum(Til) | 4 | 200.00 | 825.00 | 948.64 | 284.59 | 1897.28 | 105.00 | 0.00 | 136.57 | 4397.08 |
| **IV) Rabi Crops** | | | | | | | | | | | |
| 10 | Wheat | 25 | 1375.00 | 1619.00 | 1767.01 | 530.10 | 3534.01 | 1584.00 | 660.00 | 473.09 | 11542.21 |
| 11 | Gram | 15 | 1250.00 | 1296.00 | 1592.11 | 477.63 | 3184.23 | 675.00 | 660.00 | 341.62 | 9476.60 |
| 12 | Groundnut | 15 | 1800.00 | 548.00 | 1748.34 | 524.50 | 3496.68 | 445.50 | 660.00 | 394.19 | 9617.21 |
| 13 | Sunflower | 15 | 1750.00 | 2189.00 | 1733.14 | 519.94 | 3466.28 | 549.00 | 660.00 | 387.06 | 11254.43 |
| 14 | Cauliflower | 200 | 600.00 | 2169.00 | 15642.00 | 4692.60 | 31284.00 | 940.50 | 660.00 | 2093.80 | 58081.90 |

Table 11

**The Net Benefits from the Irrigated Agricultural Produces per 1000 Hec. in the Projected Scenario of the Study Area of JDP**

| Sl. No. | Crop | Net Benefit per ha in Rs. | Percentage Area of Cultivation | Area of the Crop in ha/ 1000 ha of cultivated area | Benefit in Rs. for 1000 Hec. |
|---|---|---|---|---|---|
| **I) Two Seasonal Crops** | | | | | |
| 1 | LS Cotton H2 | 8713.98 | 10.00 | 100.00 | 871398.1008 |
| 2 | Chillies | 9204.59 | 8.00 | 80.00 | 736367.4759 |
| 3 | Turmeric | 58741.86 | 2.00 | 20.00 | 1174837.151 |
| **II) Irrigated Kharif Crops** | | | | | |
| 4 | Sunflower | 7600.45 | 10.00 | 100.00 | 760045.1549 |
| 5 | Hy.Jowar | 4312.08 | 20.00 | 200.00 | 862415.1437 |
| 6 | Tomatoes | 25058.95 | 10.00 | 100.00 | 2505894.707 |
| **III) Unirrigated Kharif Crops** | | | | | |
| 7 | Groundnut | 3510.12 | 20.00 | 200.00 | 702023.4081 |
| 8 | Tur | 5803.35 | 10.00 | 100.00 | 580334.9014 |
| 9 | Sesamum (Til) | 5373.92 | 10.00 | 100.00 | 537391.6221 |
| **IV) Rabi Crops** | | | | | |
| 10 | Wheat | 8424.95 | 15.00 | 150.00 | 1263742.553 |
| 11 | Gram | 6922.17 | 15.00 | 150.00 | 1038325.83 |
| 12 | Groundnut | 10139.03 | 17.00 | 170.00 | 1723635.196 |
| 13 | Sunflower | 6596.94 | 3.00 | 30.00 | 197908.1766 |
| 14 | Cauliflower | 103030.70 | 10.00 | 100.00 | 10303069.85 |
| | Total | | | | 23257389.27 |

## References

ASCE 1995, ASCE Manual and Reports on Engineering Practice No. 28, 2nd ed., American Society of Civil Engineers.

Brunt D. 1952, Physical and dynamical meteorology, 2nd ed. University Press, Cambridge. 428 pp.

Burman R.D., Jensen M.E. and Allen R.G. 1987, Thermodynamic factors in evapotranspiration. In: Proc. Irrig. and Drain. Spec. Conf., James L.G. and English M.J. (eds). ASCE, Portland, Ore., July. pp. 28-30.

Campbell, D. E. 1995, Design and Operation of Smallholder Irrigation in South Asia, Irrigation and Drainage Series (World Bank Technical Paper Number 256), The World Bank, Washington, D.C.

Craig J.C. 1984, Basic routines for the Casio computer. Wayne Green Books, Peterborough, NH 03458. 131 pp.

Doorenbos J, Pruitt, W.O. 1975, Guidelines for predicting crop water requirements. Irrigation and Drainage Paper, 24, Food and Agricultural Organisation of United Nations, Rome, Italy.

Doorenbos J, Pruitt, W.O. 1977, Crop water requirements. Irrigation and Drainage Paper, 24, Food and Agricultural Organisation of United Nations, Rome, Italy.

Duffie J.A. and Beckman W.A. 1991, Solar engineering of thermal processes.,2nd ed., John WIley and Sons, New York. pp. 1-146.

FAO 1990, Food and Agricultural Organisation of United Nations, Rome, Italy 28-31 May 1990 Expert Consultatation on Revision of FAO Methodologies for Crop Water Requirement.

Gommes R. A. 1983, Pocket computers in agrometeorology. FAO Plant Production and Protection Paper 45, Rome.

Guyat G 1997, Physics of the environment and climate, John Wiley & Sons

Harrison L.P. 1963, Fundamental concepts and definitions relating to humidity. In: Humidity and Moisture. Vol. 3. Wexler A. (ed). Reinhold Publishing Company, New York.

IMD, Climatological Tables of observatories of India (1931-1960), Indian Metrological Department, New Delhi, India.

Jensen, M.E., Burman R.D., and Allen R.G. 1990, Evapotranspiration and Irrigation Water Requirements. ASCE Manual and Reports of Engineering Practices No 70, 350p.

Murray F.W. 1967, On the computation of saturation vapor pressure. J. Appl. Meteor. 6:203-204.

London, J and Fröhlich, C 1982, Extended abstracts presented at the symposium of the solar constants and the spectral distribution of solar irradiance. IAMAP Third Scientific Assembly, Aug 17-28, 1981, Hamburg. Federal Republic of Germany.

PAU 1999, Agricultural Directory and Dairy – Punjabrao Agricultural Universty, Akola, Maharashtra, India

Tetens O. 1930, Uber einige meteorologische Begriffe. Z. Geophys. 6:297-309.

USDA 1969, Engineering field manual for Conservation Practices, United States Department of Agriculture, Soil Conservation Services.

# 6

# Resource Conservation through Land Treatment of Municipal Wastewater

*P. R. Thawale, Asha. A. Juwarkar,*
*S. K. Singh and A.V. Kulkarni*

Land application of domestic sewage effluent is a cost-effective method of treatment and disposal. Rapid infiltration of sewage into sandy soils can result in improvement of its physical, chemical and microbiological quality (Iskandar, 1981). Reduction in nitrates, BOD and bacteria may be achieved after the passage of sewage effluent through a few meters of soil (Bouwer, 1985). Land treatment, especially rapid infiltration of municipal wastewater into relatively permeable soils, is becoming an important treatment method due to the increasing need for water reuse and protection of groundwater in many areas of the world (Asano, 1985). Land application of domestic and industrial wastewater is gaining momentum owing to the fact that land application provides primary, secondary and tertiary treatment to the waste, all in a single operation with recycling and reuse benefit of wastewater and nutrients for biomass production (Idlovich and Michail, 1984; Smith and Schroeder, 1985; Witherow and Bledsoe, 1986; Juwarkar *et al.*, 1995) besides preventing the pollution of streams and lakes. In view of the fact that fiscal resources available with the municipal agencies are inadequate, which pose constraints in putting up the conventional facilities, land treatment in conjunction with conventional treatment, or alone in worst case, is an ideal proposition. The in-plant treatment of both domestic and industrial wastewaters, using sophisticated treatment technologies like trickling filters, activated sludge process, aerated lagoons, bio-methanation and incineration etc. are energy and cost intensive and beyond the reach of developing countries. Further, the discharge of treated wastewater enriched with nutrients and with other pollutants causes eutrophication of lakes/rivers, besides creating aesthetic problem. This warrants adoption of an appropriate wastewater management system wherein twin benefits of treatment and recycling and reuse can be achieved, and eco-development can be promoted.

Presently, 6351 million cubic meter of wastewater is being generated per year from 212 class I cities and 242 class II towns in the country of which only 36% in class I cities and 14% in class II towns are collected due to limited treatment facilities. Hence conventional approach of wastewater treatment and its disposal in water bodies need to be reviewed. The existing wastewater treatment facilities are mostly confined to primary level, providing treatment to the extent of 20% and 45% of wastewater respectively in class I cities and class II towns (CPCB Report, 1988). The nutrient value of the available wastewater is estimated to be 317,000 tonnes nitrogen, 140,000 tonnes phosphate and 191,000 tonnes of potash per year. The economic value of the available wastewater is estimated to be Rs. 2592 lakhs. Further, it can also increase the irrigated command area by 1,70,400 hectares in the country. The present practice of land application is disposal oriented and hence the hydraulic and nutrient loading ranges from 22.1 - 336.8 $m^3ha^{-1}day^{-1}$ and 403 - 6062 $kgha^{-1}year^{-1}$ respectively which is 2-15 times higher than the normal loading (Juwarkar *et al.*, 1991). The excessive hydraulic and nutrient loading regardless of soil-plant assimilative capacity results in soil sickness, salinity and sodicity problem. Moreover, imbalance in C:N (Carbon: Nitrogen) ratio in plants induced more foliage and less fruiting particularly in case of vegetables and flowering plants (Juwarkar, 1989). The application of wastewater to agricultural crop and forestry plantation with appropriate assimilative capacity based design provide treatment to wastewater through the physico-chemical renovating potential of dynamic multicomponent soil ecosystem. It also promotes recycling and reuse of water, nutrients for biomass generation, replenishment of natural resource and recharge of ground water as (Fig. 1). The treatment efficiency of land application and conventional treatment systems given in Table1, further confirmed the fact that land application provide the highest degree of renovation (Juwarkar *et al.*, 1998).

In disposal of wastewater through crop irrigation, soil works as a living treatment system. Soil and its associated ecosystem components act as physico-bio-chemical reactor capable of treating or stabilizing the pollutants of solid and liquid origin through degradation, adsorption, precipitation and utilization by crop. The wastewater while passing through the soil matrix provides filtration on the soil surface leading to removal of coarse particles (Primary clarifier). The degradation of soluble organic pollutants in soil profile by microbial action and mixing and aeration extended by macro soil habitant (earthworms and macrofauna) represent the waste treatment process occurring in the aeration tank. The wastewater from the soil reactor is subjected to final polishing /

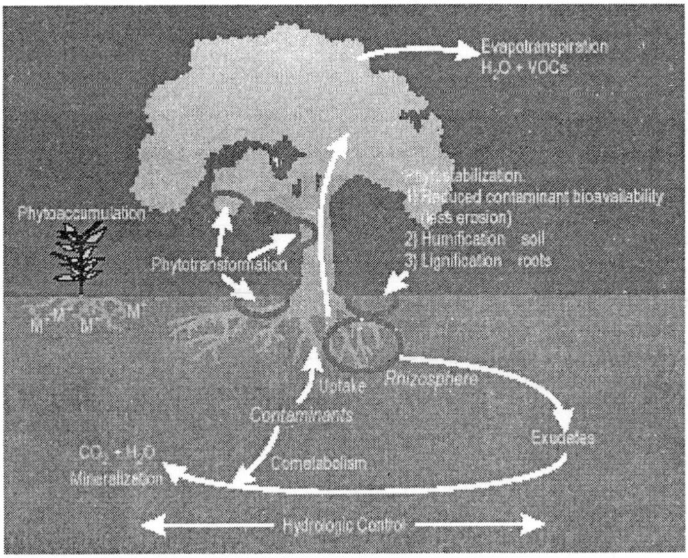


**Fig. 1. Phytoremediation: (i) in the rhizosphre, exudates stimulate the growth and activity of contaminant-degrading microorganisms, (ii) contaminats are transformed and degraded by plant specific enzymes in the roots, stems, and leaves, and may be volatilized via evapotranspiration, (iii) a zone of hydrologic control is established that can control the migration of contaminants in the vodose zone and may control and remediate contaminant plumes, (iv) by stabilizing the soil, plants minimize erosion and decrease contaminant mobility, and (v) hyperaccumulating plants take up and immobilize heavymetals.**

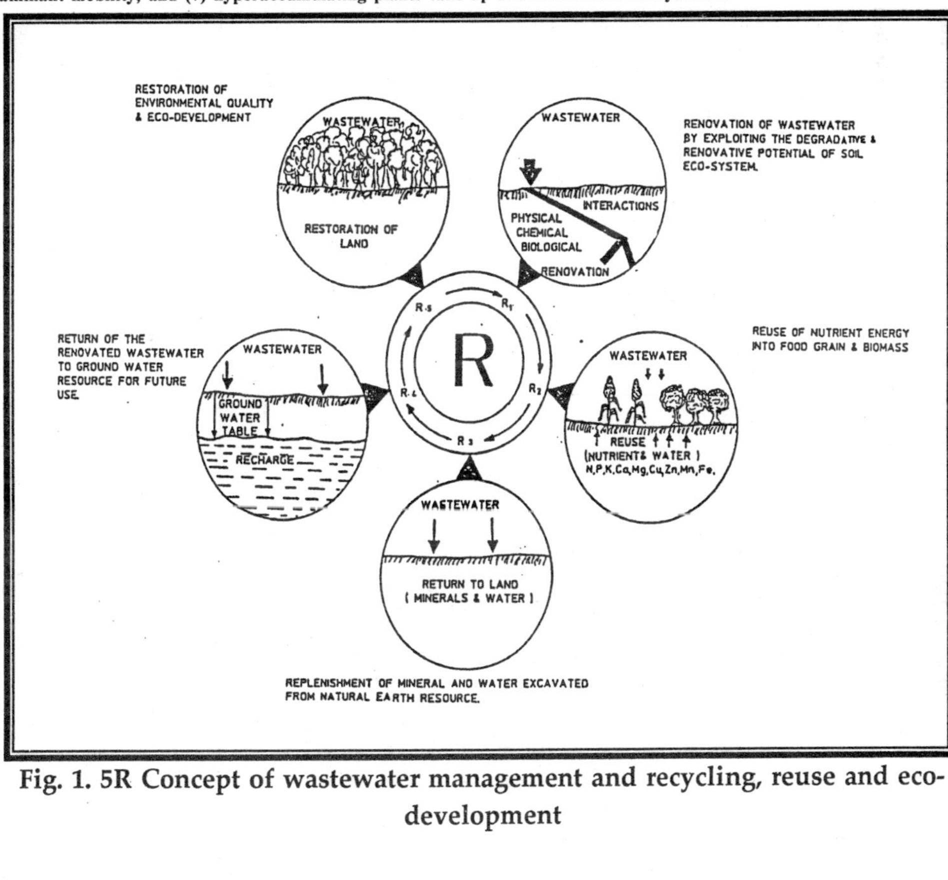


**Fig. 1. 5R Concept of wastewater management and recycling, reuse and eco-development**

**Table 1**

**Typical characteristics of treated wastewater from conventional, advanced wastewater treatment and land treatment systems**

| System | Residual Effluent constituent, $mgL^{-1}$ | | | | | | Pathogens removal, % | |
|---|---|---|---|---|---|---|---|---|
| | BOD | SS | $NH_3$-N | $NO_3$-N | Total N | Total P | Bacteria | Viruses |
| *Conventional treatment* | | | | | | | | |
| Aerated lagoon | 35 | 40 | 10 | 20 | 30 | 8 | 90-95 | 16-94 |
| Activated sludge | 20 | 25 | 20 | 10 | 30 | 8 | 85-99 | 91-98 |
| *Advanced Wastewater treatment* | | | | | | | | |
| Biological nitrification | 12 | 15 | 1 | 29 | 30 | 8 | -- | -- |
| Biological nitrification denitrification | 15 | 16 | -- | -- | 3 | 8 | -- | -- |
| Tertiary, two stage lime coagulation & filtration | 5 | 5 | 20 | 10 | 30 | 0.5 | 99-99.9 | 90-99.7 |
| Tertiary, two stage lime coagulation, filtration, selective ion exchange | 5 | 5 | -- | -- | 3 | 0.5 | 99-99.9 | 99-99.9 |
| *Land Treatment Systems* | | | | | | | | |
| Slow rate | 1 | 1 | 0.5 | 2.5 | 3 | 0.1 | 99-100 | 97-99 |
| Overland flow | 5 | 5 | 0.5 | 2.5 | 3 | 5.0 | 90-100 | 89-99 |
| Rapid infiltration | 5 | 1 | -- | 10 | 10 | 2.0 | 90-99 | 90-99 |

renovation for metals removal through adsorption and ion exchange. The suspended solids and bacterial biomass removal through adsorption, ion exchange and precipitation with hydroxides and carbonate indicates reaction processes as occurring in secondary clarifier. Removal of assailable macro and micronutrients through plant utilization resembles the tertiary waste treatment. All the renovative processes are natures own treatment process and thus land application is an eco-friendly treatment and disposal system.

The results that are based on limited data of actual systems and designs where alternatives are compared in USA (Overcash and Pal, 1979) substantiates the economic advantages of land treatment. After secondary treatment in a conventional sewage treatment plant, normally some 25-75% of phosphorus, 90% of nitrogen and about 10% of both organic matter and suspended solids remain in the effluent. This treated effluent when discharged to water bodies becomes a potential source of pollution there. To remove nutrients and other pollutants effectively from wastewater, some form of tertiary treatment, which is economical and easy to operate is required. The land treatment systems like HRTS can match with this requirement. The high transpiration capacity of plants grown on soil matrix enables the system to serve as a biopump. The plants such as bamboo (Bambusa arundinacea), acacia (Acacia mangium), neem (Azadirachta indica), shishum (Dalbergia sissoo) and eucalyptus (Eucalyptus hybrid) transpire water equivalent of 7 to 13 times of potential evapo-transpiration from the soil matrix alone. The HRTS enables the disposal of 350-450 $m^3$ of wastewater per hectare of land area per day (Juwarkar *et al.*, 1994; 1998 and 2003). As all the wastewater is utilized in this process, the ground water pollution problem is minimized. The nutrients present in the wastewater are used by the plants and partly retained in the soil matrix without affecting the soil ecosystem.

## Advantages

- The methods work on natural treatment processes and could thus be termed as eco-friendly.
- Facilitates development of forests and green belts in urban agglomerations.
- Generates sink potential for air pollutants.
- Low-energy and low-cost compared to conventional wastewater treatment systems.
- Ease of installation and simplicity of operations.
- Biomass generation and revenue returns.

## Environmental Benefits

- Reuse of effluent and its nutrient contents for biomass production.

- Stabilization of organic matter and renovation of effluent.
- Prevention of ground and surface water pollution.
- Green belt development with concomitant advantages to the environment.
- HRTS is an ecotechnology that harnesses nature's cleansing processes, and offers excellent opportunity for low cost, operationally simple, and energy efficient treatment and utilization of industrial effluent with rich dividends through the development of luxurious forests.

HRTS is a low cost, simple to operate technology and more suitable for treatment of domestic wastewater. However, insufficient information is available on the design and operation of HRTS in the country. Therefore, laboratory investigations were carried out to ascertain the efficacy of HRTS in the renovation of primary treated wastewater using *Dendrocalamus strictus* and *Casurina equisitifolia* species of plants. NEERI uses many methods to clean up pollution at metal contaminated sites in India. Some of them includes phytoremediation, bioremediation and biodegradation are considered new or *innovative*. Such methods can be quicker and cheaper than more common methods. The synergistic action of the plant and the microbes degrade or contain the pollutant of wastewater including heavy metals (Fig 4).

## Materials and Methods

The laboratory studies were carried out at the National Environmental Engineering Research Institute (NEERI), Nagpur in Maharashtra State of India representing hot semi arid climate. Cement pipes having a diameter of 40 cm placed at a depth of 80 cm filled with sandy loam soil were employed in the study. The plants of high transpiration potential, *Dendrocalamus strictus* and *Casurina equisitifolia* were grown in one area whereas another area was without plants. Four weeks after planting, primary treated wastewater was applied at the hydraulic loading rate of 5 cm $ha^{-1}$ $day^{-1}$ over the surface of the HRTS adopting a down flow system. The treated effluent was collected from a vertical tube, draining from the bottom of the cement pipe.

HRTS was installed at sea beach, Puri in Orissa state of India. Wastewater was applied in the specially designed field layouts that consist of ridges with plants and furrows laid with filter media. Impact network for treatment and disposal of wastewater of the system is given in Fig. 2. The furrows represent the treatment system resembling shallow oxidation pond. The design of ridges and furrows depend on plants to be grown. In case of *Dendrocalamus strictus* plants ridges were 1.5 m to 2.0 m wide and 0.5 m in height. The furrows were 3 m – 4 m wide and 0.5 m deep. The HRTS with ridges of 0.8 to 1.0 m wide are more suitable for plants like neem, shishum, acacia and *Casurina equisitifolia* etc. Details of ridge and furrow dimensions are given in Table 2 and schematic diagram of HRTS is depicted in Fig. 3. Length of the ridge depends on texture and permeability of soil and it varies from 5-20 m. Chabra (1989) Recommended ridges of 1 m width and 0.5 m height for Eucalyptus Hybrid. The design details of HRTS for some plants were given in Table-3.

**Table 2**

**Comparative cost of industrial waste management alternatives**

| **Type of Wastewater** | **Land Treatment Alternative** | | **Conventional Treatment Alternative** |
|---|---|---|---|
| | **Size (ha)** | **Investment cost (US$)** | **Investment cost (US$)** |
| Pharmaceutical | 49 | 490,000 | 745,000 |
| Poultry | | 220,000 | 720,000 |
| Potato | 67 | 350,000 | 780,000 |
| Nylon & Polyester | 12 | 140,000 | 300,000 |
| Refinery | 10 | 6.5/wet ton | 11.5/wet ton (Incineration) |
| Papermill | 10 | 20000 $/ m^3ha^{-1}$ | 10-1000 $/ m^3ha^{-1}$ |
| Muncipal wastewater | 10 | 10-100 $/ m^3ha^{-1}$ | 20000 $/ m^3ha^{-1}$ |

Influent and effluent samples were collected regularly and total nitrogen and phosphate were analysed as per Standard Methods for the Examination of Water and Wastewater [17]. Plants height was measured periodically and soil samples collected from HRTS were analysed by standard techniques (Black, 1965 and Jackson 1973)..

## Results and Discussions

### *Microbial Activity and Organic Decomposition in the Soil*

The microbiological processes of the soil are complex, and it is neither possible nor practical to do more than pint out some principles concerning microbial decomposition in the soil in this bulletin. It should be noted that purification of

**Table 3**

**Design Details of HRTs for some plants**

| Type of Plants | Ridges | | Furrows | | Plant to Plants |
|---|---|---|---|---|---|
| | Width, m | Height, m | Width, m | Depth, m | Distance on Ridges, m |
| **Bamboo** | | | | | |
| *Dendrocalamus strictus* | 1.5 | 0.5 | 3.0 | 0.5 | 3.0 |
| *Bambusa arundincea* | 2.0 | 0.5 | 4.0 | 0.5 | 2.0 |
| **Neem** | | | | | |
| *Azadiractha indica* | 0.8 | 0.5 | 1.6 | 0.5 | 2.0 |
| **Casurina** | | | | | |
| *Casurina equisitifolia* | 0.8 | 0.5 | 1.6 | 0.5 | 2.0 |
| ***Shishum*** | | | | | |
| *Delbergia sissoo* | 0.8 | 0.5 | 1.6 | 0.5 | 2.0 |
| ***Acacia*** | | | | | |
| *Acacia mengium* | 0.8 | 0.5 | 1.6 | 0.5 | 2.0 |
| ***Eucalyptus*** | | | | | |
| *Eucalyptus hybrid* | 0.8 | 0.5 | 1.6 | 0.5 | 2.0 |
| *Eucalyptus robusta* | 1.0 | 0.5 | 2.0 | 0.5 | 1.5 |

*Source*: *Juwarkar et al.,* Management of Wastewater through crop irrigation – An ecofriendly approach, Ecotechnology for Pollution Control & Environment Management. Enviro Media, 1998,

organic effluents applied to the soil, or organic sludge incorporated in the soil, is accomplished by the interrelated activities of the microbial population acting in conjunction with surface adsorption phenomena in much the same manner as in activated sludge units, trickling filters and sewage sand filters. Recent order of magnitude estimates of soil microbial numbers indicate the presence of about 100 million bacteria, 10 million actinomycetes, 1 million fungi, and 100 thousand algae per gram of soil, with a total live weight of 2400 pounds per acre of plow depth. Their activity is shown by the rapid decomposition of crop residues and the differences in composition of such residues and the differences in; composition of such residues and the organic matter of the soil. Of the latter, from 1/3 to ½ is estimated to be microbial in origin.

### *Irrigation Requirements of Various Crops*

Blaney and Criddle proposed a method for calculating irrigation requirements

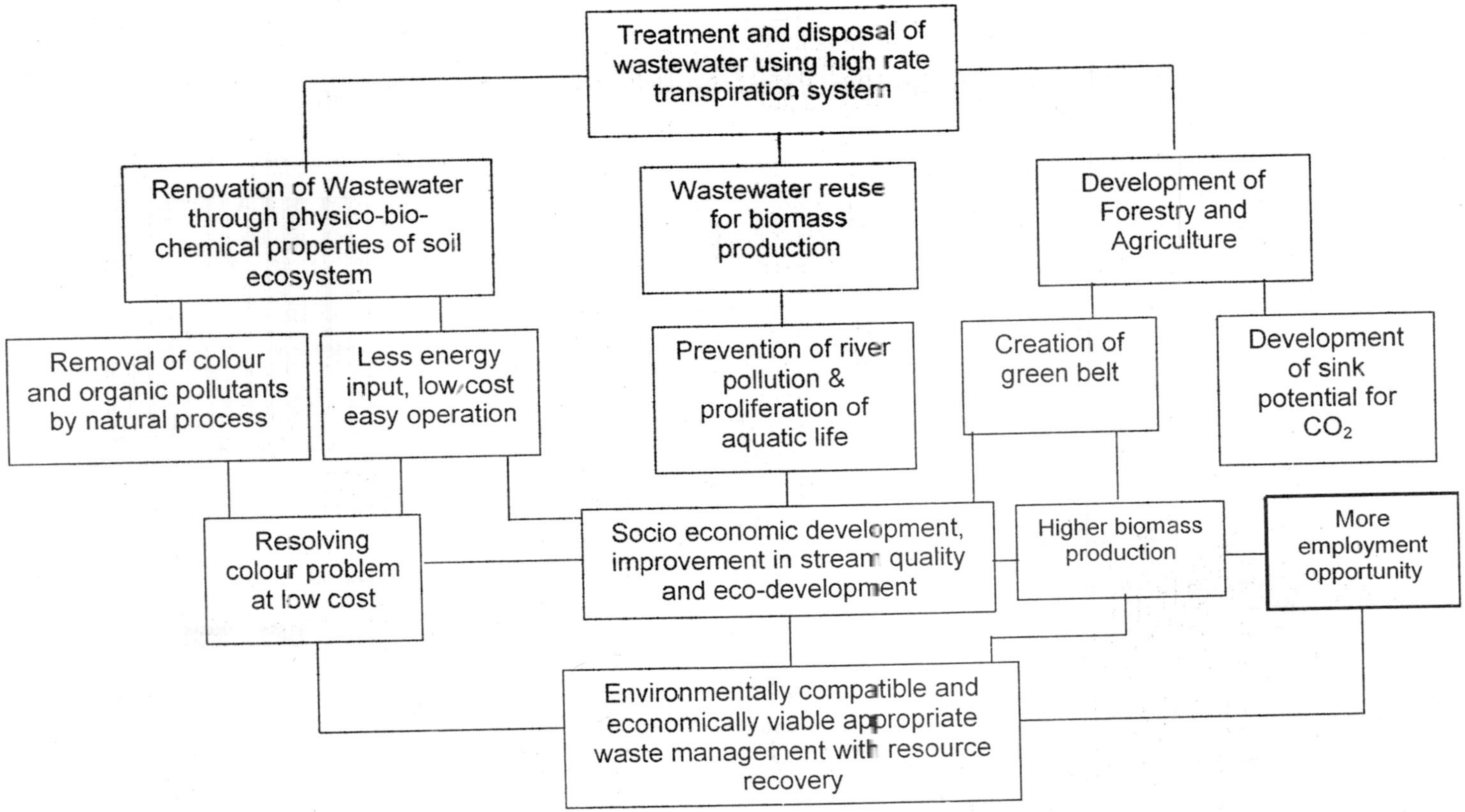


**Fig. 2. Impact network for treatment and disposal of wastewater using high rate transpiration system**

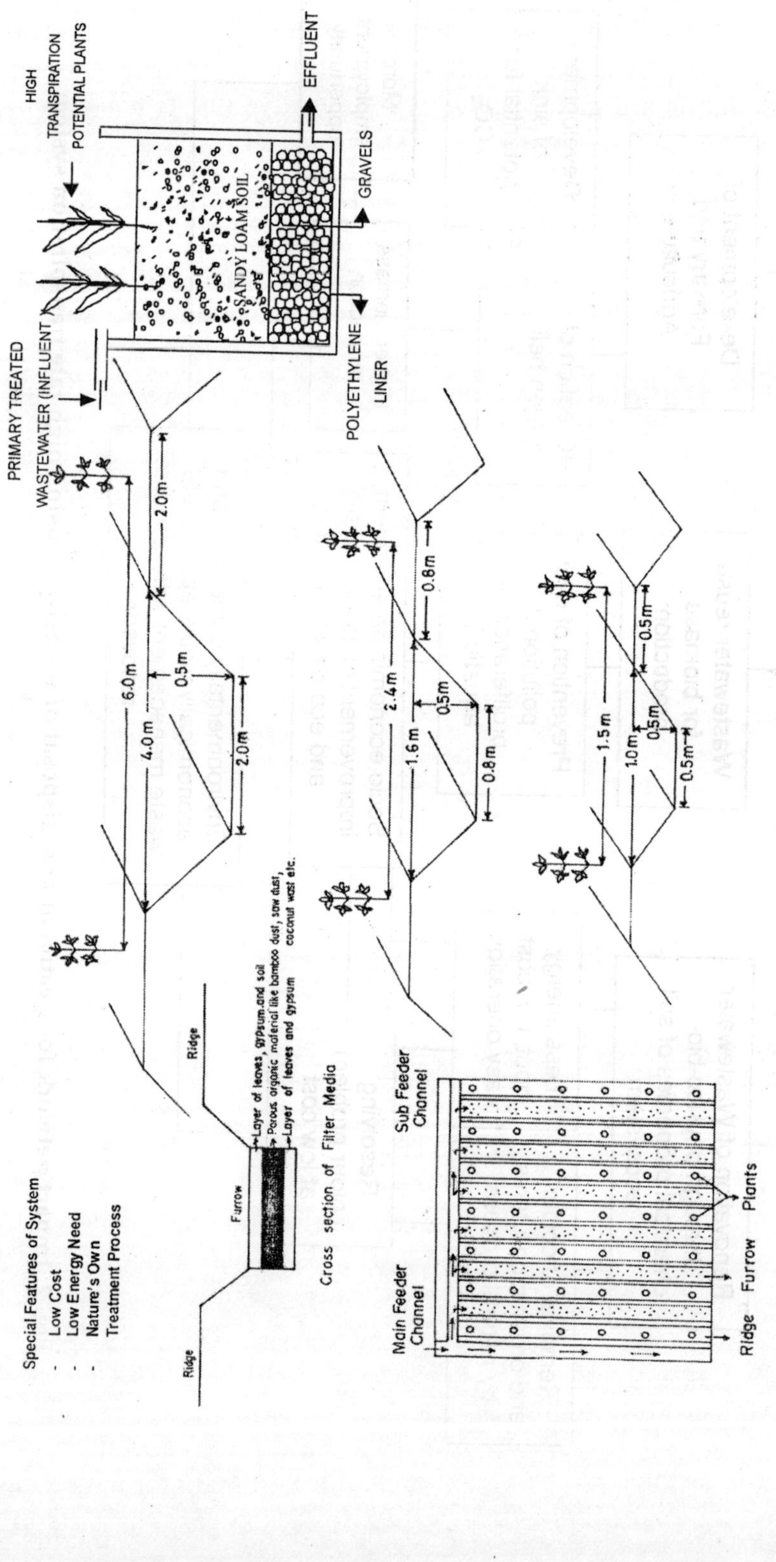


**Fig. 3.** Schematics of high rate transpiration system for treatment and disposal of wastewater

based on the premised that consumptive use varies with temperature and number of daylight hours. Empirical consumptive use coefficient K were derived from previous investigations are given below. Consumptive use U is equal to KÓf, where f = tp/100

t = mean monthly temp °F

p = monthly percent of yearly daylight hours

High transpiration capacity of plants render the system to consume more water and thus minimises the leaching problem. It has been reported that *Casurina equisitifolia* plants can transpire water equivalent to 8-12 times of potential evapotranspiration and thus very useful to grow using domestic wastewater (Chabra, 1989; Juwarar *et al.*, 1994 and Juwarkar *et al.*, 2003).

The mean concentrations of nitrogen, phosphate and BOD in the inflow and the outflow from the HRTS which received primary treated wastewater at the rate 5 cm $day^{-1}$ are presented in Table 3. The results showed that the HRTS removed N by 60 to 76.2 % and concentration of N in the outflow from HRTS varied from 8.3 to 9.5 $mgL^{-1}$ with respect to *Dendrocalamus strictus* and 8.0 to 9.0 $mgL^{-1}$ in the case of *Casurina equisitifolia*. Further, it was observed that the removal efficiency of N was greater in the case of *Casurina equisitifolia* due to profuse growth as compared to *Dendrocalamus strictus*. Removal of phosphate in the HRTS was comparatively less than nitrogen and it ranged from 17.7 to 70.3 %. Bouwer, *et al* (1980) also reported nitrogen removal to the extent of 30 to 65 % and phosphorus removal 40 to 80 % in a soil aquifer treatment system. However, nitrogen and phosphate removals by 70 and 95% respectively through soil are reported by Gupta and Nema (1991). The removal of nitrogen is mostly attributed to nitrification denitrification followed by plant removal. Coprecepitation of phosphate along with plant were are considered to be the major processes for phosphate removal. The BOD removal efficiency in the present study was observed to be 80.0 to 94.3%. The BOD of the outflow effluent ranged from 10.0 to 43.3 mg $L^{-1}$ as against the inflow BOD of 160 to 235 mg $L^{-1}$. These results are well supported by the findings reported by Sepp (1971) who observed 85 to 92% BOD removal in a soil aquifer treatment system. The comparative study of nitrogen and phosphate removal from vegetated and unvegetated HRTS shows that the presence of plants makes a difference in nutrient removal as shown in Table 4. The nitrogen removal in the unvegetated HRTS was 25.0 to 32.0 % as compared to 60.7 to 76.2 % in the vegetated HRTS system. Similarly phosphate removal in vegetated system was more than the unvegetated. Lance *et al.*, (1978) found that nitrogen removal was increased by 10 – 18.5% when vegetation was used. Levine (1980) repoted that there was 30% reduction in infiltration rates during second 34 weeks of flooding indicating some cultivation is necessary for maintenance of

**Table 4**

**Mean concentration of BOD, nitrogen and phosphate in influent and effluent from HRTS planted with *Dendrocalamus strictus* and *Casurina equisitifolia***

| Months | | Parameters, $mgL^{-1}$ | | | | | |
|---|---|---|---|---|---|---|---|
| | | Total Nitrogen | | Total Phosphate | | BOD, 20°C | |
| | Treatments | B | C | B | C | B | C |
| 3 | I | 27.4 | 27.4 | 11.6 | 11.6 | 183.3 | 183.3 |
| | E | 9.1 | 9.0 | 8.1 | 7.8 | 34.2 | 34.3 |
| | R | 66.8 | 67.2 | 30.2 | 32.8 | 81.3 | 81.3 |
| 6 | I | 30.3 | 30.3 | 12.4 | 12.4 | 205 | 205 |
| | E | 9.2 | 9.0 | 7.8 | 10.2 | 41 | 42 |
| | R | 69.6 | 70.3 | 37.1 | 17.7 | 80.0 | 79.5 |
| 9 | I | 24.3 | 24.3 | 11.8 | 11.8 | 160 | 160 |
| | E | 9.2 | 8.4 | 7.2 | 7.5 | 10.1 | 10 |
| | R | 62.1 | 65.4 | 39.0 | 36.4 | 93.7 | 93.8 |
| 12 | I | 33.6 | 33.6 | 16.9 | 16.9 | 232 | 232 |
| | E | 8.3 | 8 | 9.7 | 9 | 20.5 | 20.6 |
| | R | 75.3 | 76.2 | 42.6 | 46.7 | 91.2 | 91.1 |
| 15 | I | 33.5 | 33.5 | 15.8 | 15.8 | 211 | 211 |
| | E | 8.8 | 8 | 9.3 | 8.2 | 20.3 | 18.8 |
| | R | 73.7 | 76.1 | 41.1 | 48.1 | 90.4 | 91.1 |
| 18 | I | 24.2 | 24.2 | 16.2 | 16.2 | 235 | 235 |
| | E | 9.5 | 8.8 | 9.8 | 9.1 | 43.3 | 44.6 |
| | R | 60.7 | 63.6 | 39.5 | 43.8 | 94.3 | 81.0 |
| 21 | I | 26.7 | 26.7 | 13.5 | 13.5 | 216 | 216 |
| | E | 9.3 | 8.6 | 6.8 | 6.3 | 33 | 31.8 |
| | R | 65.2 | 67.8 | 49.6 | 53.3 | 84.7 | 85.3 |
| 24 | I | 32.5 | 32.5 | 15.8 | 15.6 | 233 | 233 |
| | E | 9.2 | 8.7 | 9.7 | 8.8 | 19.6 | 18.5 |
| | R | 71.7 | 73.2 | 70.3 | 43.6 | 91.6 | 92.1 |

B : *Dendrocalamus strictus*; C : *Casurina equisitifolia*;

I : Influent; E : Effluent; R : Removal Percent

infiltration rates. Kaur and Singh [25], recommended that for long term use of soil aquifer treatment system some cultivation is necessary to maintain through infiltration rate and enhance pollutant/contaminant removal efficiency of system. Similar type of results was also reported by the other workers (Sepp, 1971; Igbounamba, 1972; Bose *et al.*, 1973; Patra and Behara, 1974; Hansen *et al.*, 1980; Thawale *et al.*, 1999 and Hongve *et al.*, 2000).

## Conclusions

The HRTS found to be a cost effective and reuse alternative to conventional treatment processes which involve huge energy and cost. The experimental results conducted at NEERI showed that HRTS at a hydraulic loading of 5 cm $day^{-1}$ produced an effluent with BOD, nitrogen and phosphate content of 10 to 44.6 $mgL^{-1}$, 8 to 9.5 $mgL^{-1}$ and 6.3 to 10.2 $mgL^{-1}$ respectively. *Casurina equisitifolia* plant was more efficient in nitrogen removal compared to Bomboo. The plants irrigated with wastewater show increase in growth than irrigated with well water without affecting soil productivity and fertility. Further the plants got established quickly and grows profusely. Being simple in installation and operation HRTS for wastewater treatment can be adopted in small towns and villages. HRTS performance is affected by rainfall, temperature etc. Since it is not site specific, the system can be implemented near the wastewater source. The evaluation of the hazards associated to this kind of contamination would require a good understanding of the different phenomena involved not only the chemical equilibria, but also the hydrodynamics and the contact between the soil and the contaminant solution.

## References

Apha, (1995). Standard methods for the examination of water and wastewater, 19th ed. Publ. American Public Health Association, Washington, DC, USA.

Asano, T. (1985). Overview: Artificial recharge of groundwater, in artificial recharge of groundwater, T. Asano, ed.,: Butterworth Publ., Boston, MA 3-19.

Black, C.A. (1965) Methods for soil analysis, Am. Soc. Agro. Inc.,Publ. Madison, Wsieconsin, USA.

Blaney, H.F. and W. D. Cridle (1950). "determining water requiremens in irrigated areas from climatological and irrigation data." U.s. dept. Of agric., soil conservation service tech. Paper 96.

Bose, T.K., Basu, S. and Basu, R.N. (1973). Changes in rooting on cutting of Bongainvielea and Hibiscus, Indian J. Pl. Physol, 16, 127-139.

Bouwer, H. (1985). Renovation of wastewater with rapid infiltration land treatment systems. In Artificial Recharge of Groundwater, T. Asano, ed., Butterworth Publ., Boston, MA. 249-282.

Bouwer, H., Rice R.C. Lance J.C., Gilbert, R.G. (1980). Rapid in filtration research at Flushing Meadows Project. Arizona, J.W., PCF52, 2457- 2470.

Chabra, R.S. (1989). Sewage water utilization through forestry. Central Soil Salinity Research Institute, Karnal.

CPCB, (1988). Status of water supply and wastewater collection treatment and disposal in class I & II cities, CUPS/31/1988-90, New Delhi.

Crites, R.W. (1985). Nitrogen Removal in Rapid Infiltration System, J. Environ. Eng., 111(6), 865-873.

Gideon, O., Demalach, Y., Hoffman, Z., and Manor, Y. (1985) Effluent reuse by trickle irrigation wet Sci. Tech., x , 24, No. 9, 103 – 108.

Gupta, S.K. and Nema, P. (1997). Conservation and renovation of wastewater, National meet on S&T Issue to water Resource Management held at IIT, New Delhi, 133 – 144.

Hansen, E.A., Dawson, H.D. and Tolsted, N.D. (1980). Irrigation of intensively cultured plantation with paper mill effluent, Tappi, 63(11), 139.

Hongve, D., Van Hees P.A.W. and Lundstrom U.S. (2000). Dissolved components in precipitation water percolated through forest litter, European Journal of Soil Science, 51, 667-677.

Idelovitch E. and Michael, M. (1984). A new approach to an old method of wastewater reuse, J. Water Pollut. Control Fed, 56(8), 9300.

Igbounamba, O. (1972). 'Exchangeable reserve potassium and other cation reserves in agodi and ado soils', Soil Sci. 113, 394-409.

Iskandar, I.K. (1981). Introduction, in Modeling Wastewater Renovation : Land Treatment, I.K. Iskandar, ed., John Wiley & Sons, New York, 3 – 19.

Jackson, M. L. (1973). *Soil Chemical Analysis*, publ. Prantice Hall of India Ltd, New Delhi,.

Juwarkar, A. S., Moghe, M. and Thawale P. R. (1994). Sewage treatment and disposal through high rate transpiration system. National Seminar on Ecotechnological Wastewater Management through Service System of Nature, Puri, Lance, J. C. and Whisler, F.D., 1972Nitrogen balance in soil columns intermittently flooded with secondary effluent, J. Environ. Qual., 1(2), 180-186.

Juwarkar, A.S. (1987). 'A case study on the use of sewage for crop irrigation - Indian experience', Food and Agriculture Organization Rome.

Juwarkar, A.S., Asha Juwarkar, Deshbratar, P.B. and Bal, A.S. (1991). 'Exploitation of nutrient potential of sewage and sludge through land application', RAPA 1991/7, Asian Experience in Integrated Plant Nutrition. Report, of the Expert consultation of the Asian Network on Bio and Organic Fertilizers, Bangkok, Thailand.

Juwarkar, A.S., Thawale P.R. Juwarkar, A.A. and Khanna, P. (1995). A case study of environmental problems in the use of wastewater for crop irrigation. National Workshop on Health, Agriculture and Environmental Aspects of Wastewater use, UNEP-WHO-NEERI.

Juwarkar, A.S., Thawale P.R. Juwarkar, A.A. Singh, S.K. (2003). An ecofriendly approach treatment and disposal of pulp and paper mill wastewater through land management : A case study, Souvenier & Abstract, IEAM National Conference, New Delhi, 21.

Juwarkar, A.S., Thawale P.R., Jambulkar H.P and Juwarkar, A.A. (1998). Management of Wastewater through crop irrigation – An ecofriendly approach, Ecotechnology for Pollution Control & Environment Management. Enviro Media, 25 – 48.

Kaur, S. and Singh, M. (2002). Soil aquifer treatment system : A Case Study. Indian J. Environ. Health, 44 (3), 244 – 246.

Lance, J..C. and Whisler, F.D. (1976). Stimulation of denitrification in soil columns by adding organic carbon to wastewater, J. Water Pollut. Control Fed., 48(2), 346-356.

Lance, J.C., and Whisler, F.D. and Rice R.C. (1976). Maximising denitrification during soil filtration of sewage water, J. Environ. Qual., , 5(1), 102-107.

Lance, J.C., Rice., R.C. and Whisler, F.D. (1978). Effect of vegetation on denitrification and phosphate movement during rapid infiltration on soil columns J. WPCF, 50,. 2183 – 2187

Leach, L.E. and Enfield, C.G. (1983). Nitrogen control in domestic wastewater rapid infiltration Systems, J. Water Pollut. Control Fed., 55(9), 1150-1157.

Levine, P.E., Crites, R.W., and Olson, J.V. (1980). Soil chemistry changes at rapid infiltration site, J. Environmental Engg. Division, 106,. 869 – 883.

Overcash, M. R. and Pal, D. (1979). Design of land treatment systems for industrial waste, Theory and Practices, Ann. Arber Science Publication,.

Patra, S.K. and Behera, S. (1974). 'Effect of paper mill sludge on mineralization of soil nitrogen', Agra Univ. J. Res., 4, 111-113.

Sepp, (1971), 'The use of sewage for irrigation', A literature review, California State Department of Public Health, Bureau of Sanitary Engineering,.

Smith, R.G. and Schroeder, E.D. (1985). Field studies of the overland flow process for the treatment of raw and primary treated municipal wastewater, J. Water Pollut. Control Fed., 57(7), 785.

Smith, R.G. and Schroeder, E.D. (1985). Field studies of the overland flow process for the treatment of raw and primary treated municipal wastewater, J. Water Pollut. Control Fed., 57(7), 785.

Thawale, P.R., Juwarkar, A.S., Kulkarni, A.B. and Juwarkar, A.A. (1999). Lysimeter studies for Evaluation of Changes in Soil Properties and Crop Yield using Wastewater, Intern. J. Trop. Agric., 17 (1-4), 231-244.

U.S. EPA (1981). The process design manual for *Land Treatment of Municipal Wastewater*, October 1981, or latest edition, U.S. EPA.

Wilson, I.G., Amy, G. L., Gerba, C.P., Gordon, H., Johnson B. and Miller, J., Water (1986 ). Quality changes during soil aquifer treatment of tertiary effluent. Water Environment Research, 67, 371 – 376.

Witherow, J.L. and Bledsoe, B.E. (1986). *Design model for overland flow process*, J. Water Pollut. Control Fed. 58(5), 381.

# 7

# Profiles of Toxic Metals and Ambient Respirable Particulate Matter in Urban Area of Hyderabad City

*P.R. Thawale, D.G. Gajghate, P. Nema and A.V. Kulkarni*

Acute air pollution problem is being faced in urban agglomeration due to growth in urban centers, industrial activities and automobiles. The air pollution from these sources is imposing threat to health of population residing in urban areas. Recent concern about healthy effects of air pollution has focused on Respirable Suspended Particulate Matter concentration due to increased mortality and morbidity (Dockery and Papes, 1994) and asthma (Anderson et al., 1994). It is experienced that the toxic metals are found to be associated with fine dust of urban cities and their harmful effects are well documented. During last two decades, several studies have been reported on elemental composition of suspended particulate matter and their effect on exposure population,.

Status of airborne metals in air environment in India and various strategic approaches for air man agent of air pollution has been reported (Gajghate and Hasan 1995; 1996; 1999). Recently, it is observed that RSPM concentration in air environment of urban centers in India is found to be exceeding the WHO guideline. This experienced the serious atmospheric pollution problem of faster growing urban cities and indicates significant health problem (NEERI, 1997-98). The fine dust associated with toxic metals remains in urban air for longer duration and therefore it causes severe health problems to urban population.

A study was undertaken for the city of Hyderabad, one of the major industrial urban growth center of India. The population and traffic density of the city has grown several folds during the past decade resulting in overall deterioration in air quality. According to a study conducted by the Andhra Pradesh Pollution Control Board in seven government hospitals here between 1994 and 1997, air pollution related deaths were found to be steadily rising. It is reported that heart ailments which resulted in 6,450 deaths in 1994 rose to 9,624 in 1997, Likewise, deaths on account of bronchitis doubled during this period, while cerebro-vascular ailments increased alarmingly from 17 to 649. The study also revealed that 29% of the traffic constables working at high traffic junctions were beset with respiratory problems. Among them, the incidence was higher in respect of policemen posted at Punjagutta and points according to air quality measurements conducted by the pollution board. Interestingly, 88% of the vehicular pollution was caused by two and three wheelers, which accounted for 90% of the 13 lakh vehicles in the city. Cars and buses were responsible for 3% each and the goods vehicles and others for 4%. The findings said, "People, especially women and children, waiting at the bus stops and cross-roads are more exposed to these emissions" (Indian Express 2002). It is therefore, need is felt to carry out the ambient air quality study for monitoring the levels of SPM, RSPM and toxic metals in the city of Hyderabad. The study addressed the profile of SPM, RSPM and toxic metal in the ambient air of Hyderabad city during the two consecutive years i.e. 1997 & 1998.

## Methodology

A network spreaded over Hyderabad City was operated activity zone wise to get entire coverage of ambient Suspended Particulate Mater (SPM) and Respirable Suspended Particulate Matter (RSPM) and metals concentration. For Hyderabad city, three sampling sites representing residential, commercial and industrial activity were selected. SPM and RSPM were collected from the locations using High Volume Samplers and Respirable Dust Sampler operated at a rate of 1.5 $m^3$ /min for 24 hrs. on pre-weighed glass fiber filter of 20 × 25cm size and reweighed after sampling in order to determine the mass concentration of the particles collected. The concentrations of particulate matter in ambient air were then computed on the net mass collected divided by the volume of sampled. For trace metals analysis, known circles of 1′ diameter were punched out from the filter paper and digested in concentrated nitric acid. The content was filtered through whatman filter paper No. 42 and final volume made-up to 100ml by double distilled water. The filtrate was used to determine the metals include Cr, Cd, Fe, Pb, Zn and Ni by GBC 900 Atomic Absorption Spectrophotometer (AAS). The details, of sampling prodedure are given in elsewhere (Katz, 1997) Student's t-test is used to test the null hypothesis that the univariate sample populations are equal. The possible differences between the concentrations at pair of years can be examined with t-test. The student's t- statistics is given as.

$$t = \frac{X_1 - X_2}{\delta \times (1/n_1 + 1/n_2)^{½}}$$

Where $X_1$ and $X_2$ are sample mean of respective years

$\delta = (\delta_1^{\,2} + \delta_2^{\,2})^{1/2}$, $\delta_1$ and $\delta_2$ are the sample variances of respective years

The Spearman correlation coefficient between the concentration levels during two years was calculated from the data to support the results of t-test.

**Results and Discussion**

Three sampling sites representing residential, commercial and industrial activity zones were selected for ambient air quality monitoring. The prime pollutants include SPM & RSPM and the study was conducted during 1997 & 1998. The month wise concentration of SPSSM & RSPM for the years of 1997 and 1998 is presented in Figures 1 and 2. The profile of the air quality during these study periods revealed that all pollutants at industrial site were well below the CPCB limits for industrial areas. The annual average SPM concentrations at the commercial site were 143 & 172 µg/m$^3$ and RSPM levels were 105 & 107 µg/m$^3$ in years 1997 & 1998 respectively. The SPM levels at the residential site were 152 and 159 µg/m$^3$ in 1997 & 1998 which exceeded the CPCB standards. The RSPM levels was 110 & 107 µg/m$^3$ respectively. Additional toxic constituents were also selected because of their implication on health and their importance in crafting city-specific air emission. The cause of human health damage depends on concentration, nature and composition of RSPM. The characterization of RSPM for toxic elemental composition was performed. The monthly averaged concentrations of six metals are delineated in Figures 3 and 4. Lead in the atmosphere derived principally from the combustion of petrol containing lead activities and from smelting. Mean concentrations of Pb throughout the year ranged from 0.002 to 0.593 and 0.004 to 0.638 µg/m$^3$ during 1997 & 1998 respectively. Mean concentrations of Ni varied from 0.002 to 0.055 and 0.001 to 0.042 µg/m$^3$ at two consecutive years respectively. This may be due to the combustion of fossil fuels, smelting, crustal sources and volcanoes. Mean concentrations of Cr varied from 0.001 to 0.139 and 0.001to 0.50 µg/m$^3$ and concentrations of Cd varied from 0.001 to 0.023 and 0.001 to 0.027 µg/m$^3$ during 1997 & 1998 respectively. Cadmium and chromium originates from smelters, fuel combustion, paint and plastics manufacturing and volcanoes. The mean concentrations of Zn varied from 0.021 to 0.430 and 0.068 to 3.781 µg/m$^3$ respectively. The high concentration of zinc might be attributed to high rate of

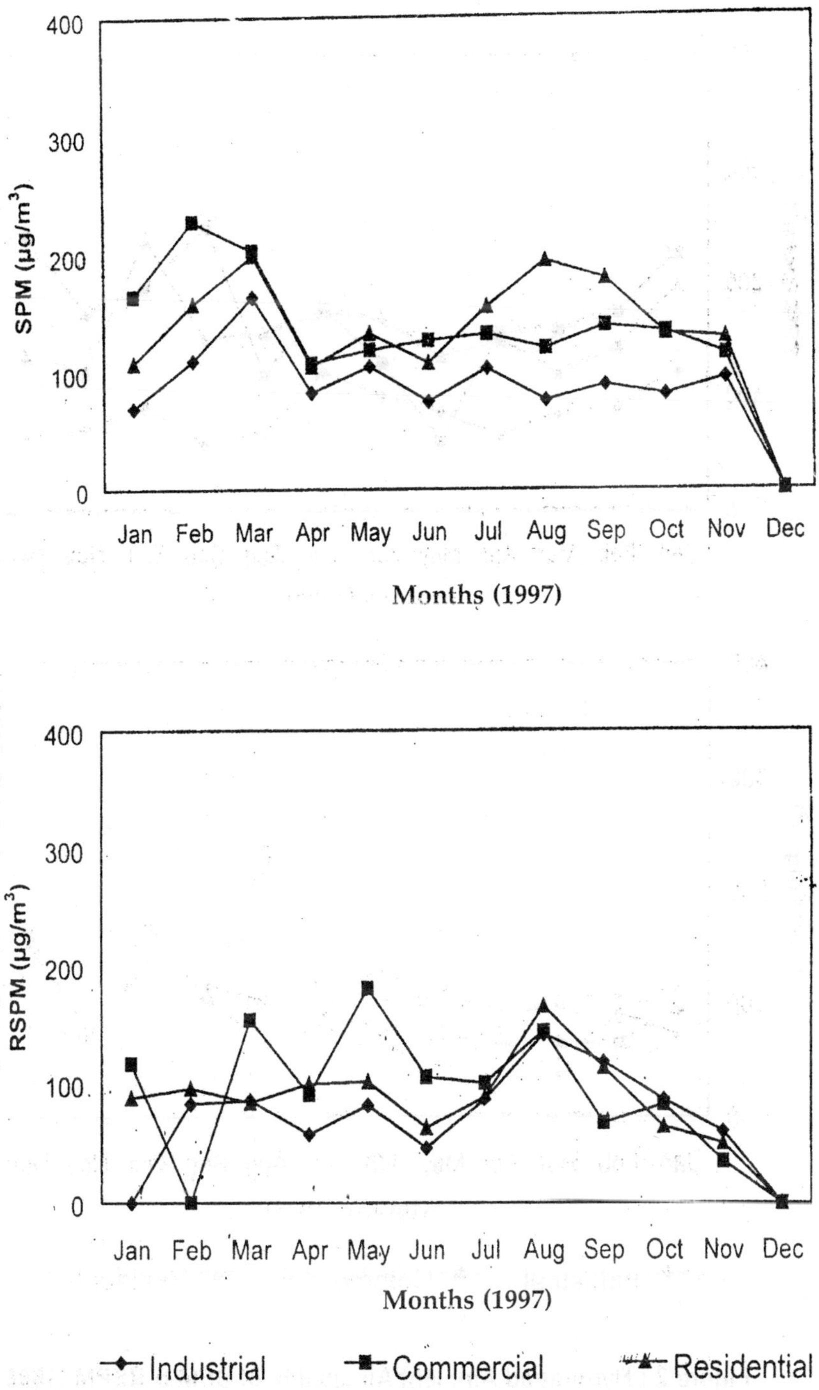


**Figure 1 : Hyderabad Ambient Air Quality of SPM & RSPM :1997**

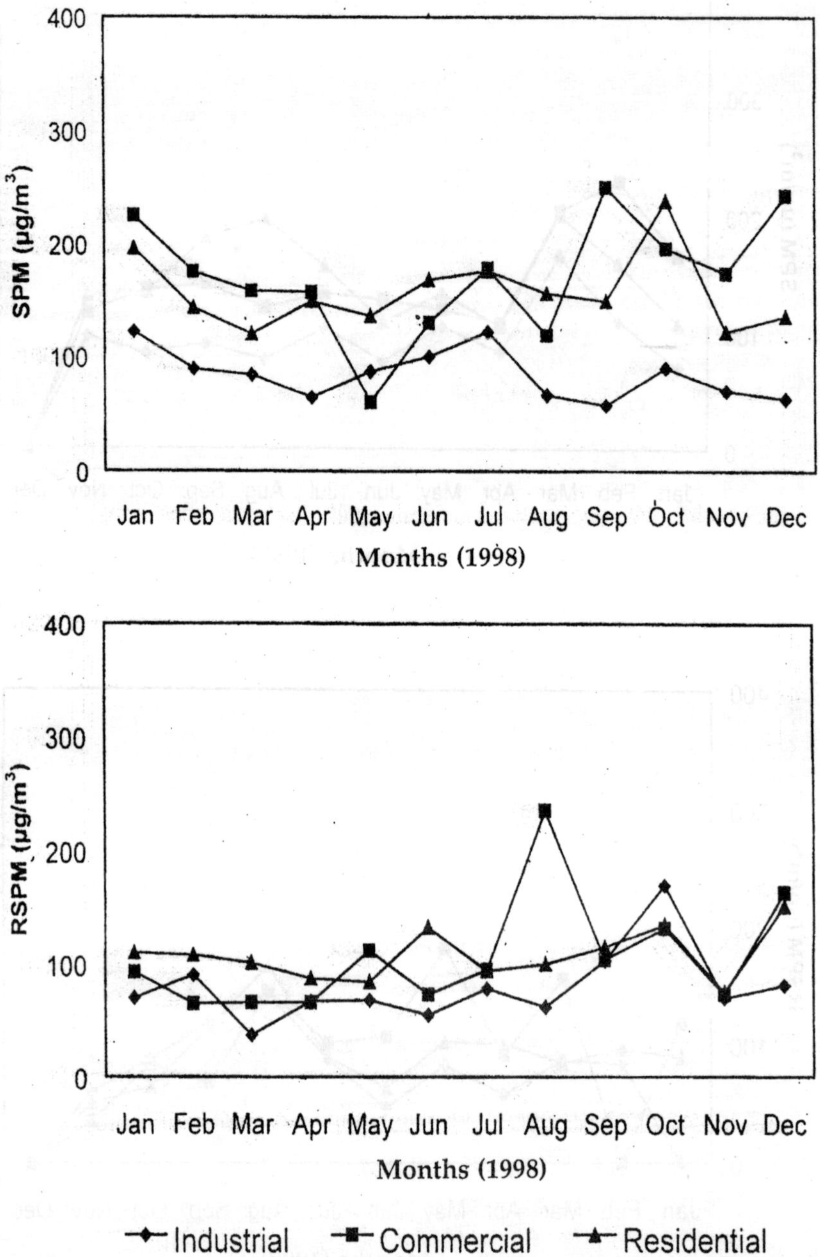


**Figure 2 : Hyderabad Ambient Air Quality of SPM & RSPM :1998**

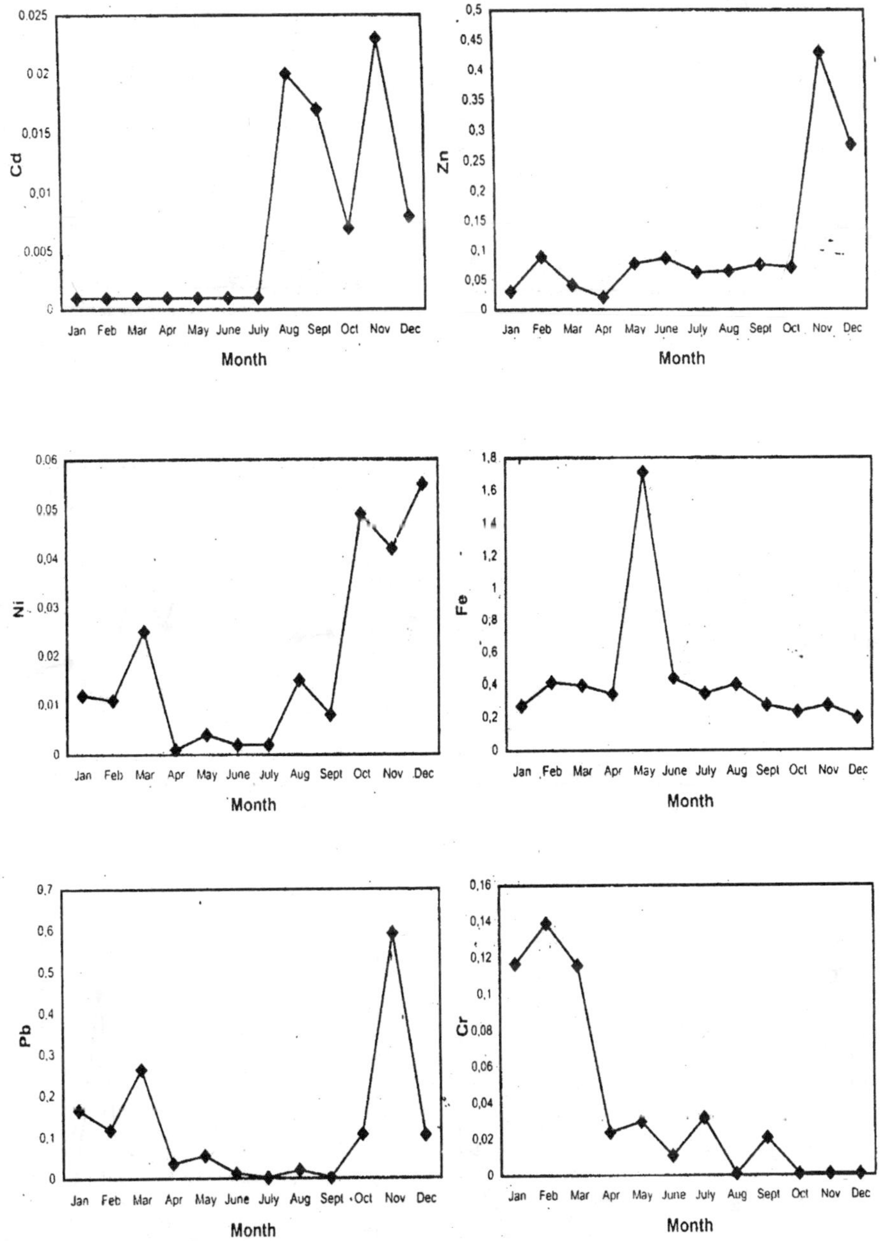


**Fig. 3. Trace Metal concentrations in RSPM at Hyderabad during 1997 (24 hrly avg. in ug/m$^3$)**

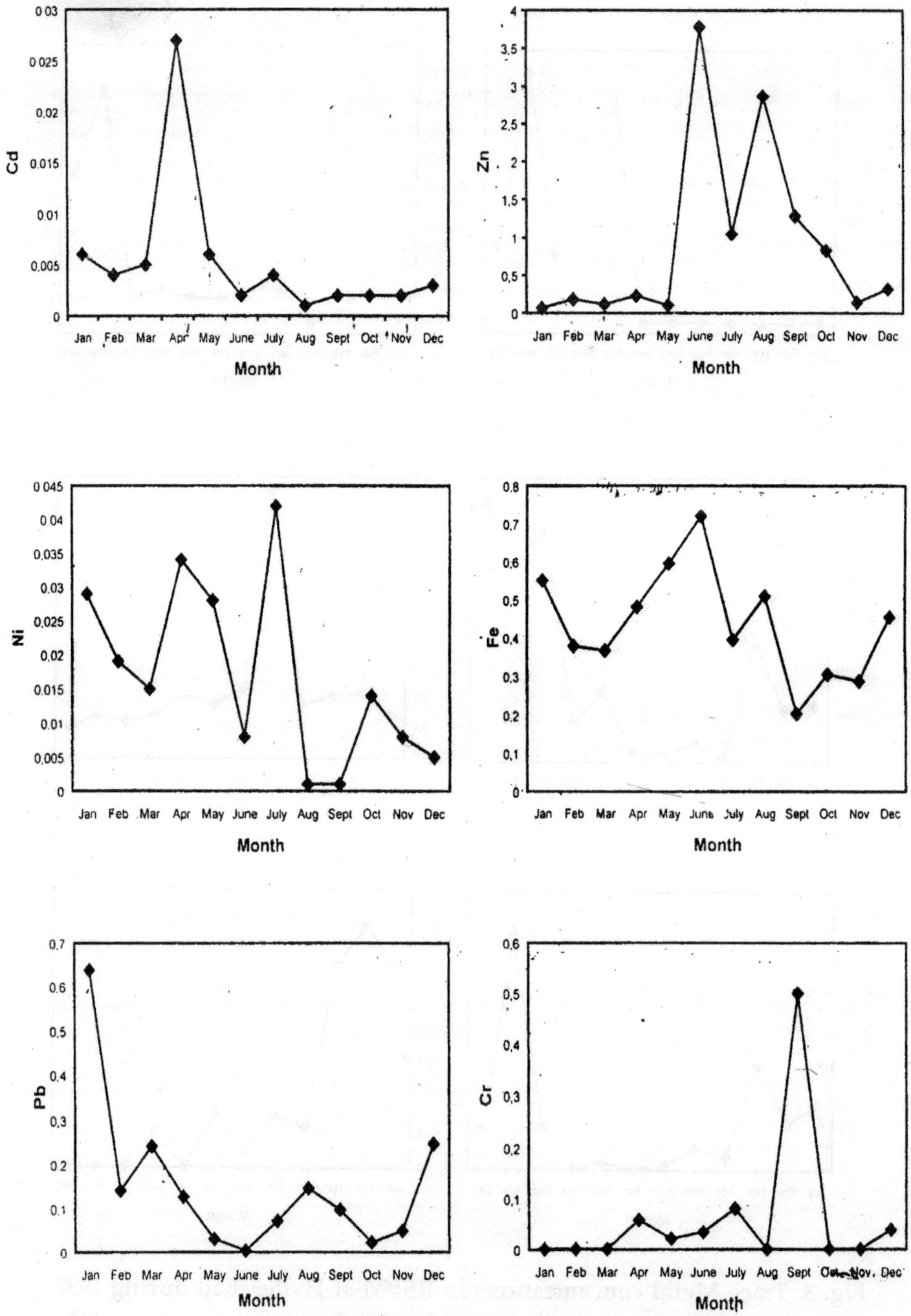


**Fig. 4. Trace Metal concentrations in RSPM at Hyderabad during 1998 (24 hrly avg. in ug/m$^3$)**

soil erosion in the months of June to September. Mean concentrations of Fe varied from 0.197 to 1.713 and 0.203 to 1.394 μg/m$^3$ respectively. While observing the seasonal variations in the trace elements, high levels of cadmium, chromium, lead, nicket, iron and zinc were observed in winter followed by summer and monsoon. The high concentrations of iron and zinc were observed among the other toxic metals during the study period. The yearly average concentrations of toxic metals is shown in Table 1. The average concentrations of Cd, Zn, Ni, Fe, Pb & Cr during 1997 and 1998 were 0.007, 0.111, 0.019, 0.442, 0.124, 0.041, and 0.005, 0.915, 0.017, 0.437, 0.151, 0.062 μg/m$^3$ respectively.

The t-test determines whether the mean of populations of concentration data for each individual element differ significantly between the years. The significant difference between pairs of years suggested that urban air quality with regard to those trace elements varied between two years. The differences between those two years cannot be explained in terms of any known large fixed sources. But exposure to vehicle emissions, re-suspended dust and industrial emission might have played an important part in the differences between two years. The results from the student's t-test are given in Table 2. significant difference were observed

**Table 1**
**Yearly average and Minimum, Maximum concentration of metals**

| Range | Trance Metals (Micro gram/m3) | | | | | |
|---|---|---|---|---|---|---|
| | Cd | Zn | Ni | Fe | Pb | Cr |
| Minimum | 0.001 | 0.021 | 0.001 | 0.197 | 0.002 | 0.001 |
| Maximum | 0.023 | 0.43 | 0.055 | 1.713 | 0.593 | 0.139 |
| Average | 0.007 | 0.111 | 0.019 | 0.442 | 0.124 | 0.041 |
| Minimum | 0.001 | 0.068 | 0.001 | 0.203 | 0.004 | 0.001 |
| Maximum | 0.027 | 3.781 | 0.042 | 1.394 | 0.638 | 0.5 |
| Average | 0.005 | 0.915 | 0.017 | 0.437 | 0.151 | 0.062 |

**Table 2**
**The statistical analysis of toxic metals during 1997 & 1998**

| | T-Statistics | Speraman Correlation |
|---|---|---|
| Cd | 0.47 (0.640) | -0.37 |
| **Metal** | 2.29 (0.032) | -0.18 |
| Ni | 0.27 (0.791) | -0.46 |
| Fe | 0.04 (0.966) | 0.42 |
| Pb | 0.40 (0.695) | 0.09 |
| Cr | 0.48 (0.636) | -0.18 |

between Zn levels in 1997 and 1998. Significant differences between crustal elements i.e. Zn may be attributed to local abundance in the atmosphere from a wide range of sources. From time to time no significant differences for other metals were observed except zinc. The spearman correlation coefficient between two years for the toxic elements concentration (Table 2) illustrates that the correlation between concentration levels during the two years are not significant for all metals.

The micrometeorological data was collected from IMD, Hyderabad for two years. Data were is further analyzed for wind direction and velocity. It revealed that the prevailing wind pattern during these years was from West with calm percentage of 38 and 24% in 1997 and 1998 respectively. The calm percentage in winter during the two years was 31 and 35% respectively with the prevailing wind direction being from E-SE sector.

## Conclusion

It my be concluded that there may be an existing source, which contribute to the reglar emission of RSPM and other toxic elements. Auto exhaust is the main source contributes to RSPM and lead in the ambient air. The study revealed that commercial and residential sites had particulate matter pollution problem, and the levels of SPM and RSPM marginally exceeded air quality standards at these sites. The toxic metals has moderate problem as compared to other mega cities of India. However, swelling of urban population, increased vehicular traffic and unplanned development of industrial activities in and around the city will lead to severe air pollution reported by earlier studies. Morbidity and mortality caused by air pollution reported by earlier studies, demands health care services and increase health care resource utilization and economic burden on society. An approach for designing policies and implementation of air management system is needed to identify major sources of air pollution and subsequently restore air quality by corrective measures. Therefore it is advocated that urban air quality management is to be prepared immediately to improve the future air quality in Hyderabad city.

## Acknowledgements

The authors are grateful to the Director, National Environmental Engineering research Institute, Nagpur for according permission to publish the paper. They are thankful to Scientist & Head, Air Pollution Control Division for able guidance and other colleagues who generated primary data on ambient air quality.

## References

Anderson, K.R., Avol, E.L., Edward, S.A., Shamoo, D.A., Peng, R.C., Linn, W.S. and Hackny, J.S. (1992). Controlled exposures of volunteers to respirable carbon and sulphuric acid aerosols. J. Air and Waste Manag. Asso., 42: 771 – 776.

Chow, J.C., Watson, J.G., Fujita, E.M. and Lowson, D.R. (1994). Temporal and spatial variations of PM 2.5 and PM10 aerosol in the Southern California air quality. Atmos. Environ., 28: 2061-2080.

Dockery, D.W., and Papes, C.A. (1994). Acute respiratory effects of particulate air pollutión, Ann. Rev. public hlth., 15: 107-132

Gajghate, D.G., Hasan, M.Z. (1999). Ambient lead levels in urban areas, Bull, Environ. Cantam. Toxicol., 62: 403-408.

Gajghate, D.G. and Hasan, M.Z. (1996). Approaches for management of air pollution and control strategy. 3rd international conference on env. Planning and manag. Nagpur. Proceedings, pp 169-175.

Gajghate, D.G., Hasan, M.Z. (1995). Status of aerosol with specific reference to toxic trace metals constituents in urban air environment. J. Chem. and Environ. Sci., 4: 67-74.

Garzon, L., Perez, J.M. Fernandas, J.M. Fernandaz, R. and Arganzer, B. (1986). Lead concentrations and their strength in the atmosphere of an Urban site. Atmos. Environ.'20: 1441-1446.

Indian Express (2002) (April 30) pollution-related deaths on rise in clean Hyderabad.

Katz, M. (1977). Methods for air sampling and analysis. 2nd Edition.

NEERI Report (1997-1998). Ambient Air Quality Status in ten cities in India, National Environmental Engineering Research Institute, Nagpur, India.

Simpson, R.W. and Hongchang, Xu. (1994). Atmospheric Lead pollution in an Urban Area-Brisbane, Austria. Atmos. Environ, 28 (19): 3073-7082.

# 8

# Conservation of Endangered Nitrogen Fixing Plants of Cold Desert/Alpine Region of Garhwal Himalaya

*Pankaj Kumar, Monika Tyagi and T.C. Pokhriyal*

A continuous forest growth depends upon the maintenance of the available N pool and its efficient use. Forest growth can be enhanced either by addition of N or by increasing the rate of N cycling within the ecosystem and one of the greatest potential for increased forest productivity lies within enhanced nitrogen fixation on forest sites (Davey and Walleen 1984, Gordon et al., 1979). It is well known that initially, NFPs play not only cater themselves for their N requirement through biological nitrogen fixation by bacteria living in their root nodules, but they are generally also hardier, more tolerant to forest, resistant to pests and disease and have potential to grow in low moisture regime. Biological nitrogen fixation is a quite cheap, renewable source essential for sustainable productivity of forest soils. Gordon and Wheeler (1983) reported that the use of NFPs in some circumstances may be cheaper and environmentally sound. Multipurpose uses, plus nitrogen fixation ability and rapid growth, often have preference of nitrogen fixing trees for commercial and agroforestry plantation over other species (Mac Dicken, 1994). Soil conservation is not only the protection against erosion but also conservation of soil fertility or the productive capacity of the land. Considering the increase of productive land, reclamation of wastelands area where NFPs have an important role to play.

Garhwal Himalaya is very fragile ecosystem. Large areas of which have become degraded due to increased biotic interference over a period of time. Top soil is lost every year due to erosion, accelerated and floods occur an a regular basis from times immemorial. This has invited a serious drought on the status of these mountains, instability of hill slopes, their geological aspects and the forest cover.

Although denudation of the hill slopes have accelerated the process of soil erosion, landslides and silting-up of streams but the geological processes have also enhanced the stability of the slopes in the recent years.

Re-clothing/afforesting these areas is the greatest challenge in front of us. As our soils are deficient in nitrogen, the growth of plants is not good. To over come this problem we have to find the best answer- afforesting these areas with such plants that can fix atmosphere nitrogen and grow under the prevailing extreme conditions of moisture and temperature. Hence, the need for identification of NFPs in this region.

Only some sporadic work has been done in the identification of indigenous legumes and non-legumes nitrogen fixing flora in the Himalayas. Himmat Singh (1997) has done work on nodular behaviour of nitrogen fixing plants from Nepal Himalays. Pokhriyal et al., (1992, 93) have examine and identified nitrogen plants of Garhwal Himalayas for nodular behaviour. No attempt has been made for the survey of indigenous legumes and non-legumes nitrogen fixing plants in cold desert areas of Garhwal Himalayas. In the present paper an attempt has been made to survey, identification, nodular behaviour, nitrogenase activity of neglected nitrogen fixing plants of cold desert region of Garhwal Himalayas.

## Study Area

Garhwal Himalayas covers about 7% area of the total Indian Himalayas. It lies between the latitude 29°31′ 9″ N and 31°26′ 5″ N and longitude 77° 33′ 5″ E and 80° 6′ 0′ E a total geographical area of 29089 $km^2$ in the newly formed state. Cold desert or alpine zone lies above 3000m and are highly in hospitable areas, which remain covered under snow for the most part of the year. The temperature is extremely in these areas and vegetation is scanty but highly specialized and adapted to theses areas. Nelong valley area of Uttaranchal have been surveyed for this study.

## Methodology

The transect survey was conducted mainly on the cold desert of Uttaranchal (above 3000m altitude) covering Nilong valley and valley of flower along different foot traits in Garhwal region in the month of August to September, 2002 which is most ideal season for nodule formation, collection, uprooting of plant materials and estimation of nitrogen fixation activity similarly as earlier suggested by Holliday (1984).

The nodules of the individual plants were collected and samples were kept for the estimation of nitrogenase activity. Nitrogenase activity was determine by studying the reduction of acetylene as described by Hardy et al., (1968) and procedure similar to that described earlier by Pokhriyal et al., (2001).

**Results and Discussions**

Total eight genera and 17 species of leguminous and non-leguminous NFP were screened and identified under this region. Out of the 17 species recorded, 14 were leguminous and 3 were non-leguminous (i.e., *Hippophophae tibetiana, H. salicifolia* and *Sibbaldia pareviflora*). In survey 82% herbs and 18% NF shrub were examined. No NF trees and climbers were found from this region. All leguminous NFP were herbs except one shrub i.e., Indigofera belongs to sub-family Papilionoideae, whereas two genera of non-leguminous NF (i.e., *Hippophophae* and *Sibbaldia*) belongs to family Elaeagnaceae and Rosaceae respectively. Genera Astragalus has maximum (7) number of species followed by Lespedeza (2), Trigonella (2) and Hippophophae (2). However, remaining four genenra i.e., *Cicer, Oxytropis, Indigofera* and *Sibbaldia* were identified with single species each. Out of total herbs, 79% showed presence of nodules whereas 21% did not possess nodules. All the 21% shrubs were unnodulated.

It was observed that species that are normally known to have nodules in other areas did not possess nodules in harsh edaphic and environmental conditions of the cold deserts. But nodules were abundant in species that were found growing close to some water source indicating thereby that not only stress but some amount of moisture is also very important for the symbiotic association of plants.

Maximum nodule 2.95 g fresh weight, 0.50 g dry weight and 31 nodules per plant were observed in *Astragalous coluleocarpa* and minimum 0.63g fresh and 0.08 g dry weight and 3 nodules per plant in *Asragalous himalayansis* (Table 1).

Whereas, maximum nitrogenase (49365.7 and 9705.3 n moles acetylene reduced g-1h-1 and p1-1h-1) activity was observed in *A. amberstions*, followed by Astragalous coluleocarpa, Trigonella pubescens and minimum in A. chlorostachy. Nitrogen fixation activity is influenced by the bioclimatic zones of distribution patter of individual species, availability of favourable temperature, light, soil moisture and harsh weather at higher altitudes as compared to the lower. Nitrogen fixation is influenced by several environmental factors. It is positively correlated within soil moisture and temperature and inhibited by dry conditions and extreme cold (Jordon et al., 1978; Stewart et al., 1978; Jager and Werner, 1981). Nodule number, its biomass and Nitrogenase activity markedly reduced at higher altitude may due to the prevailing harsh environmental conditions as already reported by Roughley (1970); Huang (1975); Gibson (1976), Sundstrom and Huss Danell (1987) and Himmat Singh (1997, 1998).

**Table 1**

**Showing name of the species, altitude, plant status, place of collection, family/sub-family, nodule biomass, nodule per plant nodule status and nitrogenase activity in endangered nitrogen fixing plants collected from different parts of cold desert/alpine region of Garhal Himalaya.**

| S. No. | Name of the Species | Altitudes (m) | Plants Status | Place of Collection | Family/ Sub-family | Nodule Fresh wt. (g) | Nodule Dry wt. (g) | Nodule per Plant | Nodule Status | Nitogenase Acticity (n mole $C_2H_2$ reduced) g-1 fresh wt. $h^{-1}$ | Nitogenase Acticity (n mole $C_2H_2$ reduced) $P1^{-1}h^{-1}$ |
|---|---|---|---|---|---|---|---|---|---|---|---|
| | **Leguminous** | | | | | | | | | | |
| 1. | *Astragalus amberstianus* Royle. | 3500 | Herb | Between Naga to Nelong (T.G.) | Papilionoideae | 2.16 | 1.02 | 28 | +ve | 49365.7 | 9705.3 |
| 2. | *Astragalus caluleocarpa* Boiss. | 3800 | " | Between Naga to Nelong (T.G.) | " | 2.92 | 0.50 | 31 | +ve | 44825.3 | 8813.7 |
| 3. | *Astragalus multiceps* Wall. | 3300 | " | Between Naga to Nelong (T.G.) | " | 1.97 | 0.97 | 12 | +ve | 28108.4 | 6792.2 |
| 4. | *Astragalus peduncularis* Royale. | 3500 | " | Between Naga to Nelong (T.G.) | " | 1.87 | 1.04 | 16 | +ve | 29132.5 | 7032.8 |
| 5. | *Astragalus candolleanus* | 3200 | " | Between Naga | " | 0.95 | 0.15 | 7 | +ve | 5339.0 | 4216.3 |
| 6. | *Astragalus chlorostachys* Lindl. | 3200 | " | Balley of flowers | " | 1.00 | 0.10 | 3 | +ve | 993.8 | 582.4 |
| 7. | *Astragalus himalayansis* Klotz. | 3200 | " | Valley of flowers | " | 0.63 | 0.08 | 3 | +ve | 1082.5 | 617.3 |
| 8. | *Cicer microphyllum* Benth. | 3750 | " | Between Naga to Nelong (T.G.) | " | * | * | * | -ve | - | - |
| 9. | *Lespedeza gerardiana* Grah. | 3200 | " | Between Naga to Nelong (T.G.) | " | ** | ** | ** | +ve | - | - |
| 10. | *Lespedeza juncea* (L.f.) Pers. | 3000 | " | Ganga - Yamuna Velley | " | ** | ** | ** | +ve | - | - |
| 11. | *Oxytropsis duthleana* Ali. | 4500 | " | Between Naga to Nelong (T.G.) | " | * | * | * | -ve | - | - |
| 12. | *Indigofera heterantha* Wall. ex Brandis | 3000 | Shrub | Above Govind Chat | " | * | * | * | -ve | - | - |
| 13. | *Trigonella emodi Benth* | 3500 | Herb | Between Naga to Nelong (T.G.) | " | ** | ** | ** | +ve | - | - |
| 14. | *Trigonella pubescens* Edgw. ex Bak. | 3000 | " | Between Naga to Nelong (T.G.) | " | 1.84 | 0.483 | 16 | +ve | 30268.2 | 7203.8 |
| | **Non-leguminous** | | | | | | | | | | |
| 15. | *Hippophophae tibetinan* | 3300 | Shrub | Between Naga to Nelong (T.G.) | Elaeagnaceae | * | * | * | -ve | - | - |
| 16. | *Hippophophae salicifolia* Linn. | 3000 | " | Ganga - Yamuna Valley | " | * | * | * | -ve | - | - |
| 17. | *Sibbaldia salicifolia* Sensu. Hook f. | 3200 | Herb | Valley of flowers | Rosaceae | * | * | * | -ve | - | - |

*No nodules and **nodule seen but because of biomass and NA activity not taken.

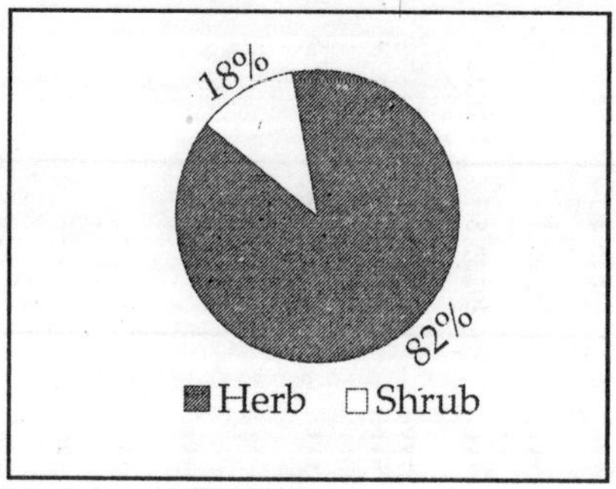


Nodule development is the summation of a number of physiological and biochemical processes and the time for nodule to appearance have been used as measure of environmental effects on the establishment of symbiosis (Gibson, 1977). Therefore, there is an urgent need to understand the nodule behaviour and nitrogen fixation activity of these neglected NFPs under cold deserts region of Garhwal Himalayas of Uttaranchal.

## Conclusions

A survey was conduced for the identification of leguminous and non-leguminous nitrogen fixing plant species (NFPs) distributed under cold desert area (above 3000m) in the Garhwal Himalayan Region (GHR) or Uttaranchal. Cold deserts are highly inhospitable areas, which remain covered under snow for the most part of the year. The temperatures are extremely low in these areas and vegetation is scanty but highly specialized and adapted to these areas. Total 8 genera and 17 species of NFPs were screened out in which 82% were herb and 18% were shrubs. However, no tree and climber species were found from this region. Out of total herb species 79% showed presence of nodules whereas 21% did not posses nodules. All the shrubs species were non-nodulated. All leguminous NFPs were belongs to sub family Papilionoideae except three species (i.e., *Hippophophae tibetiana, H. salicifolia* and *Sibbaldia pareviflora*) belongs (non-leguminous) to family Elaeagnaceae and Rosaceae. Overall, the nodule number its biomass and nitrogenase (n mole $C_2H_2$ reduced $g^{-1}$ fresh $h^{-1}$ and $P1^{-1}h^{-1}$) activity was found maximum in *Astragalus coluleocarpa* Boiss, whereas, minimum in *Astragalus himalayanais*. This is a high time to conserve these neglected NFPs of cold desert areas of Garhwal Himalaya in order to protect the fragile Himalayan ecosystem.

Conservation of neglected NFPs in cold desert area by chemical analysis of nitrogen fixing plants for their active ingredients; establishment of Nitrogen Fixing Plants Conservation Areas (NFPCAs) have been suggested and recommendation of some multipurpose NFPs for fulfillment of the local residents, preservation of biodiversity of the entire Garhwal Himalayan region.

## References


Davey, C.B. and Walleen, A.F. (1984). Nitrogen fixation system in forest plantation. In: Nutrients of plantation forests (eds. G.D. Bowen and E. K.S. Namblr), Academic Press, pp. 361-377.

Gibson, A.H. (1976). Consideration of growing legumes as a symbiotic association. *Proc. Indan Natl. Sci. Acad.* 40: 741-767.

Gibson, A.H. (1977). The influence of the environment and managerial practices on the *legume-Rhizobium symbiosis*. In: A Treatise on Dinitrogen fixation IV, Agronomy and Ecology (eds. R.W. Hardy and A.H. Gibson), John Willey Sons. New York, 394-450.

Gordon, J.C. and Wheeler, C.T. (eds.) (1983). Biological Nitrogen fixation in some *Coniferous* forest ecosystem. Foundations and applications. The Hague Netherland; Martins Hijhoff J/Dr. W. Junk, pp. 342.

Gordon, J.C., Wheeler, C.T. and Perry, D.A. (eds.) (1979). Symbiotic nitrogen fixation in the management of temperature forests, Forest Research Labouratory Oregon state University, Corvallis USA, pp. 501.

Hardy, R.W.F., Holslin, R.D., Jackson, E.K. and Burns, R.C. (1968). The acetylene-ethylene assay for nitrogen fixation laboratory and filed evaluation. *Plant Physiol.* 43: 1185-1207.

Himmat, K.C. and Pokhriyal, T.C. (1997). Effect of some leguminous and non-leguminous nitrogen fixing herbs, shrubs, climbers and trees species on soil nitrogen accretion. *Ann. For.* 6: 119-122.

Himmat, K.C. and Pokhriyal, T.C. (1998). Distribution of leguminous and non-leguminous nitrogen fixing plants in relation to altitudinal variation in eastern part of Nepal, *Indian Forester.* 124(7) 544-55.

Holliday, J. (1984). Register of nodulation reports for legumeinous trees and other arboreal genera with nitrogen fixing members. Nitrogen fixing tree research reports 2: 38-45.

Huang, C.Y., Boyer, J.S. and Vandrhoof, L.N. (1975). Limitation of acetylene reduction (nitrogen fixation) by photosynthesis in Soybean having low water potentials. *Pl. Physiol.* 56: 228-32.

Jagar, D. and Werner, D. (1981). Ecology and dinitrogen fixation in soils and rhizosphere, 3. Correlation and nitrogenase activity with water contents organic material and cell number of anaerobic bacteria in soils with different root system. Archev fin Hydrobiology, 91: 15-27.

Jordon, D.C., Mc Nicol, P.J. and Marshall, A.R. (1978). Biological Nitrogen Fixation in the terrestrial environment of high arctic ecosystem (Truelove Lowland Dwon Bland, N.W.T.) *Can. J. Microb.* 24: 643-647.

Mac Dicken, Kenneth, G. (1994). Selection and management of nitrogen fixing trees. Winrock International Institute for Agricultural Development Route, 3. Box 376, Morrilton, Arkansas 27110-9537 USA.

Pokhriyal, T.C., Chaukiyal S.P., Naithani, H.B. and Gupta B.B. (1992). Identification of some Nitrogenase fixing herb & shrub species from Dehradun – Mussoorie region Indian Forester. 118(9) 681-83.

Pokhriyal T.C., Chaukiyal S.P. and Sing U. (2001). Eucalyptus, acacia Mixed Planting effects on the Nodular Nitrogenase and Nitrate reductase activity in *Acacia nilotica* seedlings *J. of Tropical Forest Science* 13(2) 290-296.

Pokhriyal, T.C. Singh S.B. Prasad, K.G. Kumar, P. and Raturi, A.S. (1992). Effect of different nitrogen sources of NRA in *Leucaena leucocephala, Ind. J. Forestry,* 11287-291.

Pokhriyal, T.C. Chaukiyal, S.P. and Naithani, H.B. (1993). Nitrogen fixation and nodulation behaviour of some nitrogen fixing species from inner and outer Himalaya, *Indian Forester* 119(4): 310-320.

Roughley, R.J. (1970). The influence of root temperature *Rhizobium* strain and host selection on the structure and nitrogen fixing efficiency of the root nodules of *Trifoloium subterraneunll Ann. Bot.* (London), 34: 631-646.

Stewart, W.D. P., Sanpais, M.J., Isichin, A.O. and Sylvester, B.R. (1978). Nitrogen fixation by soil algae in temperate and tropical soil. In Berener J. Do, Buris, A.H., Hollaender, Franco, A, Negra, C.A., scotts, D.B. eds. Limitations and Potentials for Biological Nitrogen Fixation in Tropics. Plenum Press, New York.

Sundstrom, K.R. and Huss Danell (1987). Effect of water stress on nitrogenase activity in Incana Pl. Physiol. 70: 342-348.

# 9

# Prospects of Botanical Insecticides in Forest Pest Management

– *B. Deepa*

In India, forests constitute one of the principal natural resources and occupy a land area of 64 million hectares. However, forests in India have suffered considerable degradations and effective forest cover considerable degradations and effective forest cover constitutes less than 19% of the Indian land mass. In order to change this appalling scenario, massive tree planting programmes in degraded forests, barren, marginal, waste and mined lands and watershed areas have been taken up. Insect pests are one of the important biological determinants in the success of reforestation and afforestation programmes. Though the management of these pests is done successfully by the use of chemical pesticides the demand for alternative still exists. Though botanical pesticides are widely used and getting popularized in the agriculture sector, they are very seldom used for pest management in forestry. However, research attempts are documented with respect to the insecticidal potential of some common trees like neem, which can be used be used to control insect pests in the forestry sector. In recent times there has been lot of attention on eco-friendly measures to combat insect pests especially on high value timber yielding commercially important species like teak, sandal etc. This paper deals with the review of the plants that are extensively used for the management of forest pests.

## Major Insect Pests of Forest

Forest insect pests in India constitute defoliators xylophagous insects, shoot borers, canker formers, gall gorming pests, sap suckers and wood destroying insects. On account of high reproductive potential and short life cycle, most forest insects tend to multiply to large numbers, a fact responsible for the difficulties in management strategies.

The important forest pests include teak defoliators (*Eutectona machaeralis* and *Hyblaea puera*), gamhar defoliator (*Calopepla leayna*), Ailanthus defoliators (*Atteva fabriciella* and *Eligma narcissus*), khasi pine defoliator (*Eterusia pulchela*), poplar defoliators (*Clostera fulgurita* and *C. cupreata*), shisham defoliator (*Plecoptera reflexa*), poplar borer (*Apriona cineria*), sal heartwood borer (*Hoplocarambyx spinicornis*), Eucalyptus and babul stem borer (*Celosterna scabator*), Meliaceae shoot borer (*Hypsipyla robusta*), pine shoot borer (*Dioryctria abietella*), canker causing insects (*Dihammus cervinus* affecting teak and gamhar and *Sahydrassus malabaricus* affecting teak and Eucalyptus), tender leaf gall fly (*Trioza obsolete*), popular gall former (*Pemphigus imaicus*), sap-suckers (Pineus laevis affecting tropical pines) champ bug (*Urostylis punctigera*), termites and borers attack harvested wood (Sen-Sharma, 1992).

**Review on the Use of Botanicals against Forest Pests**

In spite of the discovery of antifeedant property of neem as early as 1962 by Pradhan et al., against the most notorious polyphagous pest *Schistocerca gregaria*, no work initiated to test other botanicals against forest pest till 1988. Credit goes to Bhandari *et al.* (1988) who tested methanolic extract of neem seed and nine chromatographic fractions of the same against a notorious poplar defoliator *Pygaera* (*Clostera*) *cupreata*. Ahmed *et al.* (1991) reported the insecticidal property of the ethanol, acetone and ether extracts of *Acorus calamus, Lantana camara, Adhatoda vasica* and *Melia azadirach* against *ailanthus webworm Atteva fabriciella.*

9 plant oils and 20 of their common constituent were examined for their insecticidal property against a notorious subabool pest *Heteropsylla cubana,* of which Acorus calamus proved to be the best (Sharma *et al.*, 1992). Joshi *et al.* (1993) tested petroleum ether extract of *Azadirachta indica* and *Jatropha curcas* for their antifeedant property against the third *instar larvae* of *Papilio demoleus* in the laboratory.

Effect of various neem products has been tested against arid forest insect pests (Sundararaj *et al.*, 1994). Methanolic extract of NSKE showed good anti-insect activity against rohida (*Tecomella undulata*) leaf defoliator, *Patialus tecomella* and *Taragama siva* of *Prosopis juliflora, P. cinerria* and *Acacia* spp. Neelusingh and Bhargava (1995) found terpenoid of neem effective as an antifeedant against third instar larvae of *Ailanthus* defoliator, *A. fabriciella.*

Pandey, et al. (1997) reported the insect feeding deterrency of camphor oil isolated from *Cinnamomum camphora* against Calopepla leayana and Eupterote geminata, to defoliators of *Gmelina arborea*. Ahmed et al. (1997) reported the antifeedant potency of *Eucalyptus hybrid, Dalbergia stiulacea* bark and *Adina cordifolia* leaves against defoliating pest of poplar, *Clostera cupreata*. A field trial was conducted to study the inefficacy of neem seed oil alone and in combination with two conventional insecticides viz., *monocrotophos* and *endosulfan* against the babul whitefly *Açaudaleyrodes rachipora* on *Acacia nilotica* seedling. The results demonstrated that the neem seed oil at 0.5% alone is good enough to control this pest and its combination with either 0.1% monocrotophos or endosulfan did not show any superiority than neem seed oil alone to *A. achipora* (Sundararaj, 1997).

In a preliminary study, conducted by Singh and Sushilkumar (1998) dilute latex and oil of *Jatropha curcas* applied topically was found highly effective to kill the third instar larvae of teak defoliators and *D. sissoo* defoliator. Insecticidal properties of neem and some plant extracts were studied against *Spodoptera litura* and *Clostera cupreata* (Kulkarni, 1998, 1999). Neem products considerably affected the growth, feeding and *oviposition* of teak defoliator, *Hyblaea puera*. Higher feeding and ovipocitional deterrency were found after the treatment of neem seed kernal extract, neem oil, neem gum, neem root extract and neem tender leaves than the other neem parts tested for H. puera. Larval and pupal mortality were also higher on the above neem product treatment. Neem seed kernal extract + neem gum extract; neem seed kernal extract + neem root extract and neem tender leaves showed synergistic activity (Murugan et al., 1999).

## Limitations for the Use of Botanical Pesticides for Forest Pest Control

Though the use of botanical pesticides appears promising in the laboratory conditions its use in the field poses serious problems due to following reasons:

1. Botanical pesticides degrade fast. Hence they must be applied frequently.
2. Obtaining the plant material and extracting the phytochemical involves long and cumbersome procedures.
3. Botanical pesticides vary greatly in their toxicity to humans as well as to non-target organisms. Use of all precautions recommended when using synthetic pesticides is necessary.
4. Cost is more.
5. Commercialization and marketing are difficult.
6. Forests have extensive area even under plantations, which are often situated in inaccessible hilly terrain, and this imposes serious restrictions on the routine surveillance and control operation.
7. Tall and larger sizes of trees demand use of specialized application equipments.

## Conclusions

Considering the impact of synthetic pesticides on the deterioration of earth's environment, use of botanical pesticides should be recommended as a component of integrated pest management in forestry sector although there are disadvantages in the use of botanical pesticides. Plants and plant products may be thoroughly researched for effectiveness and toxicity before recommending them as botanical pesticides. Detailed research has to be taken up to screen out flora possessing insecticial property against forest pests.

## References

Ahmad, M., Dayal, R. and Mishras, P.K. (1997). Antifeedant potency of some plants products against defoliating pest of poplar, *Clostera cupreata*. Indian Forester 123(9): 821-826.

Ahmad, M., Gupta, B.K. and Bhandari, R.S. (1991). Efficacy of some plant extracts against Ailanthus webworm *Atteva fabriceila*. Indian J. of forestry, 14(1): 5-7.

Bhandari, R.S., Lal, K.S. J., Ayyar and Singh, P. (1988). Effect of neem seed extractives on poplar defoliator *Pygaera cupreata* Butler. (Lepidoptera: Notodontidae) in laboratory. Indian Forester, 114(11): 790-795.

Joshi, K.C. Meshram, P.B. Roychoudhary, N. Kulkarni, N. and Kalia, S. (1993). Antifeedant activity of neem and Jatropha against the larvae of Papilio demoleus Linn. (Lepidotera: Papilionidae). Paper presented in "National Seminar on forest produce" at IFGTB, Coimbatore from 15-16th October 1993.

Kulkarni, N. (1998). Feeding deterrence property of some plant extracts against poplar defoliator *Clostera cupreata*. Insect environment, 3(4): 98.

Kulkarni, N. (1999). Antifeedant and growth inhibitory effects of *Azadirachta indica* A.Juss. against the larvae of *Spodoptera litura* (Fab.) Boursin (Noctuidae: Lepidoptera). National Academy of Science Letter, 22(1&2): 24-26.

Kulkarni, N. Joshi, K.C. and Shukla, P.K. (2003). Antifeedant acticity of *Aannona squamosa* Linn. Against Crypsiptya coclesalis Walker (Lepidoptera: Pyralidae). Entomon 28(4): 389-392.

Murugan, K. Senthil Kumar, N. Jeyabalan, D. Senthil Nathan, S. and Sivaramakrishnan, S. (1999). Neem as an effective bio-pesticide to control teak defoliator *Hyblaea puera* (Cramer) (Lepidoptera: Hyblaeidae). Journal of non timber forest products, Vol. 6(1/2): 78-82.

Murugesan, S., Ramsevak, R.S., Mattson, W.J., Nair, M.G., Narayanan, C. and Mohan, V. (2000). Antifeedant and antifungal compounds from *Dirca palustris*. The Indian Forester, 129(3): 364-370.

Meshram, P.B. (2000). Antifeedant and insecticidal activity of some medicinal plant extracts against *Dalbergia sissoo* defoliator *Plecoptera reflexa* (Lepidoptera: Noctuidae). The Indian Forester, 126(9): 961-965.

Neelusingh and Alka Bhargava (1995). Efficacy of neem constituents as feeding deterrent against the Ailanthus webworm *Atteva fabriciella*. Paper presented in Silviculturist meet held at Nagpur from 9th to 11th January 1995.

Pandey, A.K., Jacob, J.P., Das, D.K., Sharma, P.C. (1997). Insect feeding deterrency of *Cinnamomum camphora* oil against defoliators of *Gmelina arborea* Roxb. (Verbanaceae). Indian Journal of forestry, 20(1): 98-100.

Pradhan, S., Jotwani, M.G., and Rai, B.K. (1962). The neem seed deterrent to locust,. Indian farming, 12(8): 7-11.

Sen-Sharma, P.K. (1992). Forest insect pests, their management and environmental considerations. *Proc. Zool. Soc.*, Calcutta, 45(Suppl. B): 81-102.

Sharma, R.N., Tare, V., Pawar, P., and Vartak, P.H. (1992). Toxic effects of some plant oils and their common constituent on the psyllid pet, *Heteropsylela cubana* (Homoptera: Psyllidae) of social forestry tree *Leucaena leucocephala*. Applied entomology and Zoology, 27(2): 285-287.

Singh, N. and Sushilkumar (1998). Biopesticidal plants from Indian forests and their role in Insect pest management, 73-123. In Ecofriendly Pest Suppression, Gautam, R.D. and Prasad, D. (eds.) Westvill Publishing House, New Delhi.

Sundararaj, R. (1997). Evaluation of neem seed oil and some insecticides against the babul whitefly *Acaudaleyrodes rachipora* (Singh) (Alyerodidae: Homoptera) on Acacia nilotica in nursery. Pestology Vol. XXI No. 5, pp. 34-37.

Sundararaj, R. and Murugesan, S. and Ahmed, S.I. (1994). Evaluation of methanolic extract of neem seed kernal powder against the rohida defoliator *Patialus tecomella* Pajni (Curculionidea: Coleoptera). Presented in Second National Conference of AZRA on *Recent trends in plant, animal and human pet management: Impact on environment* at Madras, India, December 27-29.

# 10

# Fish Pheromones : Current Status and Potential Applications in Aquaculture, Breeding and Fishery Environment Management

— *A.K. Pandey*

Pheromones are externally secreted semiochemicals employed in intercommunications among individuals of the same species. There exist convincing evidence that such system is widely operative among the organisms ranging from protozoans to mammals. There are instances of chemical communication in aquatic plants too, especially for the attraction of gametes. Recent studies have shown that most of the vital activities in fish like alarm communication, sex attraction as well as synchronization of the reproductive processes, individual identification, group cohesion, parent-offspring recognition, territorial markings and migration are shaped by pheromones (Liley, 1982; Pfeiffer, 1982; Pandey, 1984, 1998; Chivers and Smith, 1998; Mann *et al.*, 2003). Pheromones of the terrestrial organisms are volatile having low molecular weight ranging from 50-300 while such semiochemicals in the aquatic species appears to be of larger size (high molecular weight) and soluble in water (Smith, 1992; Sorensen *et al.*, 1998; Pandey, 2003, 2005).

## Alarm Pheromones

Alarm substance cells (ASC) of the ostariophysan and other groups of fish release alarm pheromone from their skin on mechanical injury, presumably by a predator that elicits alarm reaction among the conspecifics and members of the closely-related species. Depending upon biology of the species, the alarm reaction includes dashing, c-turn, immobility, tight schooling, hiding and avoidance of the area (Pfeiffer, 1982; Smith, 1992). There are instances that the predator northern pike (*Esox lucius*) consuming the preys like *Pimephales promelas* and *Culaea* incostans containing the ASCs in skin, releases the alarm pheromone in the fecal matter which are detected by the members of the prey or closely-related species. The active space of alarm pheromone can exceed 58,000 l/cm$^2$ skin. The chemically induced alarm reaction may be transmitted visually among other members of the school/shoal (Pfeiffer, 1977; Smith, 1977, 1982a). There are reports that experience also plays a role in the alarm pheromone perception (Smith, 1982b; Mathis and Smith, 1993; Brown *et al.*, 1995; Chivers *et al.*, 1995; Wisenden *et al.*, 1995). Stabell and Lwin (1997) observed phenotypic changes like increase in body depth in crucian carp (*Carassius carassius*) exposed to the odour of northern pike fed on crucian carps. This phenotypic change appears to be due to primer effect of the alarm pheromone.

The alarm substance of minnow was suggested to be purine- or pterine-like (double-ringed compounds usually associated with pigments), non-volatile and extremely soluble in water. Histamine was also implicated to be alarm substance of certain cyprinids. Pfeiffer and Lemke (1973) reported the alarm substance of the European minnows (*Phoxinus phoxinus*) to be pterine (isoxanthopterine) that might be conjugated with protein for water solubility as well as species specificity. However, Kasumyan and Lebedeva (1977, 1979) showed that the alarm pheromone of *P. phoxinus* possesses molecular weight approximately 1100, behaves in alkaline medium (pH 8-9) as anion and breaks down when heated. They suggested it to be a carbohydrate or related compound containing amino groups. Pfeiffer (1982) found several heterocyclic compounds to induce effective alarm response. Hypoxanthine-3-N-oxide has been reported to be an active component of the ostariophysan alarm signaling system (Pfeiffer *et al.*, 1985; Brown *et al.*, 2000, 2003). Laboratory as well as field studies have revealed that the traps labeled with the fathead minnows skin extract, hypoxanthine-3-N-oxide, or pyridine-N-oxide caught significantly fewer fish than did those labeled with distilled water (Brown *et al.*, 2000) . The central bundles in both tractus olfactorii of the medial olfactory tract of the fish conduct the nervous excitation elicited by alarm substance to the higher center in brain (Pfeiffer *et al.*, 1984) resulting in the elevation of plasma cortisol, glucose, potassium ($K^+$) and sodium ($Na^+$) levels. Plasma calcium ($Ca^{2+}$) recorded an initial increase followed by a decline to the below normal level. These changes suggest that alarm pheromone functions as stressor in fish (Lebedeva *et al.*, 1994, 1999).

**Female Sex Pheromones**

Salmonid males congregate around the cages/pens of the mature females during breeding seasons. Fishermen of Mississippi usually catch large number of male channel catfish (*Ictalurus punctatus*) by placing ripe females in cages (Timms and Kleerekopper, 1972). Water from gravid females seems to increase the nesting and other activities associated with reproduction among the males in anabantids and cichlids (Crapon de Caprona, 1974). Tavolga (1956) was the first who systematically studied the sexual behaviour of an estuarine goby (*Bothygobius soporator*) and demonstrated that the spermiated (matured) male discriminates between males and females, and also between non-gravid and gravid females on the basis of chemical cues released through their vent. Ovarian fluid was sufficient to evoke courtship behaviour in males even in the absence of females. A female sex pheromone appears to facilitate the pair formation and copulation in the marine carcharhinid shark. Also, mature female sea lamprey, *Petromyzon marinus*, releases a pheromone in the ovarian fluid that attracts male conspecifics (Teeter, 1980). Later studies revealed that a definite threshold of circulating levels of androgen is essential to render the male differentially sensitive to the sex attractant released by the gravid females.

Mature females of goldfish (*Carassius auratus*), yellowfin Baical sculpin (*Cottocomephorus grewingki*), Atlantic salmon (*Salmo salar*), rainbow trout (*Salmo gairdeneri=Oncorhynchus mykiss*), Baikal cisco (*Coregonus automnalis migratorious*), common carp (*Cyprinus carpio communis*) and *Barilius bendelisis* release primer pheromone in the urine (Liley, 1982; Sorensen *et al.*, 1998; Sorensen and Stacey, 1999, 2004; Pandey, 2003) which probably enhances the plasma gonadotrophin (GtH) and testosterone (T) levels (within 15-30 minutes in case of goldfish) and milt (sperm and seminal fluid) volume as well as sperm motility among males after prolonged exposure (6 hours in case of goldfish) Besides urine, the priming pheromone has also been detected recently in bile of the female rainbow trout. There exists report showing a reduction in pituitary dopamine turnover resulting in increased gonadotropin secretion in the male goldfish exposed to the female sex pheromone. Interestingly, mature female *Oreochromis niloticus* becomes attractive to males after 17, 20-dihydroxy-4-pregnene-3-one administration. Though the exact chemical identity of the primer pheromone(s) is not known but the maturation-inducing hormone, 17, 20-dihydroxy-4-pregnene-3-one and its sulphate or glucuronide metabolites are the likely candidates because they induce electro-olfactogram (EOG) recordings in the olfactory tract even at the low concentration ranging between $10^{-9}$-$10^{-13}$mM. The primer pheromone released by the Atlantic salmon (*Salmo salar*) has been suggested to be F-type prostaglandins. Urine of the female evokes maximum response in the conspcific males emphasizing species-specificity of the pheromone.

A number of externally fertilizing fish release chemical couplin (s) during ovulation which elicits attraction (locomotor behaviour), courtship and spawning in male conspecifics (Murphy *et al.*, 2001; Laberge and Hara, 2003). Such releaser pheromones induce an immediate response (within 10-30 minutes) without altering the plasma concentration of gonadotrophin. Amouriq (1965) suggested the female sex attractant (substance dynamoge'ne) of *Lebestes reticulatus* to be an estrogen as hexestrol dipropinoate elicited courtship behaviour among males. Okada *et al.* (1978) showed that the fraction ($FP_1$) on DEAE cellulose column that evoked male courtship behaviour in the pond smelt declined in activity after heat (80°C; 5 minutes) or trypsin (30°C; 60 minutes) treatment and presumed the substance in genital fluid was either a protein or a substance involving protein in its structure. Kawabata (1993) reported amino acids as inducer of sexual behaviour in male *Rhodurus occellatus occellatus*. Recent studies have suggested the metabolites of prostaglandin $F_2a$ ($PGF_2a$) like 15-keto- and 13, 14-dihydro-15-keto-$PGF_2a$ to function as releaser pheromones in *Pimephales promelas, Carassius auratus, Salvelinus alpinus, Misgurnus anguillicaudatus, Oncorhynchus rhodurus* and *Astyanax mexicanus*. Interestingly, tetrodoxine has been reported to be the sex attractant pheromone to spermiating males in the marine puffer fish, *Fugu niphobles* (Matsumura, 1995). It has been suggested that the releaser pheromone(s) functions as reproductive isolating principle in sympatrically breeding gourami (*Trichogaster trichopterus* and *T. pectoralis*), suckers (*Catastomus catastomus* and *C. commersonii*) and masu salmon/rainbow trout (*Oncorhynchus masu* and *O. mykiss*). As per definition, the pheromones should be species-specific but there are instances of closely-related species responding to the sex pheromones in fish.

## Male Sex Pheromones

During breeding season, male become attractive to the female conspecifics. In some regions of France, male lampreys nearing sexual maturity are placed by the fishermen in the traps to attract females (Johnson *et al.*, 2006). Attraction on sexually mature female brook trout to mature male conspecifics appears to be mediated by pheromone (Young *et al.*, 2003). The odouriferous secretion of the cutaneous anal gland (rich in mesorchial interstitial cells) of male *Blennius pavo* attracts the females during the breeding season. It has been recorded that appendices of the 1$^{st}$ and 2$^{nd}$ anal fin spines of the mature male *B. pavo* produce a substance that attracts conspecific females. Numerous invaginations are found in the anal appendices of this species and production of pheromone form this organ is dependent on the circulating gonadotropin levels. The gill gland of male swardtail characin, *Corynopoma riisei* appears to be the source of female sex attractant pheromone. Sperm-duct gland of freshwater goby, *Padogobius martensi* appears to secrete sex attractant substance. Jaski (1939) was the first who reported that males of guppy (*P. reticulatus*) secrete a "chmical couplin" which synchronizes the sexual receptivity among the females. Teeter (1980) showed the presence of a pheromone in urinogenital fluid of male sea lamprey (*P. marinus*) that elicits courtship in the females. Out of the nine steroid hormones such as dehydroepiandrosterone, testosterone, dihydrotesto-sterone, progesterone, androstenedione, estrone, oestradiol, corticosterone and cortisol tested, only testosterone elicited the preference response in spawning-run female sea lamprey (Adams *et al.*, 1987). Testosterone or a closely-related structural derivatives has been suggested to function as sex pheromone in the male cyclostome. The male sexual pheromone of *Gobius jozo* has been identified as etiocholenolone glucuronide. Mature males of *S. alpinus* release metabolites of prostaglandin $F_2a$ to attract mature females and induce spawning behaviour.

Seminal vesicles of the male African catfish (*Clarias gariepinus*) secrete sex pheromone that attracts females during the breeding season (Resink, 1988; van Weerd, 1990; van den Hurk and Jenkins, 1992). Though 17 free and 8 glucuronide metabolites of sex hormones have been identified in the seminal vesicle fluid (SVF) but 5b-pregnene-3a,17a-diol-20-one glucuronide and 5b-androstene-3a,17a-diol are the two more potent olfactory stimulants with threshold value $10^{-9}$. Wild breeding male African catfish recorded 20-folds elevation in level of 5b-pregnene-3a,17a-diol-20-one glucuronide in the seminal vesicle as compared to those maintained under captivity which may probably be the reason for lack of natural spawning in the catfish kept in ponds (Schoonen, 1987; Resink, 1988).

Mature males of zebrafish (*Brachydanio rerio*) release a pheromone that induced ovulation in females. Gas chromatography-mass spectrometry (GC-MS) studies have revealed the presence of glucuronides of 5a-androstane-3a,17b-diol and cholesterol in male-holding waters. It has been suspected that these compounds, probably originating from liver, may be functioning as ovulation-inducing pheromone.

The most profound effect of males on the females seems to be its influence over the neuroendocrine (hypothalamo-hypophysial-gonad axis) of the latter. Mosher (1954) was the first to observe that introduction of a male in the group of females stimulates egg production and its continued presence led to regular oviposition cycles. In the angelfish (*Pterophyllum scalare*), blue gourami (*T. trichopterus*), Indian catfish (*Heteropneustes fossilis*) and zebrafish (*Brachydanio rerio*), paired females with males possessed ovaries filled with matured ova while ovaries of the grouped females (without males) contain almost no ova but numerous oocytes at various stages of development (Pandey *et al.*, 2000; Pandey, 2005; Gerlach, 2006). There exist reports that the male pheromone(s) of African catfish (*C. gariepinus*) accelerates the ovarian recrudescence and ovulation in females probably by enhancing plasma gonadotropin (GtH) level (van Weerd,

1990; van den Hurk and Resink, 1992). Recently, Degani and Schreibmam (1993) observed that the aquarium water in which male gourami (mature) built nests induced release of immunoreactive gonadotropin (GtH) in the females which, in turn, enhanced plasma 17b-estradiol ($E_2$), 17a,20b-dihydroxy-4-pregnene-3-one (17,20-P), 5b-pregnene-3a,17a,20b-triol and 11-ketotestosterone (11-KT) levels as well as vitellogenic activities suggesting the primer effects of male pheromone in this species.

## Intrasexual Stimulants and School Formation

Though the reports on the odouriferous secretions eliciting male-male and female-female stimulations are rare in vertebrates but the possibilities have been raised from time to time. In fishes, the male three-spined stickleback (*Gasterosteus aculeatus*) gets stimulated by smelling the odours of its own nest during breeding season. The males of *Hypsoblennius jenkinsi, H. robustus* and *H. genitilis* secrete a chemical during "high courting and mating periods" that releases the sexual appetitive behaviour and receptivity in the males. Since unisexual groupings of the females (without males) disrupt (inhibit) the ovarian development, it offers an opportunity to explore the female-female interacting pheromones in fish (Pollack *et al.*, 1978; Pandey, 2005; Gerlach, 2006).

A pheromone is found in the milt of male herring (*Clupea harengus pallasi*) that triggers spawning behaviour (papilla extension) and release of milt in the males. Sherwood *et al.* (1991) reported that this pheromone shows hydrophobic properties similar to those of polar steroids, prostaglandins or other conjugate forms, and at least one form appears to contain a sulphate or glucuronide group. The male pheromone appears to play important role in synchronization of spawning in the school of herrings. Recent studies have shown that the elevated levels of plasma 3a,17a-dihydroxy-5b–pregnen-20-one and 17a-hydroxyprogesterone coincided with responsiveness to the spawning pheromone in this species (Carolsfield, *et al.*, 1997).

McFarland and Moss (1967) speculated possibility of the involvement of pheromones in school formation and attraction of the mature conspecifics towards the spawning grounds. Presence of an aggregating pheromone has been implicated in the schooling Japanese marine catfish, *Plotosus lineatus*. Bloom and Perlmutter (1977) demonstrated the presence of male-male and female-female attractants in the zebrafish (*B. rerio*). Teeter (1980) also hypothesized involvement of pheromone(s) in the aggregation of adult lampreys (*P. marinus*) prior to upstream movement or to keep them together in night during the migration. It has been speculated that the sexual-aggregating pheromones of zebrafish are two in number-one attracting male conspecifics and other the females. These attractants appear to be contained in the cholesterol-ester fraction (R*f*-0.94) of the thin-layer chromatogram and the attraction response was not affected by food or diet.

**Individual and Species Recognition**

Varying degrees of socialism have been recorded in fish, however, such tendency is more marked among the visually-deficient species. The school of roach (*Ritulus ritulus)* does not disintegrate during night because members of the school maintain contact through chemical cues. There exist records that individuals of *Ictalurus natalis, Clupea herrangus pallasi, Caecobarbus geertsi, Haplochromis burtoni, Astyanix jordani, Salmo alpinus, Notropis lutrensis, Anguilla anguilla, Oncorhynchus kisutch, Oncorhynchus nerka, Ictalurus nebulosus, Astyanax mexicanus, Phreactichthys andruzzi, Carassius auratus, Heteropneustes fossilis* and *Danio rerio* are attracted towards the waters of the tank previously occupied by their conspcifics. Even broods and youngs are recognizable to their parents by the odours. The parents of Midas cichlid fish (*Cichlasoma citrinellum*) recognize to their youngs by means of pheromone. Barnett (1981) reported that fry of Midas cichlid fish discriminates even between mother and father on the basis of urine odour and suggested that the steroid titers in urine could provide information on sex and that of peptide chains could indicate species. It appears that secretions in the skin mucus, urine as well as metabolites of bile acids (voided in fecal matter) are involved in individual identification and group cohesion in these fishes (Todd, 1971; Selset, 1980; Barnett, 1981; Zhang, 1997). Interestingly, glucurnidation of sex hormones also takes place in the skin of the African catfish (*Clarias gariepinus*) which may probably have certain implication in individual/ sex recognition Recently, synthesis of 17a,20a-,17a,20b-dihyhroxy–4–pregnen–3-one, 11-ketotestosterone and other sex hormone conjugates have been recorded in the gill of goldfish, common carp and trout suggesting that the organ might play role in secretion of pheromones in some species.

Benficial conditioning of the waters by conspecifics and inhibition of growth under over-crowded condition appear to be mediated through pheromones (Solomon, 1977). Eels (*Anguilala anguilla*) are usually caught in the traps dripped with conspecific skin extracts (Saglio, 1982). Todd (1971) found that fighting as well as stress alter chemistry of the pheromones of the yellow bullhead (*Ictalurus natalis*) secreted in the mucus. Diets also appear to modify body odour of the catfish. Interestingly, Colyer and Jenkins (1976) indicated the presence of an aggression-suppressing pheromone(s) in the Siamese fighting fish, *Betta spendens*. Carr and Carr (1986) have observed that the aggression-suppressing pheromone of the brown bullhead (*I. nebulosus*) gets inactivated by heat and protease treatments and suggested it to be a protein with molecular weight less than 10,000.

During reproductive period, the males of some species deposit a layer of mucus on the spawning surface which may serves as chemical signals marking the breeding sites and may perhaps communicate status of the occupants. There exist reports on the substrate marking behaviour in fishes and secretions of skin (mucus), urophysis and urine have been implicated in the production/release of the territorial marking pheromones (Stabell, 1987). Possibly, these marks bear the same biological significance as marking fluid (MF) of the mammals.

## Pheromones and Migration

How migratory fishes locate their pathways from sea to freshwater and *vice versa* remains a mystery. Work done on homing of the salmonids has demonstrate that these fishes migrate from the nursery areas to their parental river streams, out to sea and then back to their birth place to spawn (basis of the home stream theory). It was suggested that each stream had unique bouquet of odours and fish has to follow its nose in detection and discrimination of the natal stream. Though the role of metablic products of a specific population (pheromones) as homing cues for the salmonids was proposed in 1924 and 1934 but this hypothesis was totally eclipsed by the concept of "olfactory imprinting" propounded by Hasler (1983). According to this theory, the youngs lock onto the odours of aquatic vegetations, soil run-off and other organic chemicals present in the parent stream in brain and they retain the olfactory memory during the interim period of homeward migration.

Nordeng (1971) realized the pheromone odour tracks help guide homing of the Arctic char (*Salmo alpinus*). Solomon (1973) demonstrated that the migrating Atlantic salmon (*S. salar*) prefers the river in which youngs of its own populations were living. After exhaustive studies on the migratory behaviour of *S. alpinus, S. trutta* and *S. salar,* Nordeng (1977) rejuvenated the pheromone hypothesis by emphasizing that homeward navigation is an inherent response to population-specific pheromone trails released from the descending smolts. The role of pheromones in upstream orientations has been confirmed in the anadromous rainbow trout (*O. mykiss*) and sea lamprey (*P. marinus*). Miles (1968) has demonstrated that elvers of the catadromous European eel (*A. rostrata*) are attracted towards the water of the streams containing the conspecific adults. Sorensen (1984) also confirmed this finding during his study at Rhode Island. Recently, Stabell (1992) remarked that the specific homing to a native site is under genetic influence and the possible genetic contamination of pheromones due to hatchery escapes or random stocking programmes might seriously interfere with the homing performance and population structure. He emphasized for the concept of chemical ecology in management of salmonid fisheries.

Our knowledge regarding the chemical nature of these navigational cues is still fragmentary. Miles (1968) found that the compound(s) involved with attraction of the elvers in *A. rostrata* to be biodegradable, unaffected by autoclaving and non-volatile. Atema *et al.* (1973) reported that the alwife (*Alosa pseudoherengu)* homing pheromones are heat-stable, non-volatile, polar and having molecular weight less than 1000. Doving *et al.* (1980) demonstrated the presence of migratory cues of *S. alpinus* in the fecal matter of the smolts and suggested it to be derivatives of bile salts like taurolithocholate, taurolithocholic sulphate, sulphotaurolithocholate or taurochenodeoxycholate which may convey the information to the brain through medial portion of the olfactory tract. By employing electrophysiological, biochemical and behavioural approaches, Zhang (1997) also confirmed the chemostimulatory roles of 9 out of 38 authentic bile acids in lake char (*Salvelinus namaychush*) and remarked that the olfactory sensitivity and specificity of bile acids are affected by (i) position and orientation of hydroxyls, (ii) hydrosulphation, (iii) side-chain length and (iv) side-chain substituents of the molecules. Li and Sorensen (1997) and Li *et al.* (2002) have identified peteromyzonol sulphate and allocholic acids as migratory pheromone in the sea lamprey, *P. marinus*.

## Chemoreception and Water Pollution

Four groups of chemicals like (i) amino acids, (ii) steroids, (iii) prostaglandins and (iv) bile acids (and their derivatives) have been found to be detected by the fish even at minute ($10^{-7}$-$10^{-13}$mM) concentration and each group of chemicals have different receptors in the olfactory system. Hara (1992) has given a hypothetical three-dimensional model to show the mode of interaction of amino acids, which serve as feeding stimulants in fish, with olfactory receptor sites. It has been observed that detergents even at the concentration of 0.5 ppm (much lower value that inflict lethal damage) impaired functioning of the receptor by causing erosion in the chemosensory organ. Bloom *et al.* (1978) showed that sublethal (0.5 ppm) exposure of zinc affected the behaviour to the sexual aggregating pheromones in the zebrafish. Linear alkylbenzene sulphate (LAS) significantly depressed chemoattraction of water conditioned by conspecifics to juvenile *Salvelinus fontinalis*. Laboratory studies have revealed that the freshwater spawning migration of salmons is being impaired by acidification of rivers. Electrophysiological recordings of the olfactory epithelium of the adult male *S. salar* to testosterone and urine of the ovulated females significantly reduced at pH 5.5-4.5 and abolished at pH 3.5. Exposures to sublethal concentrations of diazinon and carbofuran significantly reduced the ability of mature Atlantic salmon parr to priming pheromone of ovulated females. The pernicious effects of an anionic detergent (sodium lauryl sulphate), pH, mercury (Hg), silver (Ag), cadmium (Cd), copper (Cu), nickle (Ni) and zinc (Zn) on the olfactory system are now well established. By calculating $IC_{50}$ value (concentration which inhibits electro-olfactographic response by 50%), Ag, Hg, Cu and Cd are found to elicit most depressive effects on the chemoreception in salmonids. The whitefish (*Coregonus clupeaformis*) and rainbow trout (*O. mykiss*) exposed to 2.4 mm copper sulphate ($CuSO_4$) solution for two weeks exhibited loss in the phospholipid stainable granules in the receptor neurons of the olfactory mucosa, however, recovery in the granules were observed when the fishes were transferred to freshwater for 12 weeks. Phospholipids have been reported to be involved in the electrophysiological (electro-olfactogram, EOG) activities of the olfactory epithelium. There are reports that olfactory epithelial cytochrome P-450 and monooxygenase (mixed function oxidase, MFO) are activated by the pollutants like hydrocarbons and heavy metals. Since the present trends of rapid industrialization and increase in population have drastically changed the aquatic environment, it is imperative to undertake the studies regarding the interactions of various pollutants with the pheromonal communication system in fishes. Barnett (1981) remarked that a chemically polluted environment might interfere with the success of parent care.

## Conclusions

Earlier fishes were viewed as leaky bags that slowly release various chemicals which can serve as a chemical picture of the animal that others can smell and identify as to species, sex, stress level and perhaps size as well as individuality (Atema, 1980, 1996). Though the chemical nature of pheromones and sites of their biosynthesis (except alarm substances) are not clearly identified, there are growing evidences that these chemicals do have communicative roles among this class of vertebrates. Since sex pheromones are widely distributed among fish, a better understanding of their role is *sine qua non* to understand their role in reproductive physiology (Defraipont and Sorensen, 1993). Interestingly, GnRH has been identified as a potent olfactory stimulant in fish but its role in synchronization of reproductive process is not known. Since these signals are species-specific and operate at molecular levels, it is imperative to utilize them for management and conservation of fisheries (Korkum, 2004; Sorensen and Stacey, 2004; Klassen *et al.*, 2005; Wagner *et al.*, 2006). In aquacultural operations, they may be employed to advance puberty and accelerate ovarian recrudescence as well as selective stimulants to induce spawning at a time and place convenient to the managers (von Weerd, 1990; van den Hurk and Resink, 1992; Zhao *et al.*,

2003; Gerlach, 2006; Hong *et al.*, 2006; Johnson *et al.*, 2006). Since they are externally active, their delivery would eliminate the stress associated with the hormone or drug administration (Stacey, 2003). They may also be employed as artificial baits, selective attractants, growth stimulators as well as inhibitors of aggression and cannibalism. Recent studies point to the potential exploitation of semiochemicals in propagating the migratory fishes in new rivers and streams.

**Acknowledgements**

The author is grateful to Prof. A.G. Sathyanesan, FNA and Prof. C.J. Dominic, FNA for their constant encouragement and inspiration. Help extended by Prof. R.J.F. Smith, Prof. N.R. Liley, Prof. T.J. Hara, Prof. N.E. Stacey, Prof. P.W. Sorensen, Prof. A.C. Rossi, Dr. S.A. Ali and Ms. Rosalie N. Schaffer is thankfully acknowledged.

## References

Adams, M.A., Teeter, J.H., Katz, Y. and Johnson, P.B. (1987). Sex pheromones of the sea lamprey (*Petromyzon marinus*) : steroid studies. *J. Chem. Ecol.* 13 : 387-395.

Amouriq, L. (1965). Origine de la substance dynamogene emise par *Lebestes reticulatus* female (Poisson : Poeciliidae : Cyprinodontiformes). *C. R. Acad. Sci.* (*Paris*), 260 : 2334-2335.

Atema, J. (1980). Smelling and tasting underwater. *Oceanus,* 23 : 2-18.

Atema, J. (1996). Eddy chemotaxis and odour landscape: exploration of nature with animal sensor. *Biol. Bull.,* 191 : 219-138.

Atema, J., S. Jacobson, J.H. Todd and D. Baylan (1973). The importance of chemicals in stimulating behaviour of marine organisms : effects of altered environmental chemistry on animal communication. In : *Bioassay Techniques and Environmental Chemistry* (Glass, G.E., ed.). pp. 177-197. Ann Arbor Scient. Pub., New York.

Barnett, C. (1981). The role of urine in parent-offspring communication in a cichlid fish. *Z. Tierpsychol.,* 52 : 173-182.

Bloom, H.D. and Perlmutter, A. (1977). A sexual agregating pheromone system in zebrafish, *Brachydanio rerio* ( Hamilton-Buchanan). *J. Exp. Zool.,* 199 : 215-226.

Bloom, H.D., Perlmutter, A. and Seeley, R.J. (1978). Effect of sublethal concentration of zinc on an aggregating pheromone system in the zebrafish, *Brachydanio rerio* (Hamilton-Buchanan). *Environ. Pollut.* 17 : 127-131.

Brown, E.G., Chivers, D.P. and Smith, R.J.F. (1995). Fathead minnows avoid conspecific and heterospecific alarm pheromones in the feces of northern pike. *J. Fish Biol.,* 47 : 387-393.

Brown, G.E., Adrian, J.C., Smyth, E., Leet, H. and Brennan, S. (2000). Ostariophysan alarm pheromones : laboratory and field tests of the functional significance of nitrogen oxides. *J. Chem. Ecol.* 26 : 139-154.

Brown, G.E., Adrian, J.C., Naderi, T., Harvey, M.C. and Kelly, J..M. ( 2003). Nitrogen oxides elicit antipredator responses in juvenile channel catfish, but not in convict cichlids of rainbow trout : conservation of the ostariophysan alarm pheromone. *J. Chem. Ecol.,* 29 : 1781-1796.

Carolsfield, J.M., Scott, A.P. and Sherwood, N.M. (1997). Pheromone induced spawning in Pacific herring. 2. Plasma steroids distinctive to fish responsive to spawning. *Horm. Behav.,* 31 : 269-276.

Carr, M.G. and Carr, J.E. (1986). Characterization of an aggression-suppressing pheromone in the juvenile brown bullhead (*Ictalurus nebulosus*). *Copeia,* 1986 : 540-545.

Chivers, D.P. and Smith, R.J.F. (1998). Chemical alarm siganalling in aquatic predatory-prey systems : a review and prospectus. *Ecoscience,* 5 : 338-352.

Chivers, D.P., Wisenden, B.D. and Smith, R.J.F. (1995). The role of experience in the response of fathead minnows (*Pimephales promelas*) to skin extract of Iowa darters (*Etheostoma exile*). *Behaviour,* 132 : 665-674.

Colyer, S.W. and Jenkins, C. (1976). Pheromonal control of aggressive display in Siamese fighting fish (*Betta splendens*). *Percept. Motor Skills,* 42 : 47-54.

Crapon de Caprona, M.D. (1974). The effect of chemical stimuli from conspecifics on the behaviour of *Haplochromis burtoni* (Cichlidae : Pisces). *Experientia,* 30 : 1394-1395.

Defraipont, M. and Sorensen, P.W. (1993). Exposure to the pheromone 17a,20b-dihydroxy-4-pregnen-3-one enhances the behavioural spawning success, sperm production and sperm motility of male goldfish. *Anim. Behav.,* 46 : 254-256.

Degani, G. and Schreibman, M.P. (1993). Pheromone of male blue gourami and its effect on vitellogenesis, steroidogenesis and gonadotropin cells in pituitary of the female. *J. Fish Biol.,* 43 : 475-485.

Doving, K.B., Selset, R. and Thommesen, G. (1980). Olfactory sensitivity to bile acids in salmonid fish. *Acta Physiol. Scand.,* 108 : 123-131.

Gerlach, G. (2006). Pheromonal regulation of reproductive success in female zebrafish : female suppression and male enhancement. *Anim. Behav.,* 72 : 1119-1124.

Hara, T.J. (1992). *Fish Chemoreception.* Chapman & Hall, London.

Hasler, A.D. (1983). Synthetic chemicals and pheromones in homing salmon. In : *Control Processes in Fish Physiology* (Rankin, J.C., Pitcher, T.J. and Duggan, R.T., eds.). pp. 103-116. Croom-Helm, London.

Hong, W.-S., Chen, S.-X., Zhang, Q.-Y. and Zheng, W.-Y. (2006). Sex organ extracts and artificial hormonal compounds as sex pheromones to attract broodfish and to induce spawning of Chinese blacksleeper (*Bostrichthys sinensis* Lacepede). *Aquacult. Res.,* 37 : 529-534.

Jaski, C.J. (1939). Ein oestrus-zyklus bei *Lebestes reticulatus* (Peters). *Proc. Koninkl. Ned. Akad. Wetenschap.,* 42 : 201.

Johnson, N.S., Siefkes, S., Michael, J. and Li, W. (2005). Capture of ovulating female sea lampreys in traps baited with spermiating male sea lampreys. *North Am. J. Fish. Manage.,* 25 : 67-72.

Johnson, N.S., Luehring, Siefkes, S., Michael, J. and Li, W. (2006). Mating pheromone reception and induced behaviour in ovulating female sea lampreys. *North Am. J. Fish. Manage.*, 26 : 88-96.

Kasumyan, A.O. and Lebedeva, N.E. (1977). On the chemical nature of the alarm substance from skin of the European minnow (*Phoxinus phoxinus*). *Biol. Nauk.*,1 : 37-41.

Kasumyan, A.O. and Lebedeva, N.E. (1979). New data on the nature of the alarm pheromone in cyprinids. *J. Ichthyol.*, 19 : 109-114.

Kawabata, K. (1993). Induction of sexual behaviour in male fish (*Rhodeus ocellatus ocellatus*) by amino acids. *Amino Acids*, 5 : 323-327.

Klassen, W., Adams, J.V. and Twohey, M.B. (2005). Modeling the suppression of sea lamprey populations by use of the male sex pheromone. *J. Great Lake Res.*, 31 : 166-173.

Korkum, L.D. (2004). Pheromone signaling in conservation. *Aquatic Conserv. Mar. Freshw. Ecosyst.*, 14 : 327-331.

Laberge, F. and Hara, T.J. (2003). Behavioural and electrophysiological responses to F-prostaglandins, putative spawning pheromones, in three salmonid fishes. *J. Fish Biol.*, 62 : 206-221.

Lebedeva, N.Y., Lebedeva, V.I. and Tgolovkina, .V. (1994). Pheromone of the trout stress inducer. *Biophysics*, 39 : 527-530.

Lebedeva, N.Y., Tomkevich, M.S., Golovkina, N.A., Vosyliene, M.Z. and Golovkina, T.A. (1999). On hormesis of alarm pheromone of cyprinids. *Pheromones*, 6 : 65-68.

Li, W. and Sorensen, P.W. (1997). Highly independent olfactory receptor sites for naturally occurring bile acids in the sea lamprey, *Petromyzon marinus* to pheromonal bile acids. *J. Comp. Physiol.*, 180A : 429-438.

Li, W., Scott, A.P., Siefkes, M.J., Yan, H.G., Liu, Q., Yun, S.S. and Gage, D.A. (2002). Bile acids secreted by male sea lamprey that act as pheromone. *Science*, 296 : 138-141.

Liley, N.R. (1982). Chemical communication in fish. *Can. J. Fish. Aquat. Sci.*, 39 : 22-35.

Losey, G.S. (1969). Sexual pheromone in some fishes of the genus, *Hypsoblennius* Gill. *Science*, 163 : 181-183.

McFarland, W.N. and Moss, S.A. (1967). Internal behavour of fish in schools. *Science*, 156 : 260-262.

Mann, K., Turnell, E., Atema, J. and Gerlach, G. (2003). Kin recognition in juvenile zebrfish (*Danio rerio*) based on olfactory cues. *Biol. Bull.*, 205 : 224-225.

Mathis, A. and Smith, R.J.F. (1993). Intraspecific and cross-superorder responses of chemical alarm signals by brook stickleback. *Ecology*, 74 : 2395-2404.

Matsumura, K. (1995). Tetradoxin as a pheromone. *Nature*, 378 : 563-564.

Miles, S.G. (1968). Rheotaxis of elevers of the American eel (*Anguilla rostrata*) in the laboratory to water from different streams in Nova Scotia. *J. Fish. Res. Bd. Can.*, 25 : 1591-1602.

Mosher, C. (1954). Observations on the spawning behaviour and the early larval development of the sargassum fish, *Histrio histrio*. *Zoologica*, 39 : 141-156.

Murphy, C.A., Stacey, N.E. and Corkum, L.D. (2001). Putative steroidal pheromones in the round goby, *Neogobius melanostomus* : olfactory and behavioural responses. *J. Chem. Ecol.,* 27 : 443-470.

Nordeng, H. (1971). Is the local migration of anadromous fishes determined by pheromones? *Nature,* 233 : 411-413 .

Nordeng, H. (1977). A pheromone hypothesis for homeward migration in anadromous salmonids. *Oikos,* 28 : 155-159.

Okada, H., Sakai, D. K. and Sugiwaka, K. (1978). Chemical stimulus on the reproductive behaviour of the pond smelt. *Sci. Rep. Hokkaido Fish Hatch.,* 33 : 89-99.

Pandey, A.K. (1984). Chemical signals in fishes : theory and application. *Acta Hydrochim. Hydrobiol.,* 12 : 463-478.

Pandey, A.K. (1998). Chemical cimmunication in fishes : an overview. *Fishing Chimes* 18 (2) : 15-20.

Pandey, A.K. (2003). Current status and potential applications of fish pheromones in aquaculture and fishery management. In : *Aquaculture Medicine* (Bright Singh,I.S., Pai, S.S., Philip R. and Mohandas, A., eds.). pp. 271-292. School of Environmental Sciences, Cochin University of Science & Technology, Cochin

Pandey, A.K. (2005). Recent advances in fish pheromone research with emphasis on their potential applications in fisheries. *J. Appl. Zool. Res.,* 16 : 210-216.

Pandey, A.K., Mitra, A. and Sarkar, B. (2000). An instance of chemical communication in the Indian catfish, *Heteropneustes fossilis* (Bloch). In : *Fifth Indian Fisheries Forum* (January 17-20, 2000). . p. 83. Central Institute of Freshwater Aquaculture, Bhubaneswar

Pfeiffer, W. (1977). The distribution of fright reaction and alarm substance cells in fishes. *Copeia,* 1977: 653-665.

Pfeiffer, W. (1982). Chemical signals in communication. In : *Chemoreception in Fishes* (Hara, T.J., ed.). pp. 307-336. Elsevier Sci. Pub. Co., Amsterdam.

Pfeiffer, W. and Lemke, J. (1973). Untersuchungen zur Isolierung und Identifizierung des Schreckstoffes aus der Hault der Elritze, *Phoxinus phoxinus* (L.) (Cyprinidae, Ostariophysi, Pisces). *J. Comp. Physiol.,* 82 : 407-410.

Pfeiffer, W., Mangold-Wernado, U. and Neusteurer, P. (1984). Identification of nerve bundle in the tractus olfactorius of the tench, *Tinca tinca* L., which conduct the nervous excitation elicited by alarm substance. *Experientia,* 40 : 219-220.

Pfeiffer, W., Riegelbauer, G., Meier, G. and Scheibler, B. (1985). Effects of hypothanthine-3(N)-oxide and hypothanthine-1(N)-oxide on central nervous excitation of the black tetra, *Gymnocorymbus ternetzi* (Characidae, Ostariophysi, Pisces) indicated by dorsal light response. *J. Chem. Ecol.,* 11 : 507-524.

Pollack, E.I., Becker, L.R., and Haynes, K. (1978). Sensory control of mating in the blue gourami, *Trichogaster trichopterus* (Pisces: Anabantidae). *Behav. Biol.,* 22 : 2-105.

Resink, J.W. (1988). Steroid glucuronides as pheromones in the reproduction of the African catfish, *Clarias gariepinus*. *Ph. D. Thesis*. University of Utrecht, Utrecht, The Netherlands.

Saglio, P. (1982). Use of intraspecific biological extracts to trap eels (*Anguilla anguilla* L.) in the field. Demonstration of the pheromonal attractivity of the skin mucus. *Acta Oecol. Applic.*, 3 : 223-231.

Schoonen, WG.E.J. (1987). Analysis of steroid synthesis in the reproductive organs of the African catfish, *Clarias gariepinus* : a basis for the study of sex pheromones. *Ph. D. Thesis.* University of Utrecht, Utrecht, TheNetherlands.

Selset, R. (1980). Chemical methods for fractionation of odorants produced by char smolts and tentative suggestions for pheromone origins. *Acta Physiol. Scand.*, 108 : 97-103.

Sherwood, N.M., A.L. Kyle, H. Kreiberg, C.M. Warby, T.H. Magnus, J. Carolsfield and Price, W.S. (1991). Partial characterization of a spawning pheromone in the herring, *Clupea herengus pallasi. Can. J. Zool.*, 69 : 91-97.

Smith, R.J.F. (1977). Chemical communication as adaptation. In : *Chemical Signals in Vertebrates* (D. Muller-Schwarze and M. M. Mozell, eds.). pp. 303-320. Plenum Press, New York.

Smith, R.J.F. (1982a). The adaptive significance of alarm substance-fright reaction system in fish. In : *Chemoreception in Fishes* (Hara, T.J., ed.). pp.327-347. Elsevier/North-Holland, Amsterdam.

Smith, R.J.F. (1982b). Reaction of *Percina nigrofasciata, Ammocrypta beani* and *Etheostoma swaini* (Percidae, Pisces) to conspecific and intergeneric skin extracts. *Can. J. Zool.*, 60 : 1067-1072.

Smith, R.J.F. (1992). Alarm signals of fishes. *Rev. Fish Biol. Fish.*, 2 : 33-63.

Solomon, D.J. (1973). Evidence for pheromone influenced homing by migrating Atlantic salmon (*Salmo salar* L.). *Nature*, 244 : 231-232.

Solomon, D.J. (1977). A review of chemical communication in freshwater fish. *J. Fish Biol.*, 11 : 363-376.

Sorensen, P. W. (1984). Juvenile eels rely on odour cues for migration. *Maritimes*, 28 : 8-10.

Sorensen, P.W. and Stacey, N.E. (1999). Evolution and specialization of fish hormonal pheromones. In : *Advances in Chemical Signals in Vertebrates* (Johnston, R. E., Muller-Schwarze, D. and Sorensen, P.W., eds.). pp. 15-47. Plenum Press, New York.

Sorensen, P.W. and Stacey, N.E. (2004). Brief review of fish pheromones and discussion of their possible uses in control on non-indigenous teleost fishes. *New Zealand J. Mar. Freshw. Res.*, 38 : 399-417.

Sorensen, P.W., Christensen, T.A. and Stacey, N.E. (1998). Discrimination of pheromonal cues in fish : emerging parallel with insect. *Curr. Opinion Neurobiol.*, 8 : 458-467.

Stabell, O.B. (1987). Intraspecific pheromone discimination and substrate marking by Atlantic salmon parr. *J. Chem. Ecol.*, 13 : 1625-1644.

Stabell, O.B. (1992). Olfactory control of homing behaviour in salmonids. In : *Fish Chemoreception* (T. J. Hara, ed.). Chapman & Hall, London. pp. 249-270.

Stabell, O.B. and Lwin, M.S. (1997). Predator-induced phenotypic changes in crucial carp caused by chemical signals from conspecifics. *Environ. Biol. Fish.*, 49 : 145-149.

Stacey, N.E. (2003). Hormones pheromones and reproductive behaviour. *Fish Physiol. Biochem.*, 28 : 229-235.

Tavolga, W.N. (1956). Visual, chemical and sound stimuli as cues in sex descriminatory behavour of the gobid fish, *Bothygobio soporator. Zoologica*, 41 : 49-64.

Teeter, J.H. (1980). Pheromonal communication in sea lamprey, *Petromyzon marinus. Can. J. Fish. Aquat. Sci.*, 37 : 2123-2132.

Timms, A. M. and Kleerekoper, H. (1972).The locomotor responses of male *Ictalurus punctatus*, the channel catfish. *Trans. Am. Fish. Soc.*, 101 : 302-310.

Tood, J.H. (1971). The chemical languages of fishes. *Scient. Am.*, 224 (5) : 98-108.

Van den Hurk, R. and Resink, J.W. (1992). Male reproductive system as sex pheromone producer in teleost fish. *J.Exp. Zool.*, 261 : 204-213.

Van Weerd, J.H. (1990). Pheromones and ovarian growth in the African catfish, *Clarias gariepinus. Ph.D. Thesis.* University of Wageningen, Wegeningen, The Netherlands.

Wagner, C.M., Jones, M.L., Twohey, M.B. and Sorensen, P.W. (2006). A field test verifies that pheromones can be useful for sea lamprey (*Petromyzon marinus*) control in the Great Lake. *Can. J. Fish. Aquat. Sci.*, 63 : 475-479.

Wisenden, B.D., Chivers, D.P., Brown, G.E. and Smith, R.J.F. (1995). The role of experience in risk assessment : avoidance of areas chemically labelled with fathead minnow alarm pheromone by conspecifics and heterospecifics. *Ecoscience*, 2 : 116-122.

Young, M.K., Micek, B.K. and Rathbun, M. (2003). Probable pheromone attraction on sexually mature brook trout to mature male conspecifics. *North Am. J. Fish. Mange.*, 23 : 276-282.

Zhang, C. (1997). Bile acids as potential pheromones in lake char, *Salvelinus namaycush* : an electrophysiological, biochemical and behavioural study. *Ph. D. Thesis*. University of Manitoba, Winnipeg, Canada.

Zhao, W., Hong, W., Zhang, Q., Jiang, X. and Wu, D. (2003). Induction of maturation and spawning by sex pheromones in female skipper (*Boleophthalamus pectinirostris*). *Mar. Sci., Bull.*, 5 : 87-95.

# 11

# Drug Metabolism and Detoxification

*— S. Arun*

## History of Drug Metabolism

Drug metabolism is simply defined as a course of action, which the body metabolizes and converts the drug/xenobiotics into its active chemical substances. The major metabolic site of drug metabolism is the liver, the organ that plays a major role in digestion, metabolism and removal of above said active chemical substances (metabolites of drugs) from the body. The major pathways of drug metabolism are oxidation, reduction, hydrolysis, conjugation with glycine, sulfuric acid, and glucuronic acid, acetylation, mercapturic acid synthesis. The family of liver isoenzymes such as cytochrome P-450 is playing a vital role in drug metabolism. These enzymes have a catabolic action on substances, breaking them down into reactive metabolites which are known as hazardous to human health. In the beginning of the 19th century, reactions of drug/xenobiotic metabolism were itemized at a slow deliberated pace. It took a long time (another 50 years) to bring together a firm investigation on this field by the publications of R.T.Williams (See BOX-1). Williams proposed that xenobiotic compounds could be biotransformed in two phases known as Phase I (functionalization) and Phase II (conjugation) detoxification. Functionalization involves utilization of oxygen to form a reactive site, and conjugation helps addition of a water-soluble group into the reactive site. From the early 1950s through the 1960s, investigation of the authentic enzymes involved in metabolism as well as their intracellular locations came under study. In the 1980s, the x-ray structure of key enzymes was revealed and molecular biology began to explore the expression and synthesis of the individual proteins. The role of genetic polymorphisms in pharmacotherapy

**BOX-1**

**R. Tecwyn Williams**

Research findings of R.T.Williams deals with the elucidating of the ring structure of glucuronic acid which involved isolation of bornyl glucuronide from the urine of dogs fed borneol. He crystallized this conjugate and used it as a source material for his publication on the pyranoid structure of glucuronic acid in the magazine "Nature" in 1931. This work stimulated his interest in the metabolism of foreign compounds and led to a series of papers on the fate of phenols, terpenes and sulphonamides. Subsequently, he planned to write a book on the detoxification of foreign compounds. Due to war condition, he did not make it possible. Later, in 1959, he published a book entitled "Detoxication Mechanisms". It provides a systematic approach based on organic chemistry classification. Williams also extended his concepts of the principal biochemical reactions whereby drugs and other foreign compounds are metabolized in the body. He projected that foreign compounds were metabolized in two distinct phases: one including oxidations, reductions, and hydrolyses and the other comprised of conjugation reactions.

became apparent. As the 20th century began, the basics of transformation-oxidation, reduction, hydrolysis and synthesis were understood. Even the consistency of applications was unclear, but at present, the mechanism of drug metabolism is quite clear.

## Phase I Metabolism

### *Cytochrome P450*

The Phase I metabolism, composed mainly of the cytochrome P450 supergene family of enzymes, is generally the first enzymatic defense against foreign compounds such as environmental toxicants and drugs. Most of the drugs are metabolized through Phase I metabolism. About 60 years ago, Garfinkel and Klingenberg revealed Cytochrome P450 as a pigment having a novel carbon monoxide binding spectrum in liver microsomes of rat and pig. Then in 1978, Sato and Omura edited an excellent volume that summarizes the history of P450 (See Box-2). Cytochromes P450 comprise a large superfamily of heam- containing enzymes. These enzymes are of particular importance when studying drug biotransformation and drug metabolism. Cytochrome P450 has existed more than 3.5 billion years. The pivotal role of P450s is metabolism and detoxification of endogenous and exogenous compounds such as drugs. The cytochrome P-450 (CYP 450) serves as both the oxygen and substrate-binding locus for the monooxygenase reaction and undergoes cyclic oxidation / reduction of the heme iron during catalysis. It consists of single polypeptide chain with iron-protoporphyrin IX loosely bound by hydrophobic forces, electrostatic and covalent bonds. The cytochrome P450 monooxygenase function in association with several microsomal proteins, located in the endoplasmic reticulum together with the P450 isoenzymes that transfer electrons to reduce cytochrome P450 during its catalytic cycle. These components include NADPH-cytochrome c reductase, cytochrome b5, and NADH-cytochrome c reductase. Electrons from NADPH or NADH are transferred through these compounds and cytochrome P450 (CYP 450) insert one atom of oxygen into the substrate (drugs and xenobiotic

**BOX-2**

**T. Omura**

**R.Sato**

In connection with, early studies on liver pigments by Garfinkel and Klingenberg (who observed in liver microsomes an unusual carbon monoxide binding pigment with an absorbance maximum at 450 nm), in the year of 1978 this pigment was finally described as a cytochrome by Omura and Sato. Through the use of detergent solubilization of microsomes and interaction with isocyanide ligands, they demonstrated that the ensuing "P-420" was a cytochrome with typical a, b, and soret absorption bands compound) and reduces the second oxygen atom to form water. There have been two major views on the function of CYP 450 enzymes i.e., it is involved in the metabolism of endogenous compounds (steroid hormones, vitamins and bile acids) and exogenous compounds such as drugs (Jakoby, 1980). The mechanism of xenobiotic metabolism by heme protein (CYP 450) was clearly explicated by Hollebone (1986) and Hollebone *et. al.* (1995) (Fig. 1). In this model, the lipid bound protein is an oligomeric globin, capable of both rotation and translation within the lipid. As some sites of the globin protein would face the cytosol, oxygen molecules can be captured from the aqueous medium of the cell. By applying antibonding electron density to this trapped oxygen at the heme site, the protein is capable of heterolysis such as an oxygen molecule to a hydroxide ion, which is released back to the cytosol, and a positive oxygen atomic ion, which remains bound to the active site by back bonding to the iron (II) state of the heme. This activated site is protected by a tyrosine cover. After rotation of the globin and opening of the lipophilic tyrosine, this hydrophobic, activated site can encounter and adsorb hydrophobic xenobiotic compounds or drugs which dissolved in the lipid bilayer. Then, the oxygen positive ion oxidizes the drugs. After oxidation, the hydrophilic hydroxylated drug metabolite could force rotation of the protein and permit release of the drug metabolites into lumen for conjugation (phase II metabolism) and detoxification.

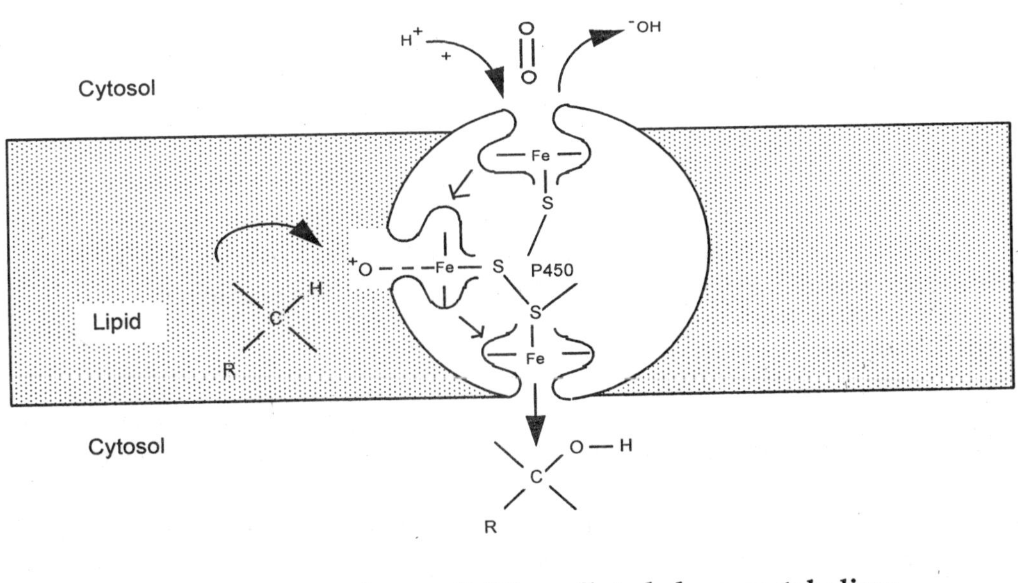


**Fig. 1 : Cytochrome P450 mediated drug metabolism (based on Hollebone *et al.* 1995).**

There are over 30 human P-450 isoenzyme systems that have been identified so far. The major ones responsible for drug metabolism are the CYP1A2, CYP2A6, CYP2B6, CYP2C8, CYP2C9, CYP2C18, CYP2C19, CYP2D6, CYP2E1, CYP3A4 and the CYP3A5-7. Among these, CYP3A-family contributes to approximately 50% of the total cytochrome P450 activity of the adult human liver. It metabolizes about 60% of all usually prescribed drugs. However, nearly 30% of all drugs are polymorphically metabolized in the human liver particularly by cytochromes P450 (CYP) 2C9, 2C19, 2D6. Fig. 2 shows few example of Phase I reaction.

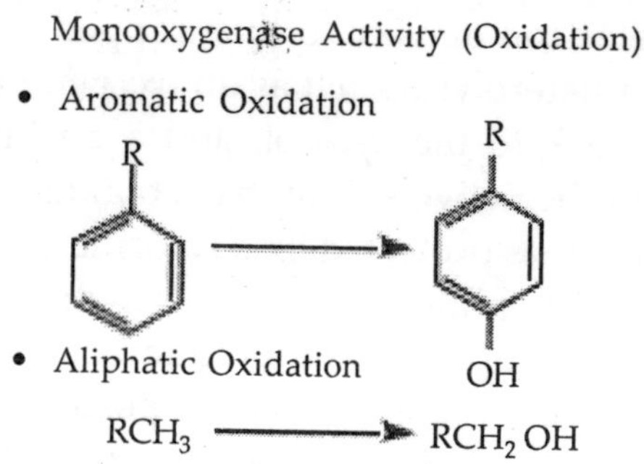


**Fig. 2. Phase I reaction**

## In Vitro Studies

For the past two decades, *in vitro* methods have been developed most extensively with the cytochrome P450 enzymes. In this period there has also been a shift of emphasis away from *in vivo* experiments with animal models to *in* vitro studies with human enzymes. However, it cannot be ruled out that human *in vivo* situation is the most appropriate. All *in vitro* systems are imperfect as models-but some are useful. On the other hand, good *in vitro* studies can be very useful in making decisions regarding further *in vivo* studies.

### *Cellular Systems*

In *in vitro* studies, human liver slices and hepatocytes has some advantage which has not only the full complement of the P450s but also other microsomal and cytosolic enzymes that contribute to metabolism. In addition, this system helps to detect or scrutinize the induction of enzymes and hepatocellular toxicity. Apart from this, there are some drawbacks in this system. Cells and liver slices cannot be easily frozen and thawed for assay and cellular systems are relatively expensive.

### *Microsomal Preparation:*

Microsomes can be prepared from frozen liver tissue and a number of different experiments can be done for drug metabolism. Outline of biotransformation of a particular P450 substrate such as drugs and or xenobiotics can be easily assessed and compared with induction of CYP profiles from experimental animals. Several techniques can be used to identify which P450s make major contributions in drug metabolism (formation of a particular product of a new drug). For example, (i) a set of microsomal samples obtained from different individuals is compared with regard to rates of reactions known to be catalyzed by individual P450s (ie., immuno detection of CYP isoenzymes by western blot and ELISA method). (ii) Chemical inhibitors have reasonably good diagnostic value for individual P450s. (iii) Several antibodies (monoclonal and polyclonal CYP antibodies) known to bind with specific P450s can be added to the microsomes and their effects on a particular reaction can be determined. In addition, now-a day, many pharmaceutical companies are developing their own systems to find out metabolism of their own product. These above said systems are having advantage that they have the normal reliability found in *in vitro* method and have accessory proteins such as NADPH-P450 reductase and cytochrome *b*5, which helps in drug metabolism. They can also use the identified metabolites to study the impact of drug metabolism and their toxicity. Thus, a number of *in vitro* approaches have been developed for the study of P450s and other enzymes in human drug metabolism. All the above said *in vitro* methods can be assisting to investigate and scrutinize the patterns of drug metabolism, i.e., which P450s are capable of catalyzing a reaction, whether metabolism generates toxicity and chemical characterization of products/drugs.

### Ah Receptor

A major step in understanding an induction mechanism of cytochrome P450 and drug metabolism occurred with the identification of aryl hydrocarbon receptor (Ah Receptor). The Ah receptor is a ubiquitous cytosolic protein that responds to planar aromatic ligands such as aromatic hydrocarbons. The AhR functions as a ligand activated transcription factor that controls the expression of several genes which includes *CYP1A*. Cytosolic AhR exists in an unliganded form and contains two molecules of 90 kDa heat shock protein (hsp90) and other proteins. Binding of ligand such as drug to the AhR leads to dissociation of hsp90 and successively bind (dimerisation) with another bHLH (basic helix -loop -helix) partner known as aryl hydrocarbon nuclear translocator (ARNT). This heterodimeric Ah receptor/ARNT complex binds to xenobiotic responsive elements (XRE) or dioxin responsive enhancer (DRE) located upstream of the CYP1A gene and promotes their transcription. These enhanced CYP enzymes further is involved in the metabolism of drug (Fig. 3).

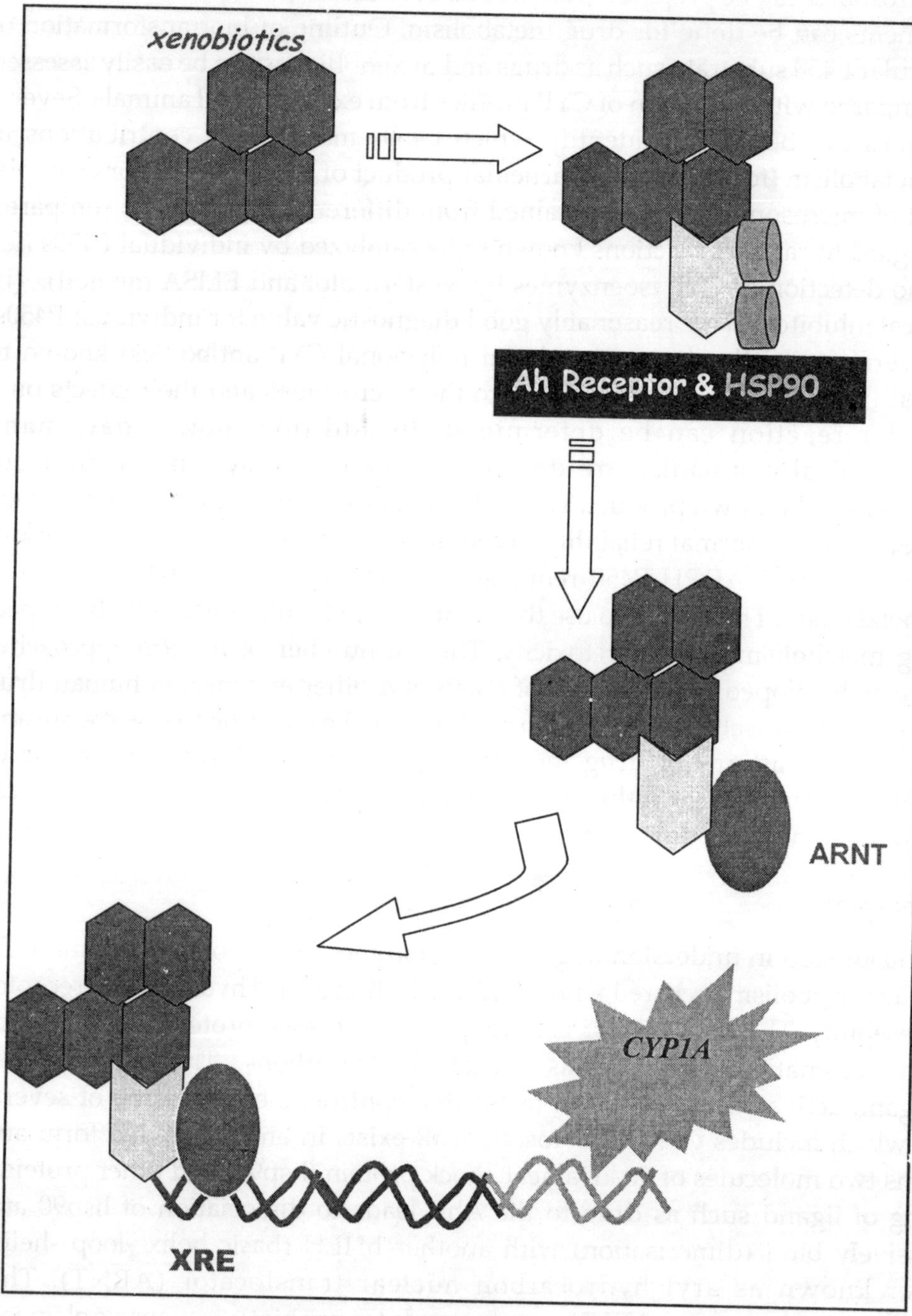


**Fig. 3. AhR mediated *CYP1A* induction.**

## *Impact of Drug Metabolites on Ah Receptor*

A benzimidazole derived antiulcer drug, omeprazole (OM), is known to be aryl hydrocarbon like inducer of CYP1A and also act as a potent activator of AhR in human hepatocytes (Diaz et al 1990) and *in vivo* in man (Rost 1992). Recently, Dr. Maurel and co-workers concentrated on structure-activity relationship study on the effect of metabolites of OM on AhR activation. It has been suspected that, phase I mediated OM metabolism may generate metabolites (i.e. OMS) that can act as an AHR ligand. Unlike OM, the metabolites of OM (OMS) reduce the activation of AhR (antagonist activity) in mouse and human AhR in Hepa-1 and HepG2 hepatoma cell lines. In order to know whether OMS affects nuclear translocation and Ah receptor, Hepa-1c1c7 cells were cultured for 2 h in the presence of TCDD (TCDD is a well known AhR activator), and in the absence or presence of OMS. Cells were fixed and analysed by immunofluorescence microscopy using an anti-AhR antibody A-1. In untreated cells, AhR was evenly distributed in the cytosol, while in TCDD (xenobiotics) treated cells it concentrated in the nucleus, (because AhR and TCDD complex dimerises with ARNT in nucleus). In contrast, when cells were treated with OMS (xenobiotics) alone or co-treated with both TCDD and OMS, AhR remained in the cytosol (i.e. OMS involves in the degradation of AhR protein or prevents dimerisation of AhR-xenobiotic complex with ARNT). Inhibition or degradation of AhR protein by OMS might affect the induction of *CYP1A* gene and its further metabolism (See Fig. 4, from Gerbal-Chaloin *et al.*, 2006). Thus, in some cases drugs become active and alter the biotransformation pathway after it has been metabolized (See Table 1, for some other drugs that produce active metabolites).

**Table.1**

**Drugs and metabolites**

| No. | Drug | Metabolites |
|---|---|---|
| 1 | Cortisone | Hydrocortisone |
| 2 | Prednisone | prednisolone |
| 3 | Enalapril | Enalaprilat |
| 4 | Diazepam | Nordiazepam |
| 5 | Morphine | Morphine 6-glucuronide |
| 6 | Paracetamol | N-Acetyl-p-benzoquinone imine |

## CYP 450 Inhibitors

The knowledge of mechanism of cytochrome P450 mediated drug metabolism

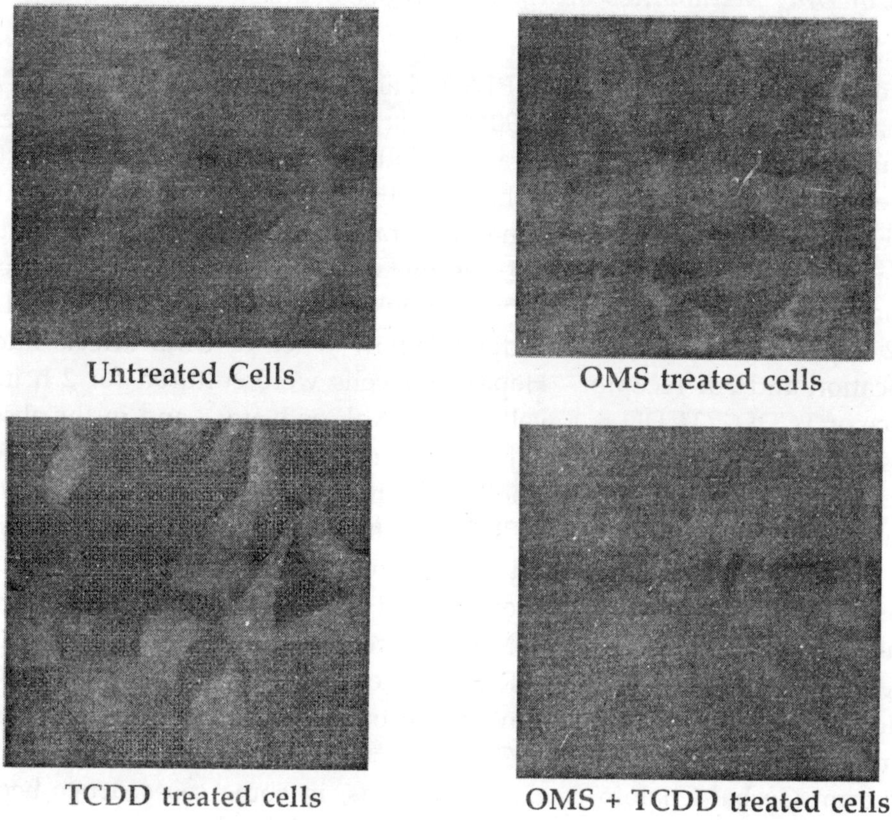


**Fig. 4. OMS blocks TCDD-mediated AhR nuclear translocation. Results are used with permission of the corresponding author Dr. Patrick Maurel, France (Gerbal-Chaloin *et al.*, 2006).**

provides an idea for predicting CYP 450 isoforms inhibition and the role of inhibitors. In brief, CYP450 inhibitors differ in their selectivity toward different isoforms of the enzyme and are classified by their mechanism of action. Some drugs can act as an active site for specific isoenzymes but are not themselves as substrates (e.g., quinidine is a potent competitive inhibitor of CYP2D6 but is not a substrate for it). No competitive inhibitors include drugs such as ketoconazole, which forms a tight complex with the Fe3+ form of the haem iron of CYP3A4 causing reversible non-competitive inhibition. So-called mechanism based inhibitors require oxidation by a P450 enzyme. Examples include gestodene (CYP3A4), diethylcarbamate (CYP2E1) and furafylline (CYP1A2). An oxidation product (e.g., a postulated epoxide intermediate of gestodene) binds covalently to the enzyme, the result of which then destroys itself (sucide inhibition). Many clinically important interactions between drugs are the result of inhibition of P450 enzymes. There are various methods available for assessing CYP inhibition. Microtiter plate-based fluorometric assays have been successfully developed for the inhibition of several principal CYP enzymes such as CYP1A2, CYP2C9, CYP2C19, CYP2D6 and CYP3A4 (Kenworthy et al 1999 and Stresser et al (2000). The assay is based on the fact that CYP enzymes convert nonfluorescent substrates into fluorescent metabolites. Enzymatic activity is determined by measuring fluorescence of enzymatic products. For a complete assessment of a drug candidate, reversible or irreversible inhibition should be distinguished. Generally, this is accomplished by determining the apparent IC50 of the drug candidate toward a standard substrate with and without a preincubation period of the enzyme with the drug candidate.

The IC50 can be calculated using the following equation:

$$\% \text{ inhibition} = \frac{\text{metabolite generated with drug}}{\text{metabolite generated without drug}} \times 100$$

Few example of common drugs act as a substrate for CYP450 isoenzymes

| CYP450 isoenzymes | Drug |
|---|---|
| CYP1A1 | Theophylline |
| CYP1A2 | Paracetamol, Tacrine and Caffeine |
| CYP2A6 | Methoxyflurane |
| CYP2C8 | Taxol |
| CYP2D6 | Clozapine, Codeine, Debrisoquine and Metaoprolol |
| CYP3A4/5 | Ciclosporin, Losartan, Nifedipine and Terfenadine |

### Phase II metabolism

As mentioned in previous section, drug metabolism is sorted out into phase I and phase II reactions. Even though, CYP mediated phase I metabolism could not make a remarkable changes in the molecular weight and hydrophilic nature of drugs, it plays a main role in oxidative reaction of drugs and generates metabolites which were subsequently eliminated by phase II reaction. In contrast, phase II metabolism/conjugation typically results in an appreciable increase in molecular weight and water solubility of drug metabolites. Most of the compounds undergo phase I oxidation prior to phase II conjugation, but molecules with sites open to conjugation may undergo phase II reactions directly [e.g. acetaminophen]. In this review, I have explicated three important reactions (sulfation, glucuronidation, and glutathione conjugation) of phase II metabolism.

### *Sulfation*

Even, sulfate conjugation is very important in Phase II reaction (sulfation is a high affinity and low capacity phase II reaction), it has been reported that it works with glucuronidation when the substrates/drug overlapped. In brief, sulfation predominates at low substrate concentration and glucuronidation predominates when sulfation has been saturated at high substrate concentration. In the cytosol, inorganic sulfate is conjugated with adenosine monophosphate by ATP sulfurylase to form adenosine 5'-phosphosulfate (APS), which is subsequently phosphorylated by APS kinase to form 3'-phosphoadenosine-5'-phosphosulfate (PAPS- Fig. 5), the donor of sulfonate groups ($SO_3^-$) in all sulfate, or more correctly sulfonate, conjugation reactions. Transfer of sulphate group from PAPS (3-phosphoroadenosine-5 phosphosulphate) to a functional group such as alcohol takes place with the help of a number of sulphokinase enzymes. The reaction is restricted to phenols and amines. The end products are highly polar and water-soluble. Generally, Sulfation of drug is mediated by sulfotransferase enzyme. The cytosolic sulfotransferase (SULT) enzymes are subdivided into three families known as (1) SULT1 as phenol sulfotransferase (PST) family, (2) SULT2 as hydroxysteroid sulfotransferase (HST) family, and (3) brain-specific SULT4. The phenol sulfotransferase family (SULT1) is further subdivided into four subfamilies known as (1) SULT1A, (2) SULT1B, (3) SULT1C, and (4) SULT1E. The SULT2 family is also subdivided into two subfamilies as (1) SULT2A and (2) SULT2B. Sulfation of drug is primarily mediated by the SULT1A subfamily, which catalyzes sulfation of hydroxyl groups and monoamine groups on phenolic-type molecules.

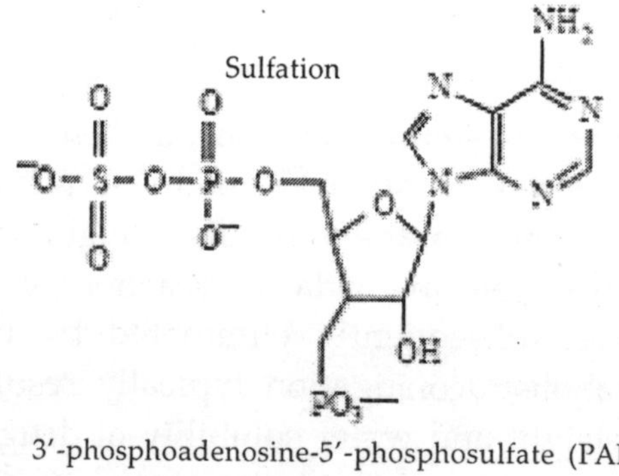


3'-phosphoadenosine-5'-phosphosulfate (PAPS)

**Fig. 5.**

## *Glucuronidation*

Formation of glucuronide is the most commonly encountered conjugation reaction in phase II metabolism. The glucuronidation reaction is multifaceted which proceed with use of the activated derivative of glucuronic acid, uridine diphosphate glucuronic acid and UDPGA. Synthesis of UDPGA occurs in two step process starting from glucose I phosphate which gets first coupled to uridine-triphosphate to give UDP-glucose. UDP glucose is further oxidized by UDPG dehydogenase enzyme which utilizes NAD as co-factor to give UDPGA (Fig-6). UDPGA helps for excretion by way of reacting with groups having active hydrogens such as alcohols, thiols, amines, acids etc., in the presence of glucuronyl transferase.

Glucuronidation

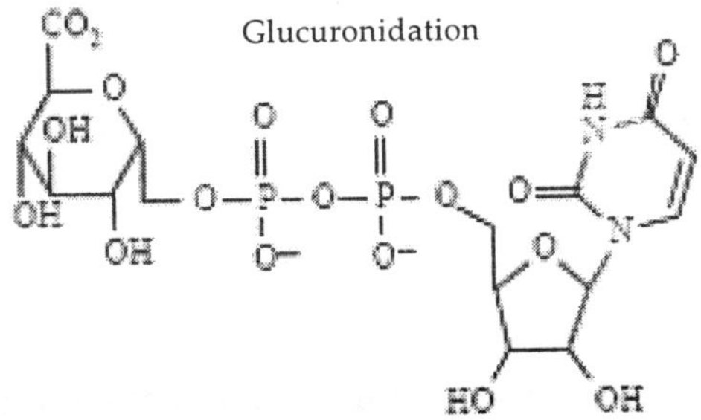

Urine 5′-diphospho-α-D-glucuronic acid (UDP-GA)

**Fig. 6.**

## *Glutathione Conjugation*

Glutathione *S*-transferases (GSTs), play a major role in phase II reaction, comprise a superfamily of ubiquitous, multifunctional enzymes, which play a key role in cellular detoxification by protecting macromolecules from the attack of reactive electrophiles. The GSTs catalyze the conjugation of the tripeptide glutathione (GSH- Fig. 7) to a wide variety of exogenous and endogenous chemicals with electrophilic functional groups (e.g. products of oxidative stress, environmental pollutants, and drug metabolites), thereby neutralizing their electrophilic sites, and rendering the products more water-soluble for excretion.

Glutathine Conjugation

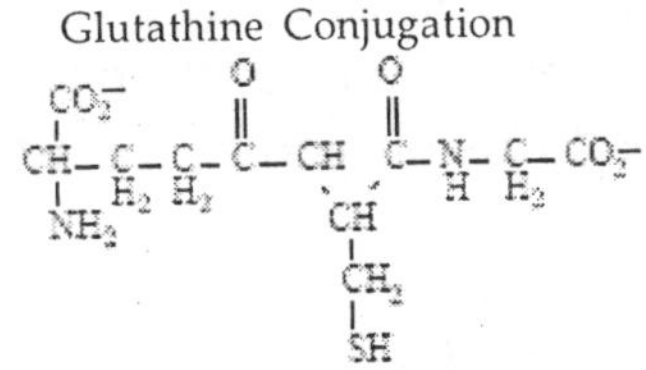

Glutathine

**Fig. 7.**

Based on sequence homology and immunological crossreactivity, human cytosolic GSTs have been grouped into seven families, designated GST Alpha, Mu, Pi, Sigma, Omega, Theta, and Zeta.

## Phase III Reaction

### *Phase III Reaction or Multi Drug Resistance*

It has been reported that antiporter activity (p-glycoprotein or multidrug resistance) has been defined as the Phase III detoxification system. The antiporter activity is nothing but an energy-dependent efflux pump, which pumps drugs out of a cell, thereby decreasing the intracellular concentration of drugs (Chin et al 1993). Antiporter activity in the intestine appears to be co-regulated with intestinal Phase I CYP3A4 enzyme. The non- metabolized drugs are pumped out from the cell and get back into the intestinal lumen which allowed more chances for Phase I reaction to metabolize the drugs before it is taken into circulation. The multi-drug resistance gene 1 (MDR1) and multi-drug resistance gene 2 (MDR2) are known as major genes that encoding antiporter activity. The MDR1 gene product is responsible for drug resistance of many cancer cells, and is normally found in epithelial cells in the liver, kidney, pancreas, small and large intestine, brain, and testes. MDR2 activity is expressed primarily in the liver, and may play a role similar to that of intestinal MDR1 for liver detoxification enzymes.

## Excretion of Drugs

Drugs are eliminated from the body by number of metabolic processes such as renal excretion, biliary excretion and fecal excretion. The kidney is the major excretory organ of drug and their metabolites.

### *Renal Excretion*

Excretion of drug metabolites in the urine takes place by three routes known as (1) glomerular filtration (most drugs, unless highly bound to plasma protein, cross the glomerular filter freely), (2) active tubular secretion (drugs which are acids and weak bases are actively secreted into the renal tubule and more quickly eliminated), and (3) passive tubular reobsorption ( drugs which are lipid soluble are passively reabsorbed by diffusion across the tubule and are not efficiently eliminated in the urine).

### *Biliary Excretion and Fecal Excretion*

In biliary excretion, liver cells involves transfer of various substances including drugs from plasma to bile by means of transport systems similar to those of renal tuble and which involve P-glycoprotein. P-glycoprotein transports lipid soluble drugs and MRP2 (multi drug resistance –associated protein –type 2) is mainly involves in the secretion of conjugated metabolites (such as glucuronides, sulfates and glutathione adducts) of drugs. In the fecal excretion, excreted substances are mainly unabsorbed, orally ingested drugs or metabolites excreted either in the bile or secreted directly into the intestinal tract and not reabsorbed.

## Conclusions

Drug metabolism describes the state of affairs where an organism, possesses a kind of characters, which allows metabolizing a particular or complex of drug. As we know, animals have evolved complex systems that eliminate foreign chemicals, including drugs. Drug elimination is the irreversible loss of drug from the body. It occurs by two processes known as metabolism and excretion. Drug metabolism involves two kinds of biochemical reactions known as Phase I and Phase II reactions. In higher organisms, induced expression of numerous proteins is involved in the protection of toxic consequences of drug metabolites. In this review, I have described the mechanism of cytochrome P450 and Ah receptor and their role in drug metabolism.

## Acknowledgement

Dr.S.Arun wishes to thank DST-India for financial support. Thanks are also due to Dr. Patrick Maurel and co-workers, Univ. Montpellier, Montpellier, France for kindly allowing me to utilize their results for this review.

## References

Chin, K.V., Pastan I, Gottesman MM. (1993). *Adv. Cancer Res.* 60:157-180.

Diaz, D. Fabre, I. Daujat, M. Saint Aubert, B. Bories, P. Michel, H. Maurel, P. Gastroenterology 99 (1990) 737.

Gerbal-Chaloin.S., Pichard-Garcia. L., Fabre J.M., Sa-Cunha. A., Poellinger.L, Maurel. P., Daujat-Chavanieu, M. (2006). *Cellular Signalling*. 18 :740 – 750

Hollebone (1986). *Drug Metab. Rev.* 17 : 93-143.

Hollebone. B.R., Brown Lee. L.J., Davis. D., Michelin. N., Purdy. D. (1995). *Envt. Toxicol. Chem.* 14: 29-41.

Jakoby, W.B. (1980). In Enzymatic Basis of Detoxification (Jakoby, W.B. Eds) Vol. 1 : pp 1-6, Academic Press, New York.

Kenworthy, K. E, Bloomer, J. C. Clarke, S. E. Houston, J. B. (1999) *Br. J. Clin. Pharmacol., 48*(5), 716- 727.

Rost, K.L. Brosicke, H. Brockmoller, J. Scheffler, M. Helge, H. Roots, I. (1992). *Clin. Pharmacol. Ther.* 52: 170.

Stresser, D.M., Blanchard, A.P., Turner, S. D., Erve, J.C.L., Dandeneau, A A.; Miller, V.P., Crespi, C.L. (2000). *Drug Metab. Dispos.*, 28(12), 1440-1448.

**Suggested Readings**

Birkett, D.J. (1998). Pharmacokinetics Made Easy. McGraw-Hill. Sydney. Australia.

Parkinson A. (1996). Biotransformation of Xenobiotics. In Casarett & Doull's Toxicology: The basic science of poisons. 5th ed. McGraw-Hill. New York.

Pratts, W.B. and Taylor P. (eds.) (1990). Principals of Drug Action: The Basis of Pharmacology, 3rd ed. Churchill Livingstone, New York.

Rowland, M. and Tozer, T.N. (eds.) (1995). Clinical Pharmacokinetics: Concepts and application. 3rd ed. Williams&Wilkins Philadelphia.

# 12

# Oxidative Stresss

*P. Krishnamoorthy and K. Muthu*

Chemical compound and reactions capable of generation of potential toxic oxygen species can be referred to as prooxidants. In a normal living cell, there is an appropriate prooxidant: antioxidant balance. However, this balance can be shifted towards the prooxidants, when production of oxygen species is increased greatly, e.g., following ingestion of certain chemical or drugs or when level of antioxidants diminished (by inactivation of enzymes involved in disposal of oxygen species and by conditions that cause low level of the antioxidant). This state is called oxidative stress and can result in serious cell damage if the stress is massive prolonged. Reactive oxygen species (ROS) are generated in oxidative metabolism and used to kill and dismantle the cells. They inflict damage on all classes of macromolecules and can ultimately lead to cell death. Indeed, ROS are implicated in a number of human diseases such as cancer, myocardial infarction, inflammation, acidosis, hemolytic anemia, atherosclerosis, hyper receptivity, Wilson's disease, Parkinson's disease and damaged tissue *in vitro*. However, a number of enzymes are involved in defense against oxidative stress. They are superoxide dismutase (SOD), glutathione peroxidase (GPx), catalase (CAT), and cyclooxygenase. The wide distribution of the enzymatic activities underscores the ever-present problem in the oxidative stress.

Several powerful oxidants are produced during the course of metabolism in the blood cells and most other cells of the body. These include lipid peroxidation (LPO), hydrogen, peroxide ($H_2O_2$), peroxy radical (ROO) and hydroxyl radical (OH). They reactive molecule and can react with protein, nucleic acids, lipids and other molecules and produce tissue damage. On the other hand, compounds and reactions disposing of these reactive species, scavenging them, suppressing their formation, or opposing their action are anti-oxidants and include components such as NDPH, GSH, vitamin-E, vitamin-A and ascorbic acid.

## Lipid Peroxidation, Fatty Acid Oxidation and Ethanol Metabolism

**Lipid peroxidation**: This is a way of generation of free radicals. Peroxidation of lipids exposed to oxygen is responsible not only for the deterioration of food, but also damage to tissues *in vivo*. The deleterious effects are initiated by free radicals (ROO, RO, OH) and produced during peroxide formation from fatty acid containing methylene-interrupted double bond (i.e., those found in naturally occurring polysaccharide and fatty acids). Lipid peroxidation is a chain of reaction providing continuous supply of free radicals that initiate further peroxidation. To control and reduce lipid peroxidation, biological system is involved in the use of antioxidants produced by its own. The propyl gallate, butylated hydroxyl anisole (BHA), butylated hydroxyl toluene (BHT) are anti oxidants used in the preparation of food.

**Fatty acid oxidation**: A modified form of β-oxidation is found in peroxisomes and lead to the formations of acetyl CoA and H2O2, which is broken down by catalase. This first dehydrogeneration is not linked directly to the phosphorylation and the generation of ATP. Acyl CoA synthase facilitates the oxidation of very long chain fatty acids. The enzyme in peroxisome does not attract short chain fatty acid. The β-oxidation sequences end at octanoyl CoA. Further, the role of β-oxidation is to shorten the side chain of cholesterol in the bile acid formation. Peroxisome also takes part in the synthesis of ether glycolipids, cholesterol, and dolichol.

**Ethanol metabolism**: Ethanol can't be excreted and must be metabolized, primarly by the liver. This metabolism occurs by two pathways. The first pathway comprises two steps. The first step is catalysed by the enzyme alcohol dehydrogenase and occurs in the cytoplasm. The second step is catalysed by aldehyde dehydrogenase takes place in the mitochondria. The second pathway for the ethanol metabolism is called the ethanol inducible microsomal ethanol oxidizing system (MEOS). This pathway generates free radical and that causes tissues damage. Moreover, because the system consumes NADPH and glutathione can't be regenerated and thus excerbating oxidative stress. The enzyme thiokinase catalyzes acetate into acetyl CoA in liver mitochondria and normally activates the short chain fatty acids. However, the further processing of the acetyl CoA by the citric acid cycle is blocked. Because NADPH inhibits two important regulatory enzymes. Isocitrate dehydrogenase and α-ketoglutarate dehydrogenase are very reactive compound forms covalent bonds with many important functional groups in proteins, impairing protein function. If ethanol is consistently consumed at high levels, the acetaldehyde can significantly damage the liver, eventually leading to cell death.

## Role of Enzymes in Antioxidation

Superoxide is formed in the blood cells by the auto-oxidation of hemoglobin to methemoglobin (approximately 3% of hemoglobin in human red blood cell has been calculated to auto oxidized per day). In other tissues, it is formed by the action of enzyme such as cytochrome-P450-reductase and xanthine oxidase. When stimulated by contact with bacteria, neutrophils exhibit a respiratory burst and produce super oxide in a reaction catalysed by NADPH reducatse. Super oxide spontaneously dismutates to form $H_2O_2$ and $O_2$. However, the rate of this reaction is speeded up tremendously by the action of enzyme super oxide dismutase. Hydrogen peroxide is subjected to a number of fates. The enzymes catalase is present in many types of cells. When $H_2O_2$ is present in excess amount catalase converts it to $H_2O$ and $O_2$. Neutrophils possess a unique enzyme myloperoxidase that uses $H_2O_2$ and halides to produce hypohalous acid. The selenium containing enzyme glutathione peroxidase will also act on reduced glotathione and $H_2O_2$ to produce oxidize glutathione and $H_2O$.

**Super Oxide dismutase (SOD):** SOD is a class of enzyme that catalyzes is the formation of H2O2 and O2 from the superoxide radical. There are two major forms of superoxide dismutase. In himan Cu-Zn isozyme occurs in cytoplasam. A Mn-containing isozyme is found in the mitochondrial matrix.

$$O_2^{\bullet\bullet} + O_2^{\bullet\bullet} \xrightarrow[SOD]{2H} H_2O_2 + O_2$$

Both the Cu-Zn SOD and Mn SOD are found on chromosome 21q 22.1 and 6q21, respectively. Moreover, SOD is considered to be a stress protein, which is synthesized in response to oxidative stress and is thought to protect the cell from damage caused by $O_2^{\bullet}$.

**Glutathione:** Glutathione (GSH) is a tripeptide with a free sulfhydryl group. It is required to combact oxidative stress and maintain the normal the helf of state in the cell. Oxidized glutathione is reduced by NADPH generated with glucose-6-phosphate dehydrogenase. It is especially sensitive to oxidative stress. The stress is most acute in blood cells, brain, microorganism and others. This stress is most acute in red blood cells because, these is no alternative means of generating reducing power due to lack of mitochondria.

**Glutathione peroxidase (GPx):** GPx can be divided into two forms viz. selenium dependent and independent forms. The selenium dependent GPx is a tetramer of MW 84000 with a very high activity towards both $H_2O_2$ and organic hydroperoxides. It is found in both cytosol (70%) and mitochondria (30%) of various tissues. Selenium dependent GPx is most responsible for controlling cellular peroxide levels. Recall that the enzyme catalyses the reduction of a variety of substances by reducing several ancillary enzyme support glutathione peroxide enzyme. Different types of oxidative stresses and their effects have been shown is Fig. 1.

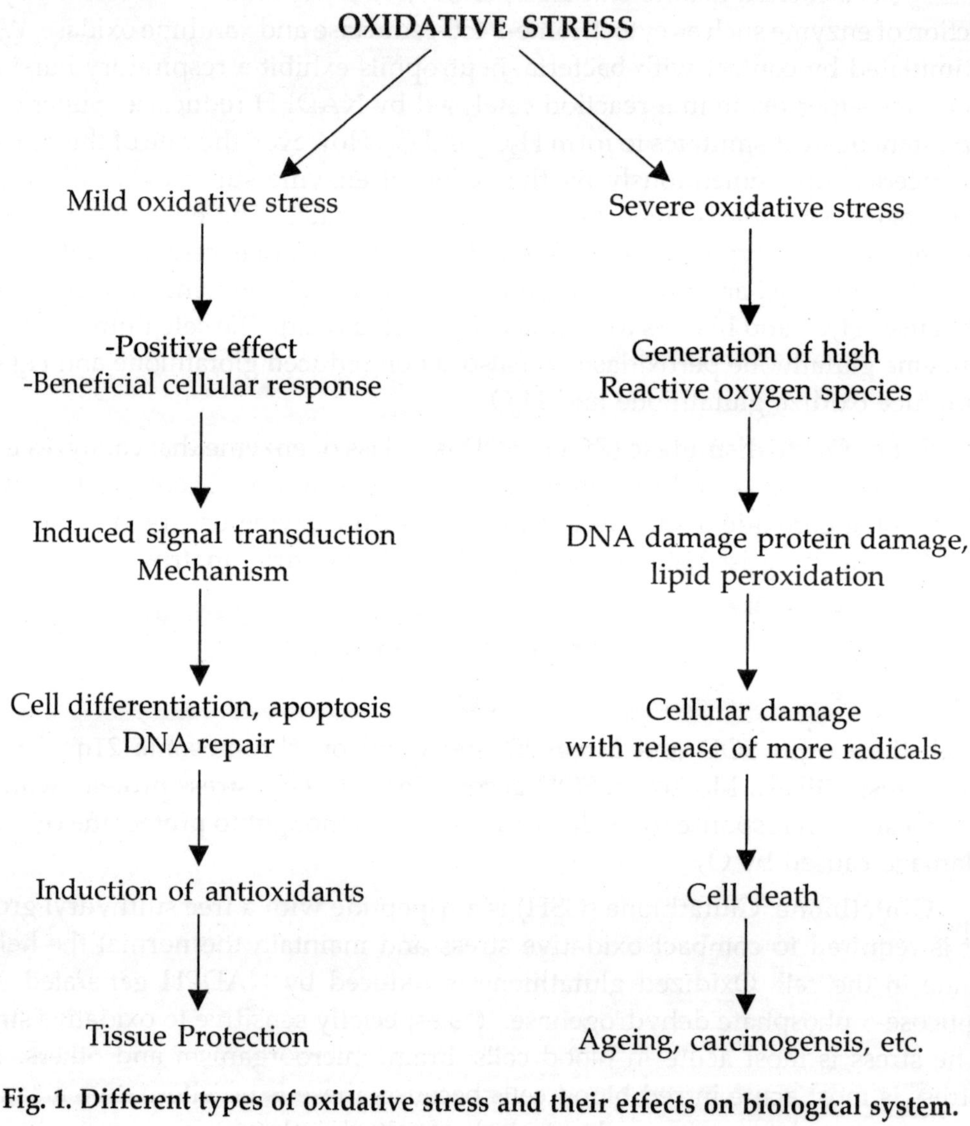


**Fig. 1. Different types of oxidative stress and their effects on biological system.**

## Oxidative Stress Initiated by Metals

**Aluminum (Al):** Aluminum may affect various oxidative processes or enzymatic properties, and even facilitates the oxidation of NADH by the superoxide anion. Aluminum has both oxidant and anti oxidant effects in mouse brain membranes and increases thiobarbituric acid-reactive substances (BARS). TBARS production and conjucated dynes formation depend upon addition of $Fe^{2+}$. The capacity of aluminum to stimulate $Fe^{2+}$ induced lipid peroxidation in liposomes was evaluated by measuring the formation of TBARS. To demonstrate aluminum – induced alterations in blood – brain barrier permeability, many experiments have been carried out so far by may researchers with various salts, using quinidine and aspirin. Similarly, aluminum chloride influences the brain kinetics of drugs. Moreover, the intraperiotoneal injections of $AlCl_3$ increase blood brain barrier permeability of iodiated N-Tyr delta sleep – inducing peptide (DSIP) and β-endophin by 60-70%. The intraperitoneal injections of aluminum chloride and lactate (but not hydroxide) can induce change in he permeability of the blood brain-barrier to ($^{14}$C) sucrose. The experiments with ($^{14}$C)-sucrose on rats intraperiotoneally injected with various aluminum salts such as lactate, acetyl lactonate, or maltolate, at pH 7.5, revealed that micromolar concentrations of aluminum increase BMEC (brain micro vessel monolayer endothelial cell cultures) permeability. Aluminum lactate in drinking water affect the barrier function of the endothelium of central cortex and blood microvessels. In aluminum-L-glutamate-treated rats, blood plasma increased and accumulation of Al was observed in the three brain regions, with higher level in the striatum. Further, a decrease in iron concentration was observed in blood plasma and in the straitum only. Involvement of aluminum in neuro degenerative processes has been clearly established. However, the ability of Al to induce neurofibrillary tangle formation, promote aggregation of β-amylopeptides or disturb neurotransmitter system are the strong arguments for the role of this metal in these diseases. Al causes alteration in blood brain barrier but the exact mechanism leading to this alteration is not known.

**Copper (Cu):** Wilson's disease is a disorder of copper accumulation which is inherited in an autosomal chromosome in recessive manner. Copper is an essential nutrient and the diet typically contains somewhat more than is required. The mechanism for getting rid of the excess copper involves hepatic excretion in bile and stool. This mechanism has been rendered ineffective in Wilson's disease due to genetic mutation. Tissue damage is related to elevated level of copper, and cessation of tissue damage is related to lowering copper level through therapeutic approaches. It has been shown that copper causes cell injury and death. However, it is also clear that certain ligands mitigate or eliminate the toxicity of high level of copper. Natural ligands include ceruloplasmin, metallothionein, and albumin. The potential toxic copper is freely dissociable from other ligands. It appears that copper toxicity is mediated through oxidant mechanisms. Copper produces $H_2O_2$, probably derived from super oxide. In red cells, level of glutathione (a major intra cellular protection against oxidation) is lowered. The critical damage may be in cell membranes or cytoskeletons.

**Magnesium (Mg):** Magnesium deficit in nervous system is the most commonly seen in clinical observation. First of all, it seems very important to discriminate between the two types of magnesium deficits viz., magnesium deficiency and magnesium depletion. Mg deficiency results in three basic effects: disturbances in cellular Ca distribution, decreased second messenger nucleotide ratio, and increased susceptibility to peroxidation. Mg deficiency in animals shows an increased susceptibility to *in vivo* oxidative stress and the tissues of these animals are more susceptible to *in vitro* peroxidation. Nervous hyper excitability (NHE) due to Mg deficit is also linked to modification in the turnover of various types of neurotransmitters such as monoamines, amino acids, nitric oxide (NO), neuropeptides, and cytokines. NO could also mediate an increase in cGMP (3′, 5′- cyclic guanosine monophosphate), whereas inflammatory neuro peptides might stimulate production of inflammatory cytokines and of free radicals.

**Managanese (Mn):** Role of endogenous antioxidant system in the striatum and striatal synaptosomes (as models or neuronal terminals) of the rat has been assessed during manganese-induced oxidative stress. Dopamine metabolism may be dissociated with an increased production of $H_2O_2$ and increased oxidative damage of the dopaminergic system. The uric acid might originate from xanthine oxidase acting on xanthine or hypoxathine, a pathway that also generates super oxide radical anion. Xanthine/hypoxanthine produce oxidative stress, suggesting a mitochondria/interference or increased energy demand during high manganese load.

Oxidative DNA damage cause mutations and cell death. L-dopa, dopamine, and 3-O-methyl-L-dopa cause extensive oxidative DNA-damage in the presence of $H_2O_2$ and traces of copper ions ($Cu^{2+}$). The question whether manganese as mangantic ions ($Mn^{3+}$) or in combination with other metal ions is able to cause DNA damage. Several mitochondrial enzymes use manganese as essential cofactors. It has been suggested that impaired oxidative energy metabolism by managanese cause additional cytotox city. Manganese neurotoxicity is possibly caused by affinity of the manganese ions to the areas with high levels of neuro melanin, e.g., the nigrotraital tract, as well as by the ability to assume multi oxidative states in areas which offer a suitable biochemical milieu for the transformation of $Mn^{2+}$ to

$Mn^{3+}$, thus favoring auto oxidation of dopamine to semiquinone and the production of toxic free radicals.

### Methyl Mercury and Iron Initiate the Oxidative Stress

**Methyl mercury (MeHg)**: It initiates membrane lipid per-oxidation and free radical-induced injury in biological tissues including brain. The brain is sensitive to oxidative injury. Vitamin E and selenium provide some proection against MeHg neurotoxicity *in vivo*. Lipid peroxidation induced by MeHg in cerebellar granule cell culture was coupled to GSH reduction and decreased viability in a close dependent manager. However, the mercurial-induced lipoperoxidation was not the direct cause of neuron cytotoxicity since melonaldehyde production could be blocked by a-tocopherol or (ethylene diamine tetra acetic acid) EDTA without significant protection. Number of pathways exist for the induction of oxidative injury by methyl mercury. Abnormalities in signal transduction and neurotransmitter function may lead to phospholipase A2 activation, increased antiodxidant production, and super oxide radical formation. Glutathione exerts major protective effects against MeHg cytotoxicity. GSH is low in neurons compared to astrocytes and glial processes, identifying a possible mechanism for neural sensitivity to the toxic compound to glia. The increase in GSH content in response to MeHg intoxication is due to transcriptional activation of the gene for a-glutamyl cysteine synthetase, a proximate enzyme in the pathway for GSH synthesis. Studies have also highlight the importance of GSH in providing cellular resistance to MeHg. Analysis of the resistant cells (PC12) revealed four fold higher levels of intracellular. GSH as compared to control cells ad increased rate of MeHg efflux. It is presumed that MeHg covalently links to GSH. In selectively transferred neuronal cell line with the anti-apoptotic gene (BCL-2) the neuronal expression of BCL-2 provided significant protection against MeHg – induced free radical production and cytotoxicity. Glutathione protects cells against MeHg – induced oxidative damage. *In vivo* cultures stress do not reveal a strict correlation between GSH content and cytoprotection.

**Iron (Fe)**: Iron attached to the transport protein transferring to the iron storage protein ferritin in though to be incapable of stimulating oxidative stress under normal physiological conditions. By contrast, non-protein – bound iron that forms the intracellular mobile iron pool is capable of stimulating oxidative stress. Hence an important function of intra cellular antioxidant defense enzymes is to scavenge $O_2$ – and $H_2O_2$ before they can come into contact with any available iron. Rapid free radical reactions may occur as a consequence of brain injury after inshemia or trauma. This problem is compounded by three additional factors: sensitivity of free radicals to membrane lipids rich in polyunsaturated fatty acid side chains, poor level of catalase and moderate level of super oxide dismutase and glutathione peroxidase and high concentration of ascorbic acid in brain. However, ascorbate/iron ascorbate/copper mixtures generate free radicals. If catalytic iron is generated in the CNS (central nervous system) as a result of injury, ascorbic acid might then stimulates OH generations within the brain and CSF (cerebro spinal fluid). Iron can play a role in production of ROS in a number of different ways. An increase in iron concentration is associated with an increase in oxidation reaction that generates free radicals. Perhaps the best-know iron-catalyzed free radical reaction is the Fenton reaction. Recycling of iron between its oxidized and reduced state allows iron to act as a catalytic agent, producing a cascade of free radicals. From a clinical perspective, the free radical – oxidative stress hypothesis in Parkinson's disease (PD) provides a scientific rationale for trails of antioxidants in the hope that they might be "neuro protective" and modify the natural progression of the disease.

### Diseases Due to Oxidative Stress

**Third line defense**: Third line antioxidants are a complex group of enzymes for repair of damaged DNA, damaged protein, oxidized lipid and peroxides and also to sop chain propagation of peroxyl lipid radical. These enzymes repair the damaged biomolecules and reconstitute the damaged cell membrane e.g., lipase, protease, DNA repair enzymes, transferase, methionine sulphoxide reductase etc.

**Cardiovascular diseases**: Epidemiological studies have indicated that vitamins C and E exert protective effect against cardiovascular diseases. Low plasma level or low dietary intake of vitamin C is associated with high blood pressure and unstable coronary syndrome. Vitamin C improves the endothelium dependant vasodilation by increasing type III NOS (eNOS) expression and ensuing NO production. Vitamin E supplementation reduced non-fatal heart attack, risk of coronary diseases and atheroclerosis by inhibiting oxidation of LDL by free radicals.

**CNS diseases**: Neuronal cells appear to be particularly vulnerable to oxidative stress due to high oxygen consumption, low glutathione (antioxidant status) content, high levels of free iron and oxidizable substances like PUFA in the central nervous system (CNS). This oxidative stress with decreased antioxidant defense does have a role in the genesis and progression of Alzheimer's disease, Parkison's disease, brain neoplasm, Down syndrome, and also some other neurodegenerative disorders.

**Carcinogensis**: A major development of carcinogensis research during the last two decades has been the discovery of significant levels of DNA damage by oxyradicals of endogenous cellular source. These oxyradicals attack DNA causing change in genomic sequences leading to mutation, deletion, gene amplification or rearrangement. In this manner, DNA damage was found important in the etiology of many human cancers.

**Inflammatory diseases**: In chronic inflammatory conditions phagocyte derived ROS have been implicated in inflammation related injury. Intracellular protection of cytoplasmic component against phagocyte derives oxidative injury is mediated predominantly by antioxidant enzymes like SOD, catalase and glutathione peroxidase. The development of mutations in p53 tomour suppressor gene and other key regulatory genes promote inflammation into chronic disease in rheumatoid arthritis and other inflammatory disorders.

**Renal diseases**: Free radical injury and oxidative stress have been implicated in many renal diseases like acute renal failure (ARF), IgA nephropathy, anaemia of chronic renal failure (CRF) and ischaemic kidney. Plasma glutathione peroxidase (an index of renal antioxidant defense) is reported to the markedly depressed in CRF and patients undergoing hemodialysis.

**Diabetes mellitus**: Different studies have given a lot of evidences of increased oxidative stress with depleted antioxidant enzymes and vitamins in diabetes. Now oxidative stress is acknowledged as a pathogenic mechanism in diabetic complications like diabetic retinopathy, nephropathy, microangiopathy.

**Respiratory diseases**: Oxidative stress is assumed to play an important role in the pathogenesis of a number of lung disease, like-chronic obstructive pulmonary diseases (COPD), bronchial asthma and acute respiratory distress syndrome, not only through direct injurious effects but also by involvement of oxidative stress in the pathogenesis of COPD include oxidative inactivation of antiproteinases, airspace epithelial injury, increased sequestration of neutrophils in the pulmonary microvasculature and gene expression of proinflammatory mediators.

**Infectious diseases**: A hallmark of HIV infection is immunodeficiency with progressive CD4+ lymphocytic cell depletion. HIV and opportunistic infections directly or indirectly lead to an oxidative stress, caused by the excess generation of ROS/RNS by activated cells of immune system. Different studies have shown that even in early stages of HIV infection, plasma/lymphocyte SOD level is decreased parallel to the decrease of CD4+ cells number.

**Aging**: Aging in human is associated with changes in physical characteristics and the decline of many physiological functions. Increased accumulation of free radicals heightens the vulnerability of older individual to a variety of oxidative insults. These radicals are capable of causing apoptosis, necrosis and cell death.

**Liver diseases**: Several studies wee conducted to find out the role of oxidant stress and AOS in alcohol induced cirrhosis, which is considered as the terminal irreversible stage of liver diseases. These studies have shown increased lipid peroxidation by HO radical and hydroperoxides in experimental acute and chronic alcoholic liver diseases. Also it is suggested that ROS and lipid peroxidation may play a role in pathogenesis of hepatic fibrosis with loss of normal liver architecture.Viral hepatitis is caused by both RNA (hepatitis A, C, D, E) viruses and DNA virus (HBV). In these infections, there is an associated enchanced production of ROSNS via long-term oxidant stress. In spite of active diseases caused by hepatitis viruses, the superoxide dismutase and total antioxidant levels in the body do not show significant changes. The behaviour of all the viruses remains similar and does not exacerbate with during viral hepatitis and protective caliber of body against oxidative stress remains intact.

## Regulation of Oxidative Stress

Generation of energy by the electron transport chain is dependent on the catalytic spin pairing of triplet oxygen. During this process, oxygen species that are toxic to DNA, protein and lipid components of the cells formed through both enzymatic and spontaneous chemical reactions. The natural by – product of aerobic metabolism are the reactive components super oxide ($O_2^-$) and hydrogenperoxide. The oxidative stress modulation includes at least 80 proteins induced during exposure to super oxide. About half are also induced by $H_2O_2$. The two known systems Oxy-R and Sox-RS are used by *E. Coli.*

The Oxy-R regulation comprises nine of the protein induced by $H_2O_2$. The positive regulator Ox-yR controls all. Oxy-R is a class I type activator that, once bound to a target DNA sequence, interacts with the α-C-terminal domain of RNA polymerase. The protein is activated by oxidation which results in the formation of a disulfide bond between cysteine residues C199 and 208. Under non-inducing conditions these systems are reduced by gluta redoxin I. For most Oxy-R dependant gene, the oxidized but not the reduced form of OxyR binds target DNA sequences. However, the reduced form of this regulator can bind to the Oxy-R, Oxy-S target operator (promoter region not activate transcription until Oxy-R itself is oxidized). Some of the enzymes whose expression is regulated by Oxy-R include catalase (Kat-G), glutathione reductase (gor A), gluta redoxin I (grx A) and alkyl hydroperoxide reductase (ahpC and ahpF). The *in vivo* function for alkyl hydroperoxide reductase would be the detoxification of lipid and other hydroperoxide produced during oxidative stress.

## Anti Oxidant Molecules

The tem antioxidant applies to the substance that delays or inhibits oxidative damage to a target molecules. Living organisms use anti oxidant molecules to protect themselves from radicals. Some prominent antioxidants are GSH, α-tocopherol (vitamin-E) ascorbic acid (vitamin-C) and β-carotene. Antioxidant fall into two classes (i) preventive antioxidant which reduces the rate of chain reaction, and (ii) chain breaking antioxidant which interferes with chain propagation. The preventive antioxidants include catalase and other peroxidase that react with ROOH and chelator's metal ion such as DTPA (diethylene triamine penta acetate) and EDTA (Ethylene diamine tetra acetate). The chain breaking anti oxidants are often phenol or aromatic amine. *In vivo*, the principal chain breaking anti oxidant is superoxide dismutase which acts in the aqueous phase to trap super oxide free radicals ($O_2$). Perhaps urate and vitamin E which acts in the lipid to trap ROO radical. Antioxidative vitamins have a number of biological activities such as immune stimulation, inhibition of nitrosamine formation and alteration of metabolic activations.

**Vitamin-E (a-tocopherol):** a-Tocopherol, a potent radical scavenger, belongs to a class of compounds phenolic antioxidants. Phenols are effective antioxidants because the radical products of these molecules are resonance stabilized and thus stable. Because vitamin-E (found in vegetables seed oils, food grains) is lipid soluble. Vitamin E plays an important role in protecting membrane from lipid peroxyl because it neutralizes reactive oxygen molecules and reduces oxidative DNA damage and genetic mutations. It is a chain breaking antioxidant and has immune competence.

**Vitamin-C (ascorbic acid):** Humans have no ability to synthesis of Vitamin C due to mutation in the gene coding for L-gluconolactone oxidase an enzyme required for the biosynthesis of vitamin via the glucouronic acid pathway. These Vitamin-C obtained through the diet. Ascorbate protects membrance through two mechanism. First ascorbate reacts with peroxyl radical formed in the cytoplasam before they can reach the membrane, thereby preventing lipid peroxidation. Second, ascorbate enhances the antioxidant activity of vitamin E by regenerating reduced α-tocopherol from the a-tocopheroxyl radical. Reacting with GSH then regenerates ascorbate.

**Vitamin-A (β-carotene):** β-Carotene is found in yellow orange and dark green fruits and vegetable such as carrots, sweet potatoes, and apricots. It is a member of a class of plant pigment molecules referred to as the carotenoids. In plants tissue the carotenoids absorb some of the light energy used to drive photosynthesis and protect against ROS. In animals, β-carotene is a precursor of

Vitamin-A and an important antioxidant in membranes (retinol is a precursor of retinal, the light absorbing pigment found in the red cell of retina).

## Conclusions

Reactive oxygen species (ROS) are referred to as Oxidative stress. These reactive oxygen species (free radicals) are generated during oxidative metabolism and used to kill the cells. Lipid peroxidation is a source of generation of free radicals. The oxidative stress are also initiated by some metals such as aluminium, copper, magnesium, manganese, methyl mercury, iron and other toxic substances. In tissue defense mechanism, a number of enzymes like super oxide dismutase, glutathione peroxidase, catalase, etc. are involved against the oxidative stress. The oxidative stress causes many diseases such as cardiovascular disease, central nervous system diseases, carcinogenesis, inflammatory diseases, renal disease, diabetes mellitus, respiratory disease and aging etc. However, some of the prominent antioxidant molecules such as vitamin E, vitamin C, and vitamin A are protecting the living cells from free radicals by scavenging/destroying.

## References

Albert G. Moat, John W. Foster, Michael, P. Specter (2002). Microbial physiology. Fourth ed., *A John Wiley and Sons, Inc., Publication*, 591-595.

Albert L. Lehinger (1982). Biochemistry. Second ed., Kalyani Publication, New Delhi, 504: 713.

David E. Metzler, Carol M. Metzler, David J. Sauke (1997). Biochemistry (The chemical reactions of living cells), Second ed., Harcourt/Academic Press, Vol. 2.

G. Banananda Ray and Syed Akhtar Husain (2002). Oxidants, antioxidants and carcinogenesis. *In. J. Ex. Boil.*, 1213-1232.

Geoffrey L., Zubay, William W. Parson, Dennis E. Vance (1995). Principles of Biochemistry, Printed in United States of America by WM.C. Brown Communications Inc., 2460 Kerper Boulevard, Dubuaque, IA 52001: 336-337.

Gerit schuuramann, Bernd Market (1998). Ecotoxicology (Ecological fundamentals, Chemical exposure, and biological effects), A John Wiley and Sons. Inc. and Sperkum akademister verlagco-publication, 541-542.

H. Robert Horton, Laurence A. Moran, Raumonds S. Ochs, J. David Rawn, K. Gray Scrimgeour (1996). Principle of Biochemistry. *Second ed., Prince Hall, Upper Saddle River*, NJ 07458, 149.

M.N. Chattergea, Rana, Shinde (1994). Texbook of Medical Biochemistry. Jaypec Brothers Medical Publishers Pvt. Ltd., New Delhi, pp. 292, 81, 821, 823.

Masayuki Yasui, Michael, J., Strong, Kilchael Ota, M. Anthony Verity (1997). Mineral and Metal neurotoxicology. *CRC Press, New York*, 93: 160-162: 204: 324: 366-485.

Rober K. Murray, Day K. Granner, Peter A. Mayes, *et al.* (2000). A language medical book Harbers Biochemistry. Twenty Fifth ed., 736-737.

Rosen GM *et al.* (1995). Free Radicals and Phagocytic Cells. *FA and EB J*, 9: 20.

Sarada Subramaniayam, K. Madhavankutty (2002). Texbook of Human Physiology, Six revised ed., Published by S. Chana and Company Ltd., New Delhi, 32.

William J. Marshall (1995). Clinical Biochemistry (Metabolic and Clinical aspects). Published by Churchili livings tone, *Medical Dividion of Pearson Professional Ltd.*, 456: 766-771.

William J. Marshall (1995). Clinical Chemistry, Third ed., Published by Mosby, London, 77-86: 179: 276

# 13

# Cancer: A Consequence of Environmental Pollution

*K. Ramalingam*

Normal cells have definite life span and controlled cell divisions for which their genome is programmed. Certain cellular lesions can transform these normal cells into phenotypes that proliferate in an uncontrolled manner. Such transformed cells are called neoplastic cells or cancer cells (Currie and Currie, 1982). Their new growth is referred to as neoplasia. Tumours are very characteristics of vertebrates, as their prolonged life span favours the slow developmental process of tumourigenesis and their full expression as neoplasia. The transformation of normal cells to neoplastic cells takes place at all stages in the life cycle. The transformed neoplastic cells are capable of invading and destroying their contiguous healthy and normal tissues and could bring death to the organisms ultimately. Neoplastic cells may persist for a long time and may go unnoticed in the body until they develop into either slow growing benign tumors or vigorously growing malignant tumours. Once the tumours growth is initiated, its development becomes autonomous and independent, depending on the genetic, hormonal and immunological status of the host to certain extent (Grant, 1978).

## Morphological Features

Both classical and modern investigations have failed to show the truly specific morphological features in malignant cancer cells. Although tumour cells present on the whole a more homogenous appearance than do the numerous varieties of normal differentiated tissue/cells, they do exhibit considerable diversity as compared to the normal homologous tissues from which they originate. Generally cancer cells tend to have a more simple structural organization and their nucleocytoplasmic ratio is often changes in favour of the nucleus. Brachet and Mirsky (1961) in their book "The Cell" have summarized the morphological characteristics and changes in the nucleus and cytoplasm of cancer cells (Fig 1).

## Nucleus

The cancer cell nucleus in interphase often has a more irregular form than normal, and its size may be considerably increased. Every neoplasm, has its own pattern of nuclear behaviour, and it is impossible to determine direct relationships

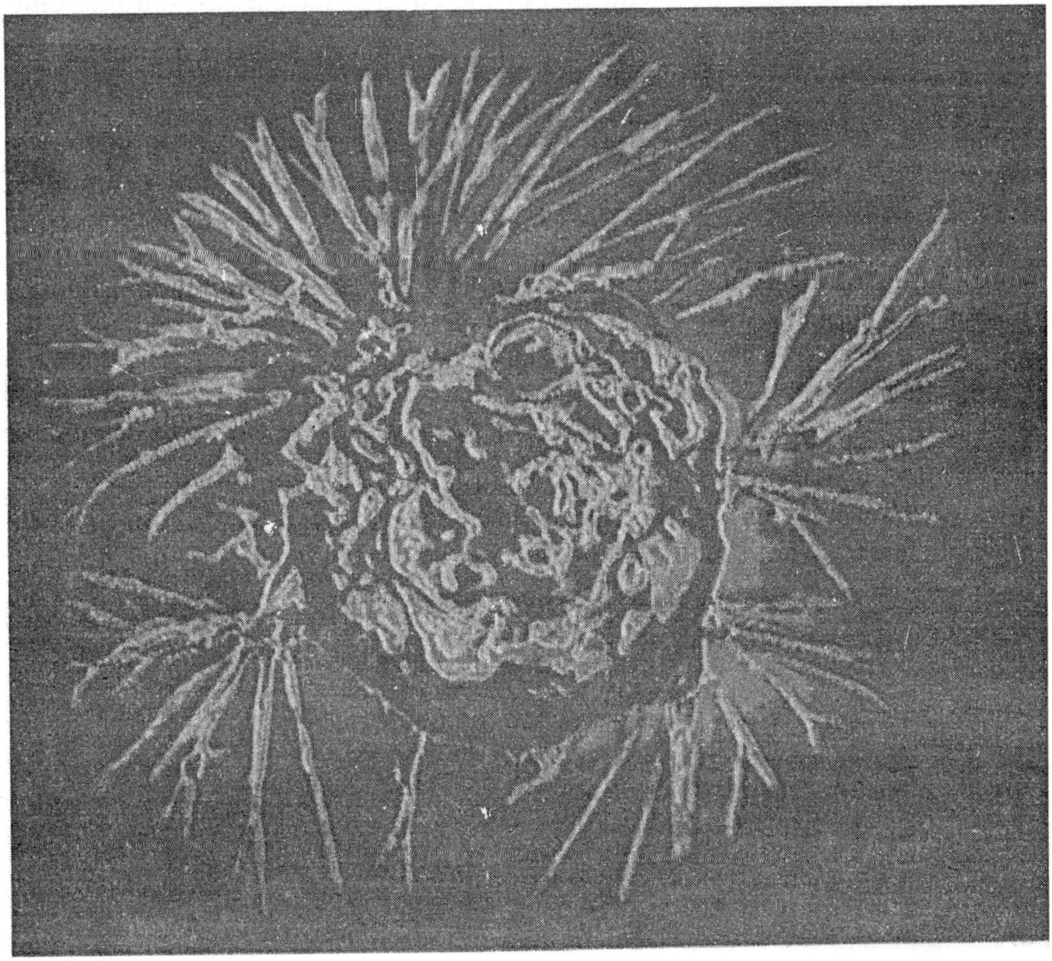

*Source*: Medicine's new vision by Howard Sochurek Page No. 112, Mack Publishing Co. Pennsylvania (1988).

**Fig. 1. A single breast cancer cell and its ultrastructural features (an electron microscopic view) with spikes/tentacles reaching out from the cell's central core.**

between neoplastic transformation of a cell and hypertrophy of its nucleus; heterochromatic anomalies in the resting nucleus are very marked in tumours but have not shown specific features, as yet; cancer cells often show an increased nuclear deoxyribonucleic acid content, but this increase varies in different tumours and even within the same tumour, alongside changes in nuclear volume and chromosome number; the nucleolus is as a rule hypertrophic. There is no constant ultrastructural difference as compared with normal nucleoli; The nuclear membrane frequently shows deep invaginations; its ultrastructure does not seem to be changed as compared with that of the normal nucleus; the chromosomes show the most marked irregularities as to number and shape. Cancer cells are often heteroploid and the chromosomes may show considerable structural changes. There has never been any demonstration, so far, that these anomalies are a cause rather than a consequence of malignant transformation; changes in the duration of mitosis are considerable. The metaphase seems to be relatively longer than in homologous normal tissues; the mitotic frequency is not increased in cancer tissue as compared with embryonic tissue in full growth; amitosis, in the classic sense of the word, does not seem to play a role in the multipliucation of cancer cells.

**Cytoplasm**

The number of mitochondria varies considerably from tumour to tumour. All the intermediate states between extreme richness and almost complete absence of chondriosomes are found; often, however, tumour cells tend to show fewer, smaller, and denser mitochondria. Phenomena of degeneration are also found, such as defective ultrastructure and swelling; Cytoplasmic basophilia is more marked, and on the ultrastructural level, this basophilia corresponds to an increase of RNP granules, freely dispersed in the cytoplasm and rarely to an increase of membranous ergastoplasm. The latter, on the contrary, tends to disappear in the cancer cells; The Golgi apparatus can be very hypertonic in certain tumours and may be hypertrophic in others. Its ultrastructure does not differ from that of the Golgi zone in normal cells; the ulstrastructure of the centriole in cancer cells does not seem to be abnormal. The position of this organelle and of the golgi apparatus determines the polarity of the cells, which may be reversed or completely lost in differentiated cancer tissues; the cytoplasm of cancer cells may contain numerous inclusions of variable origin, resulting from degenerative processes, from partly preserved functional activity, from the phagocytic properties of certain cancers, or from the presence of virus corpuscles in filterable tumours; the ground substance shows no electron microscopic alterations.

Thus the cancer cells exhibit differences from that of their normal counterparts in their nucleus and cytoplasm. The diagnostic structural features of malignant cells are increased nuclear – cytoplasmic ratio, increased nuclear - nucleolar ratio, enhanced nuclear staining and increased incidence of mitotic figures. Electron microscopic observations also could reveal relatively changes in the nuclear membrane, mitochondria and in the endoplasmic reticulum.

### Chromosomes

According to chromosomal theory of cancer proposed by Boveri malignancy was a consequence of abnormal chromosomal content of cells. Several overt structural chromosomal abnormalities are detectable in malignant cells. Clinically the most important stable chromosome abnormality associated with malignant cells is the philadelphia chromosome associated with chronic myeloid leukemia. The recent findings suggest that more than the gross structural changes in chromosomes, subtle changes in the gene expression are attributed for the genesis and behaviour of malignant cells. The specific changes in gene sequence and in genic arrangement in various cancer types are discussed later in this chapter.

### Cytoskeleton

Normal nucleated mammalian cells are in possession of an internal cytoskeleton which represents a complex intracellular network of structures, viz the microtubules and microfilaments. Microtubules are complex protein structures which when fully polymerized maintain rigidity. They also play an important role in the mitotic spindle. Attached to microtubules and also to proteins at the inner surface of the cell membrane are the microfilaments. These are the contractile elements with actin as their major constituent. These microfilaments are responsible for cell movements and overall alterations in cell shape. Fluorescent antibody techniques have revealed that normal cells are endowed with large cable of actin just inside the cell membrane and attached to it, whereas in malignant cells these cables of actin are poorly formed and attachment to the membrane proteins is defective. The cytoskeleton plays an important role in cell surface architecture, in cell shape, spreading and adhesiveness. The defective cytoskeleton of malignant cells has a correlation to their differences with the normal cells phenotype.

### Cell Membrane Structure

Changes in the cell membrane structure are very characteristic of malignant cells. They are the appearance of new surface antigens, proteoglycans, glycolipids and mucins. Glycosylation patterns of cell surface glycoproteins and glycolipids change during transformation of normal cells into malingnant ones.

Tumour associated carbohydrate antigens belong to three groups viz., 1. epitopes expressed on both glycolipids and glycoproteins, 2. epitopes expressed only on glycolipids and 3. epitopes expressed only on glycoproteins. The first group is formed in cancers such as lung, breast, colorectal, liver and pancreatic cancers. The common back bone structure for these epitopes is Ga1β1 → $3G_1$CNA cβ1 → 3-Gal (Type 1 blood group) or Gal1β1 → 3 Gal (Type 2 blood group).

The second group of epitopes expressed exclusively on glycolipids is the ganglio or globo series structures. These are expressed abundantly on cancers such as melanoma, neuroblastoma, small cell lung carcinoma and Burkitt lymphoma. The third group of epitopes seen only on glycoproteins is the multiantennary branches of N-linked carbohydrates and the alternates of O-linked carbohydrate chains seen in some mucins.

The altered glycosylation patterns result from three kinds of altered processes viz.:

1. incomplete synthesis and/or processing of normally existing carbohydrate chains and accumulation of the resulting precursor forms,
2. neosynthesis resulting from the activation of glycosyl transferases that are either absent or of low activity in normal cells and
3. organizational rearrangement of tumour cell membrane glycolipids.

## Anaplasia

The term anaplasia is used to denote the loss of structural evidence of cell differentiation. As for cancer, they represent çells which are morphologically and functionally quite undifferentiated. Clinical cancer presents however a complete spectrum of functional and morphological differentiation ranging from the well differentiated hormone producing tumours to the completely undifferentiated and apparently functionless anaplastic carcinomas. The malignant cells in a tumour mass are not morphologically uniform but present a wide range of assorted shapes and sizes of both nuclei and cytoplasm. An established tumour is a complicated ecosystem and is not able to be regarded as a mass of homogenous malignant cells alone. Solid tumours contain besides malignant cells, various amounts of host derived materials. Stromal cells, vascular endothelium and a range of inflammatory cells from the host make up the gross structure of tumour alongside the malignant cells. e.g., Scirrhous carcinomas of breast. A tumour mass often shows large areas of necrosis as well as avascular and anoxic regions.

## Malignant Stem Cells

The number of fully viable proliferating, cancer cells are known as 'clonogenic cells' or 'malignant stem cells'. But these may be very low. In a mass of malignant tissue those cells nearest the blood supply, well oxygenated and well nourished are the cells most likely to survive and proliferate. Anoxic cells which are poorly nourished are dead or dying cells. However some of these anoxic cells may survive to repopulate a tumour under circumstances of rapid changes in oxygenation and vascularity. The chlonogenic cells in many tumours are anaplastic and may give rise to progeny which undergo terminal differentiation. A substantial number of daughter cells may proceed along a proliferative dead end. Thus a considerable cell loss also occurs in malignant tumour tissues. The factors that contribute to such cell loss with in tumour tissues could be immunological, metabolic, proliferative incompetence and pressure. The immunological factor may be the infiltrating cytotoxic immune cells of the host like lymphocytes and macrophages. The metabolic factor may be due to unused nutrient requirements of malignant cells and their non-availability causing death of some tumour cells. Proliferative incompetence is unlikely to make a major contribution to cell loss since defective malignant cells with the absence of mechanics of cell division are rare to appear. When space becomes a constraint to an expanding mass of malignant tissues or trying to expand, the pressure may bring distortion and contribute to necrosis by interfering with extra cellular fluid perfusion.

## Vascular Structure

The blood supply within malignant tumours is highly disorganized. An in situ cancer presumably requires no specialized vascular supply. Diffusion of nutrients and oxygen from the extracellular fluid is sufficient to maintain the viability of very small collection of cells. However an expanding tumour cell population demands more, so much so that the passive diffusion from the extracellular portions becomes insufficient and inappropriate for the viability of cell mass. The size and surface area of a tumour mass become a determining factor for acquiring blood supply. In most tumors the critical diameter for demanding blood supply is 1-2 mm. This critical size triggers vascularisation or angiogenesis. The tumour often releases a 'tumour angiogenesis factor' (TAF) which stimulates proliferation of capillary endothelial cells and thereby promotes the development and growth of capillary sprouts towards the malignant cells. The release of TAF and angiogenesis represents an early and crucial step in the evolution of malignant neoplasms.

## Functional Attributes of Cancer

All neoplastic or cancer cells exhibit certain common phenotypic traits and functional attributes. viz.

1. Uncontrolled cell division and autonomous growth.
2. Heritability and transplantability.
3. Altered cell surface.
4. Invassiveness and metastasis.
5. Dedifferentiation and loss of specialized functions.

## 1. Uncontrolled Cell Division – Autonomous Growth

Uncontrolled cell division is the primary phenotypic trait of a neoplastic cell. Tumour cells proliferate autonomously, irrespective of the growth arresting signals of the host. The duration of cell cycle and their divisions which are regulated by a genetic mechanism in normal cells become altered in neoplastic cells such that they undergo uncontrolled proliferation. The most likely lesion that makes no control over the cell is that of DNA replication, at some point during the G1 phase. Most normal somatic cells have either a prolonged G1 phase or enter into the G0 phase after mitosis. In most tumour cells, the shift from G1 to G0 does not operate and the cell progresses naturally into the S phase of DNA replication after each mitosis. Moreover the G phase in tumour cells is comparatively shorter than their normal counterparts. The total length of the cell cycle is however is similar in both normal and neoplastic cells. A lesion may alter a cell's ability to respond to the systemic factors that regulate cell division activity in normal cells. One of the first changes in the neoplastic transformation seems to be at the cell surface suggesting that an abnormal phenotype may arise when a cell loses surface receptors sensitive to growth inhibitory signals.

## 2. Heritability and Transplantability

Once a normal cell becomes neoplastic, it transmits and inherits the neoplastic phenotype to all its progeny. Many tumours are transplantable. When a small tumour fragment is implanted into a tumour free host, it grows rapidly and kills the host. Before the host dies, another fragment of the newly developed tumour may be reintroduced into a new host, and the process may be continued indefinitely with the tumour, retaining its characteristics, morphology, growth rate and biochemical properties in each host. It is this property of the tumour cells that enabled their culture as tumour cell lines in vitro permanently. Tumour cell lines may be stored in frozen state indefinitely and thawed at any time to resume their growth either in vitro or in vivo, without loss of their tumourigenic potential.

The transplantability is usually host specific. A tumour must be genetically and immunologically compatible with the host. Otherwise the host's immune system rejects the tumour as foreign. Tumours that arise in strain A mice cannot be transplanted to C3H mice. They can be maintained only in Strain A mice. However this host specificity may be lost after many transplant generations in one strain as well in other strains. Presumably such tumours would have lost their histocompatibility antigens that induce graft rejection by other strains. Host specificity is typical of the malignant neoplasms. Phenotypic stability of transplanted tumours suggests that all tumour cell properties are due to heritable changes perhaps to mutations in somatic cells.

### 3. An altered Cell Surface

Spontaneous transformed tumour cells as well as induced tumour cells both have altered cell surfaces. Normal cells exhibit contact inhibition of proliferation and motility when grown in vitro. On the contrary tumour cells are not contact-inhibited and exhibited indiscriminate movement. Compared to their normal progenitors, tumour cells possess an increased net surface charge on their outer membrane, migrate at different rates in an electrophoretic field and exhibit different histogenic patterns when cultured in vitro.

Normal cells organize into compact parallel arrays. Tumour cells spread at random as diffuse masses and penetrate normal cells. Neoplastic cell are also less adhesive. They are more readily detached from a substratum than normal cells. Tumour cells are more motile and possess greater invasive qualities. Antigenecity of the tumour cell surface is also different. Tumour specific antigens appear in the virally transformed cell membranes(TSA). Spontaneously arising and chemically induced tumour cells also possess surface antigens. But that is absent in their normal counterparts.

**Foetal/embryonic antigens:** The reappearance of embryonic or foetal antigens over the cell surface of neoplastic tumour cells point out a recapitulation process in the transformation of normal cells to neoplasia. This also implies dedifferentiation of organised cells to embryonic conditions. Tumour cell surfaces also exhibit a different structural features from the normal counterparts. Tumour cells possess more microvilli. The surface changes in tumour cells could result from 1) the loss or addition of surface constituents or 2) reorganization of existing surface components by conformational changes in membrane proteins or by the displacement of proteins within the membrane. The surface change may relate to mechanisms that regulate DNA synthesis and cell division.

Studies have revealed that neoplastic cell surfaces, in contrast to normal cells are sensitive to plant agglutinins such as Concanavalin-A. Normal mitotic cells are also agglutinable by lectins which suggests that proliferating cells undergo a surface change that exposes Con A binding sites.

Mitotic activity correlates with intracellular cyclic AMP levels. Its concentration is lowest during mitosis and highest at most other phases of the cell cycle. Tumour cells contain less cyclic AMP. Both contact inhibition of growth and motility are restored when neoplastic cells are treated with dibuteryl derivatives of cyclic AMP.

The cyclic AMP implicates a surface enzyme Adenyl cyclase in growth regulation. It regulates gene transcription by its action on histone phosphorylation or it may regulate the assembly of microfilaments and microtubules associated with the cell surface in some manner. Neoplastic transformation is accompanied by a change in the subsurface systems of organellae, a surface modulation system that may regulate the movement of membrane constituents.

## 4. Invasiveness and Metastasis

### *(a) Cell Motility*

A relevant comparison of normal cell versus malignant one, may give an insight to the functional dynamics of cancer cells motility and invasiveness. The motility or the motile behaviour of normal and cancer cells have been investigated by several authors like Ranvier, Virchow, Waldeyer, Lambert and Michael Abercrombie using tissue culture techniques. These studies have revealed not only the differences between the normal and malignant cells but enlightened the anarchic social organization of a tumour mass with cells rushing in all directions completely devoid of normal restraints, which play a vital role in the invasion of adjacent tissues. Abercrombie demonstrated that normal cells showed mutual contact inhibition of locomotion ie., when a cell bumps into another cell it stops moving. On the contrary the malignant cells do not exhibit contact inhibition, rather show contact guidance ie., they move along the surfaces of normal cells and tissue planes. Adjacent normal epithelial cells are mutually adhesive and stick to each other. Malignant cells show decreased mutual adhesiveness. These can detach from the neighbours and can penetrate tissues. They could be carried in extracellular fluid from the tumour foci into the normal tissues. The tumour cells can also gain access to blood or lymph vessels and extend inside them. The above channels of lymph and blood enable the distant dessemination of the malignant cells. The motility behaviour of cancer cells gives way to the most lethal characteristic of malignant cancers namely the metastasis.

### *(b) Metastasis*

The process of metastasis involves several stages viz., the dissemination of cells, the arrest of cells at distant sites and the survival and proliferation of cells in such sites. There are many ways in which malignant cells may reach distant sites. The simplest mode of transferring malignant cells from one site to another is inter epithelial contact. Another simple mode of cell transfer is by direct invasion and then disseminaton within serous cavities. The latter gives rise tumour nodules formation. Ovarian cancer gives rise to extensive ascites with large numbers of malignant cells which spread out all over the peritoneum. Another less common form of serous dissemination involves the spread of tumour cells from the primary cancer of ovary to the upper part of the abdomen such as stomach and/or pancreas leading to secondary tumour formation. Of all the afore mentioned ways, the most significant mode of cell dissemination is intravascular either lymphatic or haematogenous (Mareel and Calman,1984).

### *(c) Lymphatic Dissemination*

The cells shed from a primary tumour rapidly gain access to the draining lymph nodes. Though the lymph nodes act as a barrier preventing the escape of malignant cells downstream to the next lymph node group or into the blood stream via the thoracic duct, the above lymph nodes barrier and their trapping function is abolished by factors such as steroid hormones or local irradiation etc. The above barrier disruption facilitates the entry of further secondary and tertiary deposits of malignant cells all along the lymphatic chain and their final access to the blood stream.

### *(d) Haematogenous Dissemination*

Haematogenous spread is quantitatively an inefficient mechanism. But the blood supply to a cancer is being highly disorganized and is associated with chaotic breakdown of vascular integrity, malignant invasive cells directly enter into the blood. Some forms of cancer spread exclusively via the blood stream. Choriocarcinoma, Osteogenic sarcoma, thyroid and kidney carcinomas give rise to distant metastasis through blood stream than lymph nodes. The haematogenous spread of malignant cells depends upon the vascularity of the primary tumour and the secretion TAF by the malignant cells in the primary tumour. Tumours of bone always metastasize via the blood stream.

### *(e) Fibronectin and Invasion*

Among the extracellular components in cell-substrate interactions and cell-cell interactions, fibronectin is probably the most important. It is a high moleculal weight glycoprotein present in blood plasma, extracellular matrices, basal lamina and at cell surfaces. In the quiescent normal cells it plays an important role in anchorage to substrata such as collagen. Its disappearance after malignant transformation contributes to the reduced adhesiveness and increased invasiveness of the cancer cells.

***(f) Proteolytic enzymes:***

Recent studies have demonstrated that the release of proteolytic enzymes represents an important phenotypic charactesitic of malignant cells. This enzyme converts plasminogen into the potent fibrolysis plasmin. The enzyme also cleaves many other proteins. This enzyme-mediated proteolysis by malignant cells may account for the invasive and destructive properties of a cancer and it may also play a role in the reduced adhesiveness of transformed cells. The breakdown of collagen in the normal connective tissues by collagenolytic enzymes is also thought to be involved in tumour invasion. In addition to cancer cells own battery of enzyme action, the extra-corporeal release of a range of enzymes such as neutral proteinases, collegenase, plasminogens activator, and the secretion of elastase by the infiltrated macrophages within a tumour could also contribute to its invasive and destructive properties.

**5. Dedifferentiation and Loss of Specialized Functions**

In the normal tissue differentiation, cells know by virtue of their position what to do? As a result different cell lineages give rise to their characteristic cell types. But in the event of transformation of normal cells into malignancy, cells lose their virtues and become anomalous in nature. The malignant cells after neoplastic transformation undergo dedifferentiation is a typical feature of cancers. The term dedifferentiation may seem to be wrong and misleading because such a term implies to reverse mechanism of differentiation and return to the embryonic state. The term 'disdifferentiation' has been advocated by several authors as the appropriate designation. The malignant cells in course of time become atypical cells/tissues with no analogous features of their normal counter parts. In many cases the tissue specific characteristics disappear in tumour cells. Thus they represent a mosaic and heterogenous population of cells in which certain tissue specific characters are retained or lost.

## TUMOURS AND NECROSIS

Following the invasive process of metastasis of malignant cells viz the blood stream and lymphatics, necrotic death of individual or sheets of cells in situ occur. Necrosis is a more readily recognizable feature in the histopathology of cancer. The abve process of necrosis gives way to cell loss in the tumour differentiation. Cell death or necrosis in cancer is a complex function and is related to the size of the primary tumour, the age of the primary tumour, its histopathological conditions; and the site of growth of the primary and secondary tumours. The necrosis of tumours may be classified into different types as follows.

1. Coagulative necrosis: It occurs as a result of failure in the malignant cell to maintain homeostatic regulation particularly the fluid volume in the face of strong physiological, physical or chemical stress. It is characterized by features such as blebbing, swelling of cytoplasmic organelles, notably of mitochondria followed by digestion of cell contents and formation of laminated masses.
2. Vasculature induced necrosis: Here qualitative and quantitative nature of vasculature may influence the probability of necrosis. The failure of vessel-supply to tumour thus may induce increased cell loss by necrosis.
3. Hypoxic necrosis: There is a close relationship between the distance from a vessel and the distance at which $O_2$ tension falls to zero, to the necrotic zone in tumor.
4. Nutrient gradients and necrosis: Recent studies attribute that the concentrations of nutrients such as glucose and the influence of intermediary metabolites like lactic acid, lactate/pyruvate ratio and their consequent alteration locally in the tumour microenvironment may serve as probable factors relevant to bring about cell death. Thus tumour necrosis is brought about by many parameters rather than one.

## Apoptosis and Cancer

Apoptosis is an important mechanism by which cells die in the normal differentiation of tissues. By this mechanism the number of cells is being regulated during development. It is a self regulatory process of balancing the cell divisions in the tissue differentiation process, and maintaining cell numbers through defined biochemical and morphological changes, so as to mould normal tissues. Loss of this normal apoptosis mechanism also may have a role to play in cancer or neoplasia as the latter is a disease due to a loss of control of the balance between cell proliferation and cell death. Many protooncogenes regulate cell division while several others regulate apoptosis concomitantly. The protooncogene c-myc has a dual role since it stimulates both cell division as well as apoptosis. In the absence of growth factors, c-myc triggers apoptosis. Mutation of the genes involved in apoptosis, results in either the absence or relative down-regulation of apoptotic pathway and manifests in cancer. Studies have also revealed that overexpression of the apoptotic genes such as bcl-2 may also result in cancer. Generally anticancer drugs damage DNA and trigger apoptosis in cancer cells specifically. However, the resistance to these drugs develop due to suppression of the apoptosis. This occurs in 50% of human cancers by the mutation of the tumour suppression p53 gene, nicknamed as the 'guardian' of the genome. p53 triggers both apoptosis as well as DNA repair. In the event of p53 mutation and also that when the damage is too great for repair, the derangement leads to the relapse of cancer.

### Oncogenes

By definition an oncogene is a gene whose action promotes cell proliferation. Certain viruses infect the host and induce the cells of the latter to proliferate in an uncontrolled fashion and transform them to neoplastic state. These are called tumour viruses or oncoviruses. These are endowed with an oncogene or (gene) in their genome. Both RNA viruses (Retro) and DNA viruses can cause cancer in hosts but they do so by entirely different mechanisms.

Some retroviruses in addition to their protein coding life cycle genes carry an oncogene that gives them the ability to transform cells they infect. These are the oncogenic retroviruses. Eg., Rous Sarcoma virus (RCV). These viruses may or may not have the capability to replicate. Different retroviruses may have different oncogenes viral oncogenes (V-oncs) are responsible for many different cancers. Certain bacteriophages can pick up an oncogene from the genome of the cell and transducer the gene/genes to other cells. These are called transducing retroviruses. The table 1 illustrates some of the transducing retroviruses and their viral oncogenes (Fredrick *et al.*, 1999).

### Cellular Protooncogenes

The existence of genes that induce cancer inside the normal cells has been unequivocally established in both human and animals. Moreover in the early 1980s R.A. Weinberg and M. Wigler showed independently that a variety of human tumour cells contain oncogenes. These genes when transfected to other cells growing in culture transformed these cells also into cancer cells. Human oncogenes were shown to be closely related to genes found in normally growing cells. They were also found to be very similar to the viral oncogenes. When these proto – oncogenes are mutated they become active oncogenes and induce tumour formation (Fig 2).

In normal cells, expression of protooncogenes is tightly controlled in such a way that the cell growth and division occur only at appropriate time, characteristic of the cell type involves. However when these protooncogenes are changed to oncogenes, the tight control is lost and results in unscheduled cell proliferation (Russell, 1998).

The various genetic and chromosomal mechanisms that bring tumourigenesis

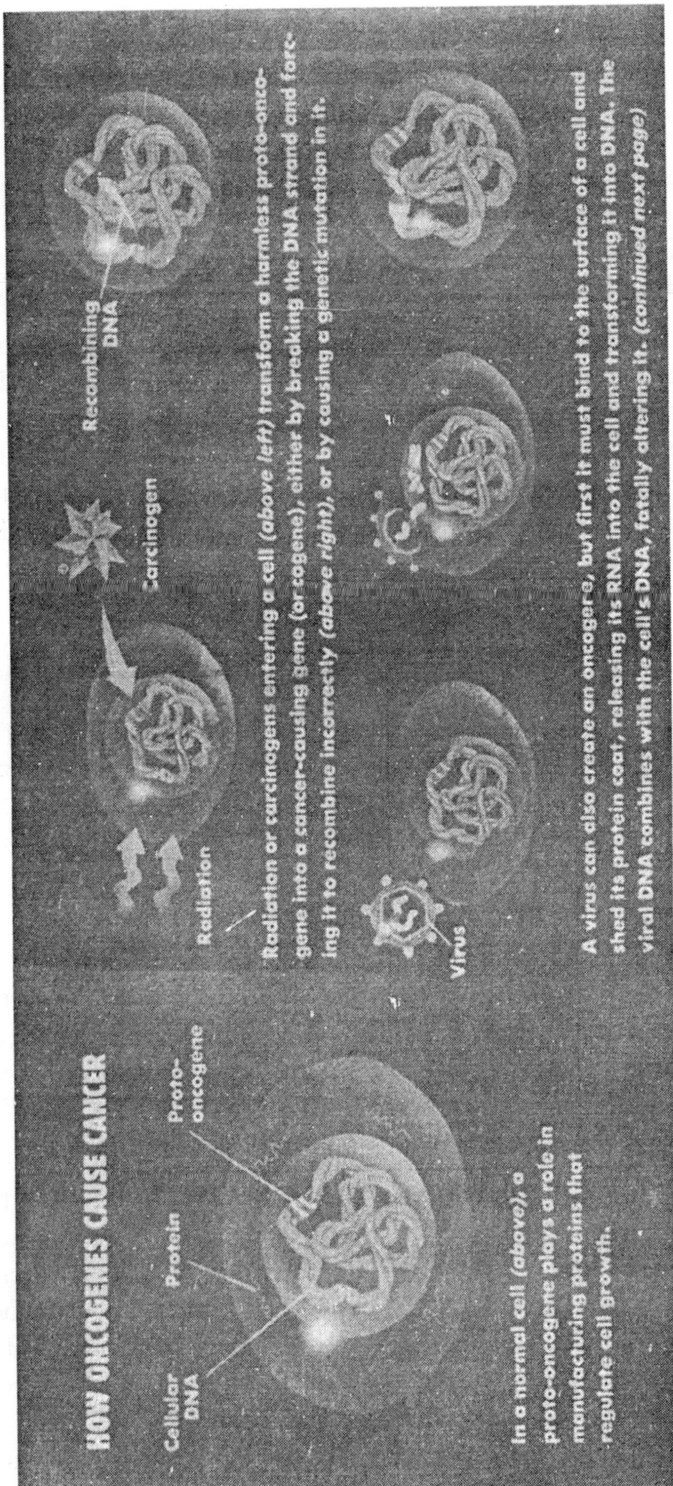


*Source:* Times Magazine (1998) Page No. 53.

**Fig. 2. Mechanism of oncogenes and tumourigenesis**

include point mutation, deletion, gene amplification, chromosomal translocation etc.

### Point Mutation

Point mutation in the coding region of a gene or in the controlling sequences viz., promoter, regulator, enhancers can change a protooncogene into an oncogene and cause the abnormal increase in the expression of the gene or the activity of the gene product leading to an eventual increase in the amount of gene product.

For instance the 'ras' genes are a family of genes encoding membrane associated G protein. A single point mutation in the ras genes may produce an abnormal (mutant) protein which can transform normal cells into malignant cells. Point mutations have been found in malignant cells from patients with bladder cancer, lung cancer, colon cancer and leukemia. The point mutations cause the G proteins to lose their ability to be regulated so that constitutive growth signals (continuous) induce the commencement of unregulated cells proliferation.

### Deletion

Deletions of part of the coding region or of part of the controlling sequences of a protooncogene were found frequently in oncogenes. For example the 'myc' oncogene arises from its protooncogene by deletion. The normal gene with three exons and two introns when becomes deleted of its first exon and intron results into myc oncogene and culminate in increased amount of myc protein.

### Gene Amplification

In tumours multiple copies of oncogenes due to random overreplication are formed. These multiple copies result in an increased amount of gene products which in turn causes unscheduled cell proliferation.

### Chromosomal translocation

A number of human tumours have been shown to be associated with such chromosomal aberrations as translocations. In chronic myelogemous leukemia (CML) reciprocal translocation between chromosomes 9 and 22 has been documented. Similarly Burkitt's lymphoma is due to reciprocal translocation between chromosomes 8 and 14. Chronic myelogenous leukemia (CML) is a fatal cancer involving uncontrolled replication of myeloblasts which are stem cells for the WBCs. In these patients leukemic cells exhibited reciprocal translocation event in which a part of the long arm of chromosome 22 is translocated to chromosome 9 and a small part from the top of the latter is translocated to chromosome 22 The above reciprocal translocation involves the shift of the gene C-abl portion of 9 and positioning of it with in the 'bcr' (breakpoint cluster region) gene of 22. This hybrid oncogene caused uncontrolled expression of both genes and produces leukemia. The above translocated chromosome was discovered in Philadelphia in leukemia patients. Hence it is called the Philadelphia chromosome ($Ph^1$). Ninety percent of CML patients have this Philadelphia chromosome (Fig.3).

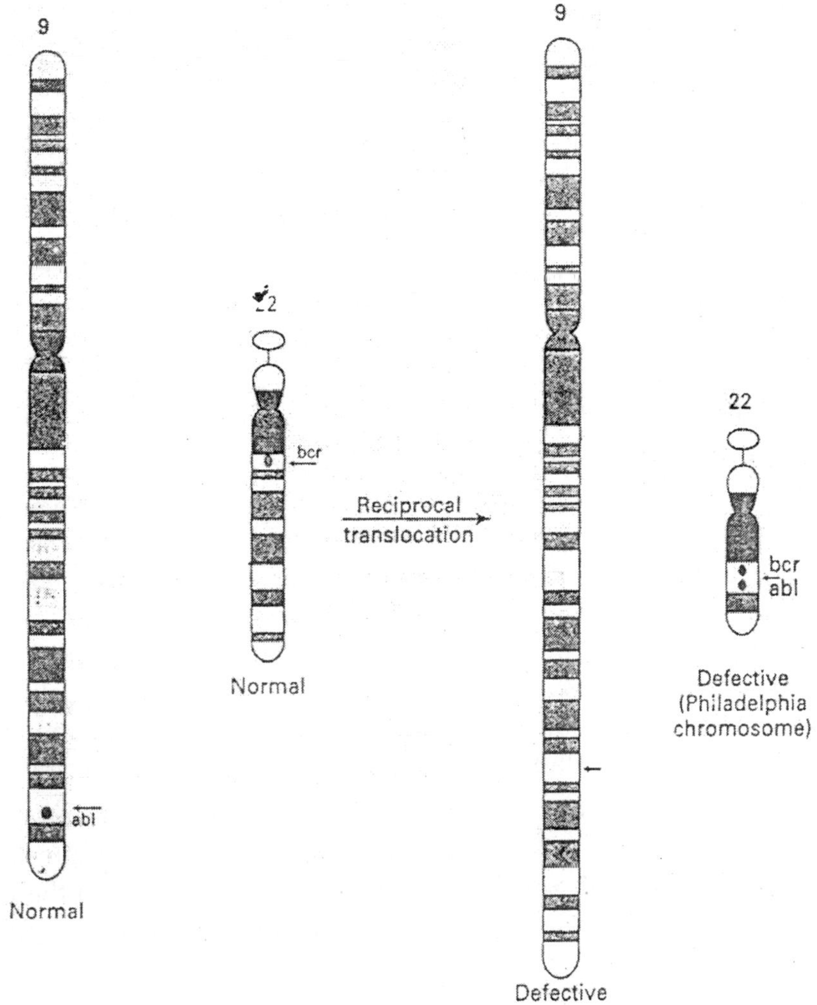


*Source*: Genetics Vol. I and II by Peter J. Russell Page No. 203, Publ. by Benjamin/Cummings Publishing Co.

**Fig. 3. Chromosomal and gene translocation in malignacy (Philadelphia chromosome).**

**Table 1**
**Some Transducing Retroviruses and Their Viral Oncogenes**

| S.No | Retrovirus Isolate | Oncogene | V-Oncorigin | Type of Cancer |
|---|---|---|---|---|
| 1. | Rous sarcoma Virus (RSV) | src | Chicken | Sarcoma |
| 2. | MC29 | myc | Chicken | Sarcoma, carcinoma, and myelocytoma |
| 3. | Avian myeloblastosis virus (AVM)<br>AVM – E26 | myb | Chicken<br>Chicken | Myeloblastosis<br>Myeloblastosis and Erythroblastosis |
| 4. | 3611-MSV | raf | Mouse | Sarcoma |
| 5. | Harvey MSV (Ha-MSV) | H-ras | Rat | Sarcoma and Erythroleukemia Sarcoma |

## Tumour Suppressor Genes

Contrary to the existence of oncogenes whose gene products have a carcinogenic role in cell growth and division, normal cells are also endowed with genes which could suppress the uncontrolled cell proliferation characteristic of cancer cells. These genes are called Tumour Supressor genes. These genes have been discovered in the middle of 1980s by Henry Harris at Oxford. Thus oncogenes cause cancer if they are switched on while tumour suppressor genes cause cancer if they are switched off. Tumourigenesis can progress well if only a cell has both a jammed on oncogene and jammed off tumour suppressor gene. That is, the expression and suppression of two antagonistic genes.

In contrast to mutations that activate oncogenes, mutations of tumour suppressor genes are recessive, that is, cell proliferation can be affected only if both alleles are inactivated. Several tumour suppressor genes and their encoded proteins have been studied in order to understand their role in the cancers with which they are involved.

## Retinoblastoma Tumour Supressor Gene (RB)

The RB gene in human is a tumour suppressor gene. Its locus is at chromosome location 13q14.1 - q14.2. The RB gene contains 180 kb of DNA and encodes a 110 kDa nuclear phosphoprotein (pRB), which is involved in regulating the cell growth.

In patients with retinoblastoma, RB gene has undergone different mutations that lead to loss of its gene function. The mutant allele is probably formed through mitotic recombination, chromosomal non-disjunction or gene conversion.

In the cell cycle phases G1-S-G2, the pRB plays a major role by regulating the passage of cells from G1 to S. The retinoblastoma protein pRB containing 92 amino acids becomes highly phophorylated near the end of G1 phase and at the beginning of S phase ending up with 11 phosphate groups on serine and threonine residues. In its hypophosphorylated or unphosphorylated form the RB associates with the transcriptional factors E2f and Drtf1 and removes them from circulation. As a result the cell will remain in G1 or enter the quiescent state viz G0. If the cell is signaled to progress, the pRB is phosphorylated by cyclin/cyclin dependent kinase (cdk) complex. As a result of pRB phosphorylation it is no longer available to bind to E2F molecules, which in turn (the free E2F) binds with the binding sites of genes and switch on transcription and enable the cell to enter from G1 into S phase.

In the case of mutant RB alleles, pRB is unstable and does not bind to E2F. The free E2F activate the genes for transcription and the cell division takes place in an unprogrammed manner.

**Tumour Suppressor p53 Gene:**

p53 gene was first discovered by David Lane in Dundee. It is so named because it encodes a protein of molecular weight 53 kDa (p53). 50% of all human cancers such as breast, brain, liver, lung, colorectal, bladder and leukemia are caused due to mutations of both alleles of p53. Both mutations and deletions have been shown to cause the inactivation of p53 gene. Infact over 3400 mutations in p53 have been recorded in databases.

The mechanism involved in the p53 gene activity in the suppression of tumourigenesis and genetic stability is illustrated in the figure.

The antitumour activity of p53 gene following cellular damage by irradiation involves the production of p53 proteins which acts as a transcription factor and activates WAF1 gene. The product of WAF1 is a 21 kDa protein called p21. p21 when overexpressed has been shown to cause cells to arrest in G1. The above arrest occurs due to binding of p21 to cyclin/cyclin dependent kinase (cdk) complexes and blocking of the kinase activity. The absence of kinase activity blocks the transition of cell from G1 to S. The cell cycle will not be resumed and the cell will die by programmed cell death (apoptosis). The induction of apoptosis is an important function of p53.

If p53 alleles (both) are inactivated in the cell or mutated, active p53 will not

**Table 2**
**Some Known Tumor Suppressor Genes**

| S.No | CANCER TYPE | GENE | MODE OF ACTION |
|---|---|---|---|
| 1. | Colon carcinoma | APC | Cell adhesion molecule |
| 2. | Breast cancer | BRCA1 | Transcription factor |
| 3. | Breast cancer | BRCA2 | Transcription factor |
| 4. | Colon carcinoma | DCC | Cell adhesion molecule |
| 5. | Melanoma | P16 | Transcription factor |
| 6. | Colon cancer; many others | P53 | Transcription factor |
| 7. | Retinoblastoma | RB | Transcription factor |

be presumed to activate WAF1 and the subsequent cascade events. As a result, the cell cycle may proceed to S. Thus in the absence of p53 protein, p21 protein and apoptosis, the unrepaired cells proliferate, accumulate and increase the probability of cancer.

In FAP, a form of colorectal cancer patients, loss of tumour suppressor gene in chromosome 5 has been established. The deletion of both copies of tumour suppressor gene DCC in chromosome 18 results in adenoma class III cancer of colon. Similarly deletion of both copies of tumour suppressor gene p53 in chromosome 17 results in colon carcinoma.

The development of most cancers involved multiple mutational events involving both activation of oncogenes and inaction of tumour suppressor genes. These cumulative mutations result in the breakdown of multiple cellular mechanisms that regulate the growth and differentiation and culminate in cancer (Fig 4).

## MUTATOR GENES AND CANCER

In addition to oncogenes and tumour suppressor genes, mutator genes also can contribute to tumourigenesis or cancer. In normal cells, the mutator genes, in unmutated forms are involved in such important activities as DNA replication and DNA repair. Mutations of these genes can significantly affect the above processes and can make the cells error prone, such that the accumulated mutations manifest in the form of cancer. In human, the hereditary nonpolyposis colon cancer (HNPCC) results due to mutator genes effect. It accounts for 5-15 percent of colorectal cancers. It is an autosomal dominant genetic disease. Gene mapping has identified four genes $hMSH_2$, hMLH1, hPMS1 and hPMS2 for the HNPCC. These mutator genes, when mutated confer a phenotype of hereditary predisposition to NHPCC. The first two genes are responsible for about 90 percent

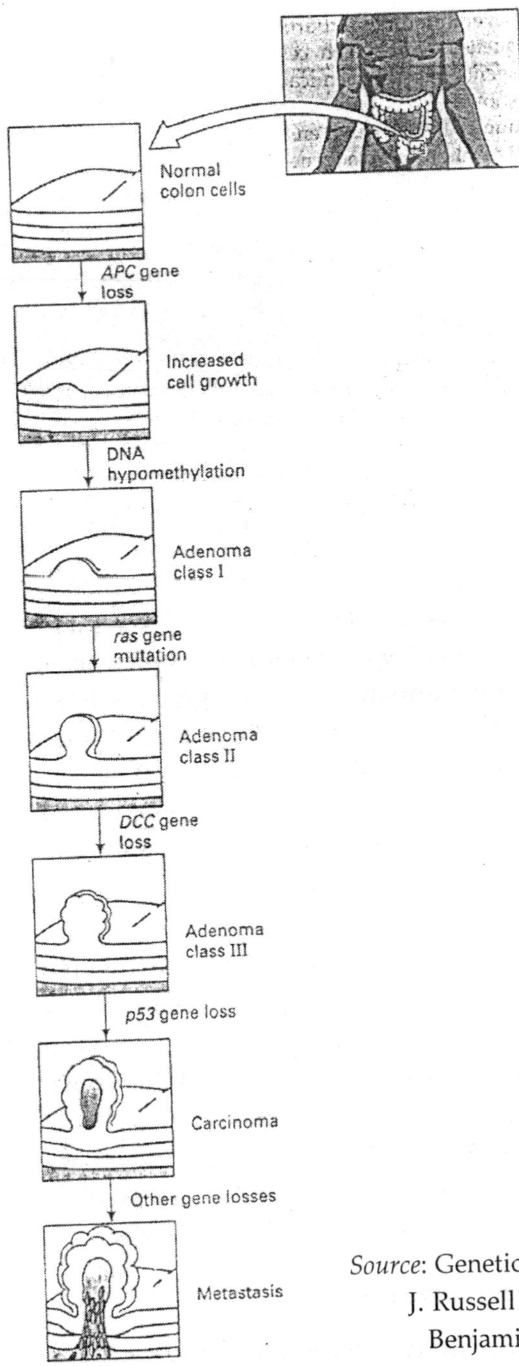


*Source*: Genetics Vol. I and II by Peter J. Russell page no. 585, Publ. by Benjamin/Cummings Pub. Co.

**Fig. 4. Bent vigelstein's Multistep molecular event model for the development of a colorectal cancer Hareditary adnomatous polyposis (FAP). APC, DCC, ras, p. 53, Tomour suppressor genes.**

of all HNPCC cases. While the other two account for 5 percent of cases each. DNA sequencing analysis revealed that all four genes have homologous counterparts in E.coli and yeast, which are known to be involved in DNA repair in them. $hMSH_2$ is homologous to E.coli mut S and the other two genes show homologies to E coli mut.L. The role of these E. coli genes in mismatch repair mechanism during DNA replication has been well documented. The yeast genes also have similar functions. Thus $hMSH_2$, hMLH1, hPMS2, mut.S and mut.L of E.coli and yeast genes are called mutator genes. Mutations of these genes contribute to the development of cancer because the normal maintenance of the integrity of the genome through accurate and perfect DNA replication and efficient DNA repair is adversely affected. It makes the cells error prone and their increased mutation rates result in cancer development.

## Immunochemistry of Cancer

The immunology of cancer/tumour, deals in different dimensions, the various aspects of cancer/tumour growth. The diagnostic characteristics and identification of various forms of cancers and their etiological factors have been thoroughly illustrated in the field of oncology. Irrespective of the various exogenous causative factors, the genetic mechanism of carcinogenesis has disclosed the role of oncogenes in conferring the neoplastic phenotypic traits of transformed cancer cells. viz., uncontrolled cell division and autonomous growth, heritability and transplantability, altered cell surface, invasiveness, metastasis, dedifferentiation and loss of specialized function etc. In recent years the development and growth of immunological field has unraveled the intricacies and mysteries of cancer/tumour.

The informations have been gained by both in vivo studies and in vitro investigations using cell lines. Recent studies have given hope that complete cure to cancer and development of cancer vaccines are within the purview of the immunological domain. In this context our understanding of its various intricacies and literature gathered on it remind us the words of Paul Ehrlich (1903) viz., We will .... "No longer find ourselves lost on a boundless sea .... We have already caught a distinct glimpse of the land which we hope nay, which we expect, will yield rich treasures for biology and therapeutics".

## Growth Promoters in Tumours

Genetic mechanisms are responsible for the transformation of normal cells into malignant cells. But the growth of a malignant tumour involves several growth factors which besides augmenting growth also promote tumour progression and metastasis. Various growth promoters are epidermal growth factors (EGF), platelet derived growth factor (PDGF), fibroblast growth factor (FGF), transforming growth factors (TGF-$\alpha$ and TGF-$\beta$) and insulin like growth factor (IGF). Mesothelioma cells express TGF and PDGF.

## Cytotoxicity and Tumour Cell Death

Cytotoxicity is a process by which certain immune cells destroy the microbial pathogens like bacteria, viruses as well as the virus transformed tumour cells and malignant cancer cells. Virtually any cell in the immune system could conceivably play a role in eliminating tumour cells. Specific types of immune cells that mediate tumour regression could be separated, cultured, cloned and employed in cancer treatment. This method is known as adoptive transfer cellular therapy (Fortner and Rhoads, 1986). The cytotoxic killing or cell mediated lysis occurs in three discrete stages viz., 1. Recognition and binding, 2. Programming for lysis and 3) Dissolution. The last stage is characterized by membrane damage and release of intracellular materials. The cytotoxic lymphocytes, macrophages, natural killer cells, have specific surface receptors which bind with the glycoprotein target structures over the tumour cells and effect the programming and lysis (Potten, 1987).

The binding of cytotoxic cells is $Mg^{2+}$ dependent while programming for lysis and death is $Ca^{2+}$ dependent. CTLs effect the above cytotoxic lysis without antibodies, whereas NK cells and macrophages and still others are antibody dependent for their cytotoxic killing.

## List of Cytotoxic Cells

The development and propagation of immune responses to pathogens foreign antigens and cancer cells is dependent upon the interaction of a diverse groups of cells. These cells may represent the probable candidates for adoptive cellular therapy. In vitro oncological studies have employed so far the following immune cells: viz., B cells/B-lymphocytes; T cells/T lymphocytes/cytotoxic T cells (CTLs), monocytes, macrophages (the activated) NK cells, K cells, natural cytotoxic cells (NCC), lymphokine activated killer cells (LAK cells), PMN (Polymorphonuclear cells – neutrophils, basophils and eosinophils), tumour infiltrating lymphocytes (TIL, $T_8^+$ CTL cells) and Langerhans cells.

## Biochemical Changes in Cytotoxic Killing

In all cell mediated lysis of pathogens or cancer/tumour cells, characteristic biochemical cascade events occur. They are enhanced glycolysis, increased $O_2$ consumption (respiratory burst), increased $H_2O_2$ generation, enhanced hexose monophosphate shunt (HMP), increased turnover of cholesterol and phospholipids, transient increase in some lysosomal enzymes, while decreased acid phosphatase, cathepsin–D and β-glucuronidase activity; collagenase production, membrane adenylate cyclase enhancement, glucosamine incorporation, LDH increase etc. The above biochemical changes in cytoxicity may ultimately enhance the phagocytosis and tumouricidal activity.

**Lysosomal Enzymes in Cytoxicity**

Lysosomes of monocytes and macrophages as well other cells have been found to show 'shoot up' in some selective enzymes such as acid hydrolases, neutral proteases (plasminogen activator, elastase, collagenase), β-hyaluronidase, acid phosphatase (isozymes), lipase, myeloperoxidase (in monocytes only), β-galactosidase, aryl sulphatase, esterase etc.

**Antitumour Factors/Growth Inhibitors:**

While transformed cancer cells exhibit a battery of growth promoting factors, for progression and metastasis, recent investigations have identified antitumour factors, which arrest the above antagonistically. Some of them are interferons (α,β,γ), prostaglandin-E, cytokines, tumour necrosis factors (TNF), cartilage antiinvasive factors, antiendothelial cell growth factors, proteinase inhibitors; antitumour angiogenic factor etc.

**Lymphokines/Lymphotoxins**

All lymphoid cells activated either by antigens or mitogens secrete lymphotoxins/lymphokines/interleukins and tumour necrosis factors. About seventeen types of lymphokines have been recognized so far. Some of their antitumour responses have been documented well.

**Cytokines and Lymphokines**

Three major cytokines which have been subjected to rDNA engineering and clinical trial are Interferons of various classes; Tumour necrosis factor (TMF) and Interleukin – (IL-2).

**Interferons**

There are three major classes of interferons viz alpha, beta and gamma (Papadimitriou, 1985). There are two types of cellular receptors for interferon viz one for alpha and beta and one for gamma. There are atleast fifteen subclasses of alpha interferon. Interferon was first described for its main antiviral properties but several functions could be attributed for it. It has now been shown that interferons inhibits cell growth; alters the state of cell differentiation and phenotype; interferon with oncogene expression; interferon can augment the activity of cytotoxic T-cells (CTL) macrophages, NK, K etc; It will either enhance or suppress the B cell responses; It can promote the differentiation of monocytes into cytotoxic macrophages; Interferon can induce the expression of certain cell surface antigens (eg. MHC-antigens) on tumour cells to make them more susceptible to immune recognition; At lower doses it shows immuno augmenting effects while at higher doses it shows direct antiproliferative effects on tumour cells; The major effect of IFNs on the cell surface is the enhancement of MHC antigens which is a vital controlling mechanism of many immune responses; Gamma interferon is an effective treatment in patients with steroid resistant-rheumatoid arthritis; Alpha interferon is active in the management of laryngeal papillomatosis, condyloma accuminatum, hepatitis-B, non A and non B hepatitis and other chronic viral illnesses; IFN binding cell surface receptor elicites a cascade effect, culminating in enhanced synthesis or inhibition of number of cellular proteins and unusual enzymic activities; Human NK cells activity is stimulated by IFNs both in vivo and in vitro; Pour cell types with cytotoxic potential are Macrophages PMN cells, T-lymphocytes and NK cells. IFN can enhance or induce such activity; It has efficacy in veneral warts and can prevent the development of carcinoma of the cervix; IFN increase macrophage phagocytic and cytotoxic capacity; Interferon has been shown to be an active against a number of tumour types viz., hairy cell leukemia, cutaneous T cell lymphoma, ovarian cancer, low grade malignant lymphoma, chronic myelogenous leukemia, melanoma, myeloma renal cell cancer, breast-lung – colon cancers etc. In leukemia patients alpha interferon enhances the hemoglobin, leucocytes count, platelet count, and monocyte count within first few weeks. The disapprearance of abnormal cells indirectly indicates its cytogenetic influence on the oncogenes.

## Tumour Necrosis Factor (TNF)

TNF is a necrosis factor with 157 amino acids residue. The TNF polypeptide is elaborated primarily by the monocytes in response to certain stimuli such as lipopolysaccharide or immune adjuvants (such as BCG). It can directly cause necrosis of animal tumour when injected directly. A similar cytotoxic proteins viz., TNF – beta is produced by the lymphocytes. The genes for both TNF proteins are located together on human chromosome 6 very near the human MH complex. TNF will be found to have some useful antitumour effects when combined with another biological agent, chemotherapy agents or radiation therapy.

## Interleukin-2

IL-2 is a 15000 dalton protein that interacts with specific receptors on a variety of haematopoietic cells. It was originally described as T-cells growth factor, because of its capacity to induce the proliferation of T cells in cultures of human bone marrow. IL-2 receptors are expressed in monocytes and B cells. IL-2 can have trophic effects on a variety of cell types including natural killer cells (NK). IL2 and its tumour responses have been demonstrated in isolation at maximum dose as well as with its combination. But IL2 does not have the capacity to exert direct antitumour effects. Hence Tumour therapy involves employing IL2 to stimulate the immune system to mount an attack on the tumour. IL2 can cause the proliferation and activation of cytotoxic T-lymphocytes, helper T cells (Th) and natural killer cells (NK). IL2 can induce the lymphokine activated killer (LAK) cell phenomenon and a number of lymphokines (e.g., $\gamma$-interferon). Bimodal dose response curve reflects that lower doses of IL2 stimulate T cells maximally, while higher doses are required to activate NK cells and LAK phenomenon.

## IFNs and Inhibition of Cell Growth

IFN induced growth inhibition has been documented in tumour cell lines such as osteosarcoma lymphoblasoid, breast carcinoma and melanoma cell lines. But tumour cells vary in their response to IFN. For e.g., lymphoblastoid cells are more sensitive to growth inhibition by a interferons, and osteosarcoma cells to b interferons.

***Clinical studies on tumour cells using IFN***: Clinical studies on tumour cells using IFN have been carried out in Osteosarcoma, Acute lymphoblastic leukemia ALL, Acute Myeloid leukemia AML, Chronic Lymphocytic leukemia CLL, Chronic Myeloid Leukemia, Lymphoma (Burkitt's), Myeloma, Melanoma, Renal carcinoma, Lung cancer (Small cell and non small carcinoma), Breast cancer, Kaposi's sarcoma, Carcinoid tumours, Gastrointerstinal cancer, Bladder Papillomas, Carcinoma of the cervix, Glioma, Renal, ovary, prostate cancer.

## Lectins and Cytoxicity

Haemagglutinins or lectins of invertebrates have been found to demonstrate the characteristics such as direct cytoxicity and cell mediated lysis by granular haemocytes, opsonisation and clearance of foreign materials from circulation similar to vertebrate immunoglobulins. Sialic acid binding lectins have been identified in various species of crustaceans and land pulmonates. They have been identified in their tissues like hepatopancreas and haemolymph. These lectins have also been isolated from the haemocytes. The molecular weight varies in different lectins derived from different tissues. The binding of metal ions like $Ca^{2+}$, $Mg^{2+}$ and $Mn^{2+}$ produces conformational changes and activates the lectins. The activated lectins bind with the cytotoxic cells such as lymphocytes and induce their multiplication (mitogenecity) (Chitra and Chowdhury, 1991, Indra *et al.*, 2000). These findings construe that the identification of pathogen specific and tumour specific lectins and their synthesis through cDNA employing such technique as PCR would have far reaching implications in immunodiagnosis and conjugate-immunotherapy.

## Signal Transduction

The discovery of a multitude of antitumour or tumouricidal factors, and their properties against isolated tumour cell lines in vitro, though hint at the feasibility of tackling the cause of cancer deaths, whatever discovered so far seems to be only the tip of an iceberg. A lot remains to be understood still, before achieving complete success. The tumour cells unlike the adult differentiated normal cells, reverse and recapitulate their stance to undifferentiated embryonic condition and evade all the hosts immune cascade of defense, by making molecular mimicry or molecular piracy. This could be understood by the process of signal transduction in normal versus cancer cells. Signal transduction is an inductive interaction mechanism in which a molecular signal is perceived and acted upon by a responding cell. The molecular signals act as a ligand for the transmembrane receptor of the responding cells. The extra cellular ligand binding domain causes a conformational change in the cytoplasmic domain and stimulates the activity of an enzyme cascade viz. protein kinase (PK). Several biomolecules act as signal transducers. The above process of signal transduction involves hormones, growth factors, second messengers, phospholipids, protein kinase cascade etc. They bring about the internalization of information from plasma membrane to nucleus and switches on the transcription process, which bring about normal growth and differentiation. The same signal tranduction operated by the oncogenes disrupts the normal growth and brings about the uncontrolled growth of the cancer cells (Fig 5).

## Lectins and Cancer

Lectins have been widely used to map cell surface carbohydrates and to observe changes occurring during malignant cells transformation and development. Lectins are also capable of immunomodulation. Human lymphocytes and monocytes have been demonstrated to undergo mitogenesis and cloning under the influence of both lectins of plant origin and animal origin.

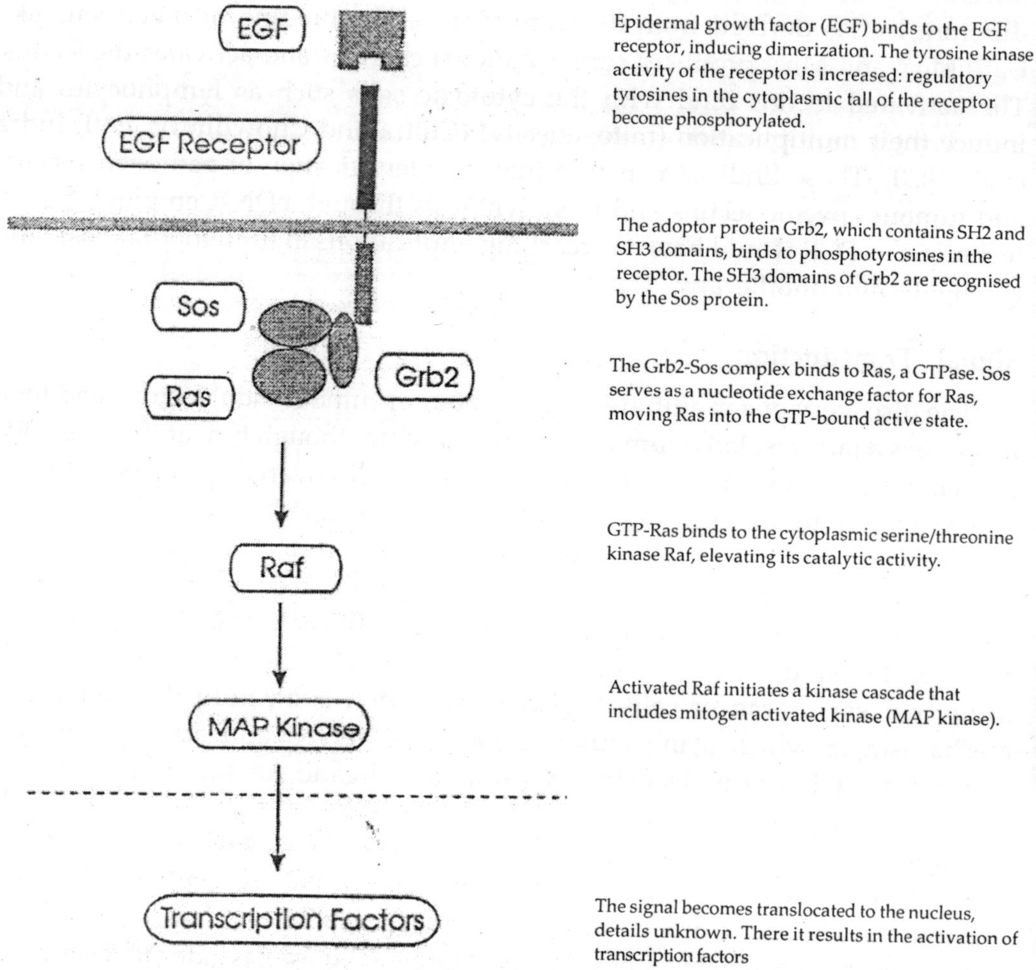


*Source*: Genetics Vol. I and II by Peter J. Russell page no. 585, Publ. by Benjamin/Cummings Pub. Co.

**Fig. 5. Multistep signal transduction pathway for cancer.**

Recent studies have revealed snail lectin (Achatinin$_H$) could arrest the $G_2$/Mphase in breast cancer cell lines MCF7 in vitro. The mechanism was attributed to the repression of $P_{53}$ tumour suppressor gene function and the consequent cell death by apoptosis (Indra *et al* 2006).

## Cancer Therapy – Future Prospects

Systematic pharmacotherapeutic drugs, are in general are not selective in action with respect to the target organs or cells. The chemical properties and the bio-distribution features of these drugs would have toxic side effects. Hence chemotherapeutic drugs used in the treatment of cancer may interfere with metabolic processes of most of cells and also on the proliferating cells.

In the case of metastatic tumours, the tumour cells migrated to other areas may have been targeted by such systemic drugs.

In cancer therapy, to avoid the side effects due to drugs, the concept of coupling a therapeutic agent to a carrier molecule with specificity for a defined target cell population has been advocated as an attractive concept.

Antibodies with affinity for antigens have been considered as a natural choice as targeting moeties for drugs.

An unconjugated antibody will show the following therapeutic effects.

1. Unconjugated antibodies can activate or mediate biological effector mechanisms.
2. They are effective agents in neutralizing toxins or infectious agents.

In cancer treatment, serotherapy involving antisera or purified imunoglobulin fractions as anti-cancer agents are of limited value only.

Immuno-conjugates ie., the toxic substances targeted by means of monoclonal antibodies fall into four major categories viz., 1. Toxins, 2. Low molecular weight drugs, 3. Biological response modifiers (BRMs) and 4, Radionucleides.

### 1. Toxins (Immuno)

Toxins are proteins of microbial and plant origin such as diptheria toxin, ricin, gelonin and poke weed anti-viral protein (PAP). Immuno-conjugates consisting of monoclonal antibodies and these protein toxins or their active fragments are termed Immunotoxins. The protein toxins (enzymes) once reach the cytosol of the cell, a single molecule may be sufficient to cause cell death.

### 2. Chemo Immunoconjugates

The chemoimmunoconjugates consist of antibodies and low molecular weight drugs such as methotrexate and adriamycin. Several drug molecules can combine with a single antibody molecule.

### 3. Biological Response Modifiers – Immunoconjugates (BRMT):

These are also called Immunomodulators, which are naturally occurring group of molecules or their synthetic analogs, which are bound to have regulatory functions. Lymphokines, growth factors and complement activating cobra venom factor, by binding to defined receptor molecules may exert antitumour activities by activating or directing biological effector systems.

### 4. Radioimmunoconjugates

The fourth group of immunoconjugates consists of a radio-nucleide coupled to an antibody. These radio immunoconjugates depending upon the radioactive isotope may become cytotoxic causing cell death through their radiation.

## Conclusion

Information derived from cancer research have so far revealed that it is a multifactorial (etiological) and polygenic disease (Smyth *et al*, 1996). The research findings in immunology so far reflect a greenish picture by tackling several of the greyish areas in oncology. Future immunobiological and cytotoxicological/ cytotoxinological investigations would certainly spot out the culprit genes and keep them at a fix and bay circumventing all the road blocks in the highway research of oncology.

## Reference

Brachet, J and Mirsky, A.E. (1961). The cell – Biochemistry, Physiology and Morphology, Academic Press.

Chitra, M. and Chowdhury, M. (1991). The polyclonal actination of lymphocytes and T-cell mitogenicity by a unique sialic acid binding lectin from the haemolymph of *Achatina fulica* snail. Immunopharmacology. 114: 63-72.

Currie, G and Currie, A. (1982). Cancer – the biology of malignant disease, Edward Arnold, Times of Castlefield Press.

Fortner, J.G. and Rhoads, J.E. (1986). Accomplishments in Cancer Research 1986, J.B. Lippincott Company, Philadelphia.

Frederick, A., Murphy, E., Gobbs, P.J., Horzinek, M.C. and Studdert, M.J. (1999). Veterinary Virology, Academic Press.

Grant, P. (1978). Biology of Developing Systems, Holt, Rinehart and Winston.

Indra, D., Ganesh, S., Ramalingam, K., Asokan, C. and Jayakumar, R (2000). Immunological significance of metal induced conformational changes in the mitogenic Achatinin$_H$ binding to carbohydrate ligands. Comp. Biochem. Physiol. (part C), 127: 177-183.

Indra, D., Nalini, R., Ramalingam, K. and Mary, B. (2006). Evaluation of anticancer properties of Achatinin$_H$ against mcf – 7 cancer cell line (communicated).

Mareel, M.M. and Calman, K.C. (1984). Invasion – Experimental and Clinical Implications, Oxford University Press.

Papadimitriou, J.T. (1985). Interferons – Their impact in Biology and Medicine, Oxford University Press.

Potten, C.S. (1987). Perspectives on Mammalian Cell Death, Oxford University Press.

Russell, P.J. (1998). Genetics, Addison, Wesley Longman, Inc.,.

Smyth, J. F., Boogaerts, M.A. and Ehmer, B.R.M. (1996). rhErythropoietin in Cancer supportive treatment, Marcel Dekker, Inc..

**Suggested**

Ahuja, M.R. (1997). Cancer – causes and prevention, UBS Published Distributors Ltd.

Akhtar, S.S. (1991). Current oncology, Matkkoff Press.

Ranga. M.M. (2003). Animal Biotechnology, Agrobios (India).

Wright, N. and Alison, M. (1984). The biology of epithelial cell populations, Oxford University Press.

# 14

# Pesticide, Hormone and Antioxidant Associated Changes in Metabolism of Fish

*G. Tripathi, J. Shasmal, H. Singh and N. Bandooni*

Toxicity of pesticides to fish has been well documented (Arora *et al.*, 1971 a,b; Verma *et al.*, 1975, 1977, 1978a,b, 1984; Ghosh and Chatterjee, 1987; Arunachalam and Palnichamy, 1980; Choudhary *et al.*,1984; Singh and Sahai, 1984, 1985; Jagdish and Sahai, 1987; Banerjee *et al.*, 1987; Hari Lal, 1989). Gupta (1989) examined the toxicity of BHC, endosulfan, malathion and carbaryl to *H. fossilis*. Mishra *et al.* (1991) studied the acute toxicity of carbaryl to a freshwater teleost *Channa punctatus*. Pesticides enter the aquatic ecosystem and adversely affect the non-target organisms such as fishes (David *et al.*, 2003). The pesticides exhibit the phenomena of biomagnification due to bioaccumulation. Biomagnification increases as the tropic level increases. Fish show high biomagnification of organochlorine, phosphate or carbamate pesticides in their vital organs. Organophosphates inhibit acetycholineesterase (AChE) activity in spleen, liver and brain of certain fishes (Hande). Insecticides have been known to induce changes in fish behaviour. Bashamohideen *et al.* (1987) and Venugopal *et al.* (1988) studied behaviour in a variety of fishes. They reported restlessness, frequent surfacing, loss of balance and irregular opercular movements in fishes on methyl parathion, malathion and monocrotophos exposure.

Effluents from pesticide industries have resulted a marked increase in the incidences of mass mortality, growth and retardation and tissue damages in fish (Konar, 1981; Dikshit *et al.*, 1990; Kumarguru, 1995; Sarkar, 1997; Lal and Pandey, 1999; Pandey *et al.*, 2000). Synthetic pesticides are commonly used in developing countries. Many pesticides are persistent and non-biodegradable in our environment. Sastry *et al.* (1982) exposed the snake headed fish *Channa punctatus* to quinolphos for 15 and 30 days and found that blood glucose, lactic acid and hemoglobin levels decreased in quinolphos exposed fish. The glycogen content of liver and muscle increased but lactic acid decreased. Hexokinase was inhibited in intestine and muscle after 30 days of exposure but increased in enzyme activity was noted in gills. However, LDH activity was inhibited in liver, kidney, intestine, brain, gill and muscle. Pyruvate dehydrogenase (PDH) activity of liver, kidney, gill and muscle was inhibited. In contrast, the PDH activity in brain was elevated. Succinate dehydrogenase (SDH) activity was elevated in intestine but inhibited in other tissues.

Yadav *et al.* (2007) studied the impact of fertilizer industry effluent on the level of protein and the activity of LDH. The fish exposed to different sublethal concentrations of fertilizer industry effluent exhibited decrease in the level of protein (8-76%) in different organs. At highest effluent concentration (7%) treatment for short (96 h) or long (15 days) duration, the liver of the fish registered a remarkable decrease (62-76%) in protein content as compared to control, whereas other organs of the fish showed only 38-52% decrease in protein. The treatment of fish with 7% effluent for 96 h caused 78% decrease in LDH activity in fish muscle, whereas after 15 days the effect was maximum in fish brain as it exhibited 86% decrease ($P<0.001$) in LDH activity as compared to control. The effect of effluent on the activity of LDH and protein contents in different tissues of the fish was dependent on concentration and duration of exposure. The significant reductions in the activity of LDH and level of protein in fish tissues due to treatment with the fertilizer industry effluent indicated the possibility of impairments in energy metabolism and protein turnover, respectively. Pandey *et al.* (2006) evaluated the effect of a single exposure of endosulfan (5 ppb) on catalase (CAT), glutathione peroxidase (GPx), glutathione S-transferase (GST) and reduced glutathione (GSH) of liver, kidney and gill of freshwater fish *C. punctatus* (bloch) after 24 h of treatment. Endosulfan showed induction in GPx and GST activity, and GSH level in tissues. However, CAT activity was found to be significantly decreased.

Sastry and Siddiqui (1984) examined the levels of hemoglobin, total plasma protein, glucose and lactic acid in the blood; glycogen and lactic acid content of liver and white skeletal muscle; and the activation of LDH, PDH and SDH in liver, kidney, intestine, brain, gill and muscle in *C. punctatus*, after exposure to sublethal concentration of quinolphos for 60 and 120 days. Hemoglobin, plasma protein, glucose and lactic acid decreased in pesticide – exposed fish. The glycogen content of liver and muscle increased but lactic acid decreased. LDH activity decreased in different tissues. PDH activity of liver, kidney, gill and muscle decreased but it was elevated in intestine and brain. SDH activity was elevated in intestine, but it was reduced in other tissues. These findings showed that aerobic oxidation of nutrients was adversely affected in quinolphos exposed fish. Ahmad *et al.* (2000) studied the activities of GPx, CAT, GST, glutathione and metallothionein in various tissues of freshwater fish *C. punctatus* in response to short term and long term exposure to paper mills effluent. Metallothionein was induced in liver after 60 days of exposure. CAT activity increased after 60 days of effluent treatment. Antioxidant pattern was different in gill and kidney showing that liver was more resistant to oxidative damage as compared to gill and kidney. Tham *et al.* (2009) reported inhibition in AChE of *C. batrachus.* exposed to insecticides. Screening of insecticide and heavy metal showed that carbofuran and carbaryl strongly inhibited *C. batrachus* AChE. It was suggested that *C. batrachus* has a potential to be used as a cheaper and readily source of AChE than other more commercially available sources.

## Impact of Organophosphate Pesticide

Tripathi and Shukla (1990) found that the activities of cMDH, mMDH and LDH reduced in the liver and skeletal muscle following methyl parathion exposure to the freshwater catfish *Clarias batrachus* after 7 days. The activities then began to recover and reached the control levels on the 28$^{th}$ day of exposure. A complete recovery occurred on the 7$^{th}$ day when the pesticide was withdrawn from the medium after an exposure for one week. The treatment of organophosphate to fishes inhibits the activities of several enzymes such as glucose-6-phosphatase, acid phosphatase, pyruvate dehydrogenase, succinic dehydrogenase and cholinesterase in brain and acetyl cholinesterase as well as cytochrome oxidase in the brain, kidney, gill, liver and muscle (Sastry and Sharma, 1980; Natarajan, 1984). It has been shown that the activity of alkaline phosphatase, ATPase and LDH remain unaltered in brain of *Ophiocepalus punctatus*, after 15 days of exposure of diazinon (Sastry and Sharma, 1980). The exposure of methyl parathion to the freshwater fish *Tilapia mossambica* for 48 h decreases the activity of SDH and LDH in gill, liver and muscle tissues (Rao and Rao, 1979). A decrease in LDH activity has also been reported in the liver of *Cyprinus carpio* exposed to organophosphorous pesticide carbetox. Malathion significantly increased the liver cholesterol level at 0.05 ppm and decreased the protein and glycogen content in *Glossogobius giuris* exposed to 0.05 to 0.5 ppm pesticide (Anandhi and Murty, 1996).

Shah and Dubale (1983) found a decline in liver protein in *Channa punctatus* exposed to malathion. *Mystus vittatus* exposed to sublethal levels of sumidon showed a significant decrease in glycogen, glucose content and the respiratory dehydrogenase in liver muscle and gill. The inhibition was concentration dependent and was relatively more in liver followed by muscle and gill (Neethirajan and Mathavan, 2004). A significant decrease in AChE activity in the brain of freshwater fish *Catla catla* has been in response to dimethoate toxicity (Kumar and Singh, 2000). Cappage *et al.* (1975) reported malathion associated inhibition in brain AChE in *Lagodon rhomboidis*. Sahib *et al.* (1980) documented impact of malathion on AChE in *Tilapia mossambica*. They found 50.7% inhibition of AChE activity in brain after 36 h. Concomitant with the observed AChE activity, the AChE content correspondingly increased. Shastri and Sharma (1981) studied the effect of sublethal concentration of diazinon on the enzyme of *Ophiocephalus punctatus*. AChE activity was inhibited from 29 to 42% in the brain of fish exposed to diazinon.

Rao and Rao (1982) studied the effect of sublethal concentration of methyl parathion on *Tilapia mossambica* and found 62% inhibition in AChE activity in brain tissue. They also observed corresponding increase in AChE content and suggested disruption in the nerve impulse conduction. Rao *et al.* (1983) reported 61.56% inhibition in AChE activity after 48 h exposure to methyl parathion in the same fish. Devraj *et al.* (1991) observed 91% inhibition of AChE in brain of *Oreochromis mossambicus* exposed to phosalane at the end of 96 h. Satyadevan *et al.* (1993) reported 36.6% inhibition of AChE activity in brain tissue of *Cyprinus carpio* exposed to sublethal concentration of dimethoate. Kumble *et al.* (1999) studied the effects of chlorpyriphos on alkaline and acid phosphatases in intestine and liver of a freshwater fish *Sarotherodon mossambicus* after acute exposure. Both sublethal (0.1 ppm) and lethal (0.14 ppm) concentrations the pesticide significantly decreased the total alkaline phosphatase activity in intestine. However, there was a significant increase in the activity of liver. Total acid phosphatase activity in intestine was increased in liver.

Sensitivities of AChE to various organophosphorous compounds have been studied by Benke *et al.* (1974), Johnson and Wallace (1987), Chattopadhyay *et al.* (1986), Wang and Murphy (1982), The inhibitory effect of parathion and paraoxon were observed in fish and other animals (Siddiqui *et al.*, 1988; Nemesok *et al.*, 1987; Jarvinen *et al.*, 1983). Methyl parathion caused 58.6 to 83.7% inhibition in AChE activity within 10 h of exposure (Gibson and Ludke, 1973). The maximum level of inhibition of AChE was reached within 2 to 4 h with methyl parathion and after 20 h with ethyl parathion (Benke *et al.*, 1974) of brook and rainbow trout. Whereas malathion decreased AChE level by 24-28% over the control. Brain AChE activity was significantly inhibited when fathead minnow *Pimephales promelas* was exposed to chlorpyriphos (Olsen and Christensen, 1980). Malathion declined glycogen and cholesterol content in the cardiac muscles of the gobiid fish, *Glossogobius giuris*. Whereas a slight fluctuation in protein and glycogen content was observed at low concentration (0.05 ppm) of malathion. The levels of protein showed a significant decrease at high concentration (0.5 ppm) when treated for a longer duration (96 h) (Venkataramana *et al.*, 2006). Quinolphos depleted the carbohydrate, protein and lipid contents in tissues of *C. mrigala* and the depletion was concentration-dependent. (Amali *et al.*, 1996). Ramakrishna and Sivakumar (1993) reported a time-dependent depletion in glycogen content of liver and muscle in *Oreochromis mossambicus* under quinolphos toxicity. Tilak *et al.* (2005) showed decrease in total glycogen and protein levels and increase in activities of selected enzymes such as lactate dehydrogenase, alanine aminotransferase, aspartate aminotransferase, DNA and RNA in tissues of *Catla catla, Labeo rohita* and *Cirrhnius mrigala* exposed to chloropyriphos. Long term exposure to quinolphos decreased the glucose content in *Channa punctatus* (Sastry and Siddiqui, 1984). They also reported decrease in protein content in liver, skeletal muscle, kidney, intestine, brain and gill. Methyl parathion and monocrotophos declined glycogen level in the same fish (Samuel and Sastry, 1989).

Khare *et al.* (2000) showed that the sublethal concentration of malathion significantly increased the protein content in kidney of *C. batrachus* during first week of exposure. Thereafter a gradual decrease in protein content was observed in the later periods. The LDH activity increased in gill, brain, skeletal muscle and liver of *C. punctatus* exposed to sublethal concentration of metasystox (Natarajan, 1984) and phosphamidon in *Clarias batrachus* (Ghosh, 1987). The activities of aspartate aminotransferase and alanine aminotransferase increased in liver, muscle and blood in acute and chronic toxicity of monocrotophos in *C. punctatus* (Sastry and Siddiqi, 1984). Malathion increased ATP and ADP hydrolysis in competitive, and an uncompetitive manner respectively in brain of Zebra fish (*Danio rerio*), but did not alter AMP hydrolysis (Senger *et al.*, 2005). Thangavel *et al.* (2005) showed reduced prolactin level and loss of selected mineral contents (Ca, P, Mg) in freshwater fish *Orechromis mossambicus* exposed to organophosphate pesticide dimecron.

Samuel and Shastry (1989) demonstrated alternations in the level of some biochemical and enzymological parameters in the blood, liver and muscle of *C. punctatus* exposed to monocrotophos for 15, 30, 60 and 120 days. Fish exposed for 96 hrs. and 15 days were hypoglycemic and hypolactemic, and a decrease was noted in the pyruvic acid level of blood and the activity of hexokinase in liver. In muscle, a decrease in the glycogen content and an increase in the lactic acid level indicated hyperactivity. After 30 days of exposure hyperglycemia, hyperlactemia, a decrease in liver glycogen was noticed. An increase in the muscle glycogen and in the activity of hexokinase and a decrease in the activities of malate dehydrogenase and succinate dehydrogenase indicated that anaerobic metabolism prevailed over aerobic metabolism. After 60 and 120 days of exposure, both aerobic and anaerobic pathways were impaired as evident by hyperglycemia, hypolactemia, decrease in pyruvic acid level of blood, increase in glycogen content of both liver and muscle and decrease in the activities of glucose 6-phosphatase, hexokinase, SDH and MDH. The activities of glutamate oxalacetate transaminase and glutamate pyruvate transaminase increased in the liver, skeletal muscle and blood in acute and chronic exposure.

Awasthi *et al.* (1984) exposed two common fish of commercial importance viz., *C. punctatus* and *H. fossilis* to sublethal concentrations of malathion and dimethoate. The exposure to malathion showed a gradual decrease in glycogen content in liver and kidney of *C. punctatus*, but thereafter a gradual increase was observed. However, in the exposure of *H. fossilis* to dimethoate, there was an increase in the glycogen content both in liver and kidney during the 2$^{nd}$ week. Thereafter a decrease was found. The specific activity of SDH showed a sharp rise initially for 2 weeks followed by steep decline in both the tissues. Rao (2006) described the effect of a novel organophosphate insecticide 2-butenoic acid-3-(diethoxyphosphino thioyl) methyl ester (RPR-II) on *Oreochromis mossambicus*. The fish was exposed for 3,7,15 and 30 days and its recovery response was observed after 7 days. AChE activity of brain, gill and muscle was inhibited by 67%, 77% and 73% respectively in 30 days. The plasma and kidney alanine aminotransferase and aspartate aminotransferase activity increased, while decreases was observed in gill and liver. Increase in acid phosphatase and alkaline phosphatase activities were observed in plasma, gill and kidney, and reduction of 20% and 61% in liver acid phosphatase and alkaline phosphatase respectively. Depletion of glycogen was observed in all tissues which indicates a typical stress related response of the fish to the pesticide. LDH activity decreased in liver and muscle, indicating tissue damage but a significant decrease in LDH activity in gill and brain was observed. Depletion of glutathione (GSH) was observed in all tissues, thereby enhancing lipid peroxidation resulting in cell damage. The induction in hepatic glutathione-s-transferase (GST) levels indicated protection against the toxicity of xenobiotic-induced lipid peroxidation. There was a significant recovery in the above biochemical parameters, in all tissues of fish after a recovery period of 7 days. These results revealed that the organophosphate insecticide RPR-II is highly toxic and affects the intermediary metabolism of *O. mossambicus*.

Durmaz *et al.* (2006) studied oxidative stress and neurotoxic potential of diazinon in the sentinel freshwater fish *Oreochromis niloticus*. Antioxidant and AChE enzyme activities and malondialdehyde (MDA) and protein levels were measured in gill, kidney, alimentary canal and muscle tissue of fish treated with sublethal diazinon concentration for 1,7,15 and 30 days. Dose-dependent inhibition of AChE was observed in all experimental fish. On the contrary, in alimentary tract the MDA level was elevated in kidney and skeletal muscle, whereas gill was not effected. AChE and MDA levels were intercorrelated in kidney and muscle tissues. Diazinon increased superoxide dimutase (SOD) activities in all tissues. Tissue-specific alternations were observed in catalase (CAT) and glutathione peroxidase (GPx) activities. However, the activities did not change in gill and muscle tissues for GPx and in gill, muscle and kidney tissues for CAT. Protein levels decreased in kidney, muscle and alimentary tract, while increased in gill and alimentary tract in 15 days. The effects of diazinon were tissue-specific and dose-dependent.

Exposure of a freshwater cyprinid (*Puntius conchonius*) to dimethoate for 24 h seriously affected carbohydrate and cholesterol metabolism (Pant and Singh, 1983). Significant hyperglycemia and glycogenolysis in liver, brain and heart of the treated fish was reported. Hypercholesterolemia and rise in liver cholesterol also occurred. Long term exposure (15-30 days) to sublethal concentration of dimethoate elicited hypoglycemia and depletion of liver glycogen, together with enhanced glycogenesis in the heart. Joshi and Desai (1988) determined the effect of exposure of monocrotophos on protein, RNA and DNA content and 5'-nucleotidase activity in the liver of *T. mossambica*. Protein content decreased by 45% after 5 days and returned to control levels at 10-30 days, and again decreased by 45 days. DNA content decreased by 2 days, returned to control by 5 days, and remained constant throughout the exposure. In contrast, RNA content was significantly lower starting from 2 days through 45 days of exposure. 5'-nucleotidase activity showed a transient increase at 5 and 30 days of monocrotophos exposure. These results indicate that its exposure altered the protein, DNA and RNA contents and the 5'-nucleotidase activity within 2-5 days.

Sancho *ét al.* (1997) showed the effects of fenitrothion (0.02 mg/l) on the energy metabolism of the European eel *Anguilla anguilla* and its recovery from intoxification. Various parameters such as glycogen, lactate, protein and glucose level were measured in different tissues of eel after 2, 8, 12, 24, 32, 48, 56, 72 and 96 h of fenitrothion exposure. Subsequently the fish were allowed for recovery period of 8, 12, 24, 48, 72, 96, 144 and 192 h in clean water and some parameters were evaluated. Liver glycogen content showed no significant changes during the exposure, while blood glucose level increased markedly. Gill, liver and lactate increased during pesticide exposure, while protein decreased in comparison to control. Hai *et al.* (1997) studied the effect of dichlorovos on antioxidant enzymes and other oxidative and redox parameters of carp (*Cyprinus carpio*) and catfish

(*Ictalurus nebulosus*). Changes in SDH, CAT, glutathione peroxidase and acetylecholinesterase activities in carp were studied. The result showed that the pesticide besides its inhibitory effect on AChE – or together with it – induced oxidative stress. Begum and Vijayaraghavan (1999) exposed *C. batrachus* to a sublethal concentration of Rogor for a period ranging from 24 to 192 h. A gradual decrease in muscle glycogen and an increase in lactate contents were observed. The activity of LDH showed a sharp rise initially for 48 h followed by a decline after 96 and 192 h.

Sastry and Malik (1982) studied the effect of acute and chronic exposure diazinon against *C. punctatus*. In acute exposure SDH activity was elevated in intestine and pyloric ceca. No alternation was noted in LDH activity but PDH was inhibited in pyloric ceca. Chronic exposure resulted in inhibition of the activities of these dehydrogenases. The studies have also been done about the effects of malathion and BHC exposure on circulating levels of plasma $T_3$ and $T_4$ in *H. fossilis*. A lower plasma $T_4$ level was found within 4 weeks of malathion and BHC exposure. High concentrations of BHC induced an increase in the plasma $T_4$ level. BHC exposed fish inhibited significantly plasma $T_3$ and $T_3/T_4$ ratio, but malathion exposure significantly increased their levels. Probably BHC acted directly on the thyroid gland reducing the synthesis and release of the thyroid hormone. Malathion increases the plasma $T_3$ level either by stimulating the extra thyroidal conversion $T_4$ to $T_3$ or by reducing $T_3$ excretion thereby increasing plasma $T_3$ and $T_3/T_4$ ratio in the fish.

Saxena *et al.* (1989) studied *in vitro* incorporation of $^{14}C$ glucose/acetate into lipids and $^{14}C$ leucine into protein by the liver of *C. punctatus* exposed to safe application rate (SAR) and sublethal (SL) concentrations of malathion for 30 days during the preparatory phase of the reproductive cycle. The decline in $^{4}C$ incorporation was more in fish exposed to sublethal concentration than to SAR concentration of the pesticide. Thangavel *et al.* (2005) investigated the endocrine response in *Sartherodon mossambichus* under dimecron toxicity by estimating the serum levels of $T_3$ and $T_4$. Exposure of fish to sublethal concentration of dimecron caused significant changes in the levels of serum hormones. Based on results, it was concluded that (i) the fish adaptability maintains a probable low metabolic rate, as indicated by the reduced level of thyroid hormone ($T_3$), which could be considered advantageous for the fish to indirectly reduce the toxic impact of the pesticide. (ii) the prolonged exposure of fish (for 5 days) to sublethal dimecron appeared to exhibit a uniform recovery response in different hormonal levels of the fish. Upnishad *et al.* (2008) studied the effects of Keltane, Metacid and Dithane and their combination at 24 h and 48 h exposure on calorific values, protein and water content in the freshwater air breathing fish *Channa punctatus*. The treatment of these pesticides increased water content and decreased protein content and calorific values to a significant level.

Luthra and Gupta (2009) studied the effects of monocrotophos on blood cells of *C. batrachus*. Fish showed reduction in the number of erythrocytes and leucocytes and hemoglobin content. AChE gene expression was significantly elevated (12.4 times) in catfish exposed to 43 mM chlorpyriphos in comparison to control groups ($P<0.05$). This indicated that AChE is very sensitive to administration of chlorpyriphos. The induction of AChE gene reflects the possibility of using this gene as biomarker for detecting defects of organophosphate insecticides in hybrid catfish (Somnuek et al., 2009).

**Impact of Pyrethroid Pesticide**

Fenvalerate intoxication has been shown to affect rate of excretion of cations ($Na^+$) in the urine of rainbow trout (Bradbury *et al.*, 1987). Total structural and soluble proteins showed decrement in gill, brain, liver and muscle of *Cyprinus carpio* exposed to sublethal concentration of cypermethrin (Reddy and Bashamohideen, 1995). Pyrethroid toxicity is highly dependent on stereochemical structure (Milam *et al.*, 2000). Effects of deltamethrin on nervous, respiratory and hematological system in fishes have been reported (Ferreira *et al.*, 1984; Golow and Godzi, 1994). Begum (2005) showed decrease in protein content in liver and gill of the freshwater catfish *C. batrachus*, cypermethrin for 10 days. Recovery in protein content was observed to a large extent in both the tissues when some fishes were released to freshwater after 10 days of exposure. She also observed increase in free amino acids *in vivo* and gill tissues throughout the treatment period and recovery response after cessation of exposure. Ammonia level was decreased in both the tissues throughout the exposure period except in liver tissue at the end of first day of exposure. A decrease in glycogen content of liver was observed during the exposure period. Gill showed decrease in glycogen content at the end of 1$^{st}$ and 5$^{th}$ day. When the fish were transferred to freshwater, recovery in glycogen content was noted. The activities of alanine, aminotransferase, aspartate aminotransaminase, glutamate dehydrognease (GDH) and glycogen phosphorylases increased in both liver and gill after cypermethrin exposure.

Decrease in protein content of fish *Labeo rohita* fingerlings exposed to cypermethrin was reported by Das and Mukherjee (2003). Tripathi *et al.* (2002) reported fenvalerate-induced macromolecular changes in *C. batrachus*. Kumar *et al.* (2007) conducted a study on acute toxicity of cypermethrin and λ-cyhalothrin for 96 h using *C. punctatus*. The behavioural pattern of *C. punctatus* got severely altered in each group due to pesticide treatment. The results suggested that even at low concentrations the pyrethroid compounds may exert toxic effect, markedly modifying their behavioural pattern. Das and Mukherjee (2003) studied the effect of exposure of sublethal concentration of cypermethrin on biochemical parameters of muscle, blood and enzyme activities in brain, liver and kidney of the Indian major carp *Labeo rohita*. RNA level decreased while DNA was elevated. Acid phosphatase was unchanged but alkaline phosphatase was depleted. Brain acetylcholinesterase decreased significantly over a period of 45 days. LDH activity in brain and liver was elevated but it was inhibited in the kidney. SDH and ATPase activities were depleted in brain, kidney and liver. There was a decrease in serum protein level over control. Philips *et al.* (1995) studied carbohydrate metabolism under *in vivo* conditions in functionally different tissues of freshwater fish *Labeo rohita*, exposed to cypermethrin for 4 days. All exposed fish exhibited a hyperglycemic condition. An increase in tissue lactate with a decrease in pyruvate, total carbohydrates and glycogen content were noted. Activity of LDH was elevated, indicating a shift towards anaerobiosis, TCA cycle enzymes, namely SDH and MDH were inhibited. Singh and Srivastava (1999) showed permethrin associated reduction in the activity of LDH and cytochrome oxidase and enhancement in SDH activity in the *Channa striatus*. The results indicated that the permethrin blocks aerobic as well as anaerobic metabolism in the exposed fish.

Sayeed *et al.* (2003) investigated the effect of deltamethrin on *C. punctatus*. A single exposure caused induction of various antioxidant enzymes and non-enzymatic antioxidants in kidney and liver. Certain antioxidants were found to be depleted in gill, enzymic activity was decreased, lipid peroxidation was induced in all tissues, gill showing the highest level. Glutathion, which is an established non-enzymatic antioxidant in fish, was significantly increased. Ascorbic acid content increased in kidney and liver, while it decreased in gills. Deltamethrin has oxidative stress inducing potential in fish and gills are the most sensitive organ. Gills are the primary sites of deltamethrin absorption and their antioxidant potential is also very poor. Total structural and soluble proteins showed decrement; whereas free amino acids and the activities of protease, aspartate aminotransferase and alanine aminotransferase significantly increased in cypermethrin exposed fish. Ammonia content decreased but urea and glutamine increased at all periods of exposure. It was observed that alternations steadily increased with the period of exposure and exhibited tissue specificity.

The effect of sublethal concentration of cypermethrin on protein content of gill, heart and kidney of *C. batrachus* was studied by Tripathi *et al.* (2006). They showed that cypermethrin significantly increased the protein content of these tissues. The maximum increase in protein content was in kidney. Tripathi *et al.* (2007) showed an increase in protein content of brain, liver and skeletal muscle

*C. batrachus* exposed to cypermethrin. Liver and skeletal muscle showed almost similar increases in protein profile. Saravanan *et al.* (2009) studied the effect of lambda cypermethrin on the functioning of endocrine glands in freshwater catfish, *Clarias batrachus*.

**Effect of Thyroid Hormone**

Tripathi and Shukla (1989) reported induction in the activities of cMDH, mMDH and LDH of the liver and the skeletal muscle of a freshwater catfish *C. batrachus*, in response to administration of $T_3$. The activities of all three enzymes in both tissues of fish exposed to thiouracil for 28 days decreased 2 to 3 times as compared to normal individuals. A single injection of $T_3$ to thiouracil exposed fish substantially increased the activities of these enzymes. Administration of thyroid hormone increases the activity of MDH in the liver of the teleost (Le Ray *et al.* 1970; Bonnet, 1970). The effects of thyroid hormones on the activities of hepatic cMDH, mMDH and LDH of thiouracil treated fish (*Anabas testudineus*) have been demonstrated (Peter and Oomen, 1989). It showed that the treatment of $T_3$ and $T_4$ reduced the activities of cMDH and mMDH but had no effect on the activity of LDH. Treatment of thyroid hormones increased the activity of cytochrome oxidase and decreased the activity of glucose 6-phosphate dehydrogenase in the liver of *Mugil auratus* (Le Ray *et al.*, 1970; Bonnet, 1970). Treatment of thyroxin to *Ophiocephalus punctatus* for 30 to 50 days increased the activity of hepatic acid and alkaline phosphatases (Ray and Medda, 1976). The effect of thyroxin on the mitochondrial respiration and phosphorylation has been reported in the trout, *Salmo trutta* (Massey and Smith, 1968).

Lynshiang and Gupta (2000) studied *in vivo* and *in vitro* effects of thyroidal hormones (Monoiodotyrosine, diiodotyrosine and triiodothyronine and thyroxine), propyl thiourea (PTU), testosterone and cyproterone. Monoiodotyrosine (MIT) and diiodotyrosine (DIT) increased the respiratory rate in a dose – dependent and temperature independent manner. $T_3$ and $T_4$ stimulated tissue respiration during summer/rainy months but not in winter. Propylthiourea decreased tissue respiration during summer/rainy season and also stimulated at low temperature. Peter and Oommen (1989) reported decreased G6PDH activity in *Anabus testudineus* 24 h after $T_3$ treatment.

Garg (2007) studied effects of oral administration of $T_4$ on growth performance, body composition and some aspects of nutritional physiology in two slow growing air breathing fish (*C. punctatus* and *H. fossilis*) under laboratory conditions. Liver glycogen levels decreased with the increase in the dietary $T_4$ level, indicating its utilization during growth. Muscle glycogen level in *H. fossilis* coincided with high growth at 100 mg of $T_4$. *H. fossilis* and *C. punctatus* feeded on low doses of $T_4$

enhanced growth, decreased excretion of metabolites and increased nitrogen retention. These observations suggested that $T_4$ supplementation of the diet may have practical utility in the culture of slow growing fish species.

For elucidation of the thyroid hormone induced responsiveness of fish brain, various doses $T_3$ were injected to a fish *H. fossilis*, for 3 consecutive days and changes in cytosolic NADP-dependent malic enzyme activity in the whole brain tissue were determined by Srabani De *et al.* (1989). The NAD-dependent cytosolic malate dehydrogenase content of brain cytosol remained unaltered with all doses of $T_3$ used. Varghese *et al.* (2001) showed that *in vivo* injection of 6-propyl thiouracil (6-PTU) did not alter the concentration of malondialdehyde and conjugative dienes in liver. The administration of $T_3$ or diiodothyronine ($T_2$) increased the peroxidation rate in hypothyroid fish (*Anabus testudineus*). However, in normal fish, only a high dose of $T_2$ caused increased malondialdehyde production rather than $T_3$. SOD activity was higher in $T_2$ treated fish in both experiments. Glutathione peroxidase activity was also high in hypothyroid fish treated with $T_2$. In normal specimens, Injection of $T_3$ and $T_2$ had no effect on GPx activity. Glutathione reductase activity was not altered by hypothyroidism, while $T_3$ and $T_2$ increased it.

To assess the role of $T_3$ in mediating short-term changes in metabolism, Ballantyne *et al.* (1992) examined the effects of intraperitoneal injection of $T_3$ on the oxidation of substrates by isolated mitochondria from liver of bowfin (*Amia calva*) and red muscle and liver of lake char, *Salvelinus namaycush*. They found that 3 h after intraperitoneal injection of $T_3$, oxidation of some substrates by mitochondria isolated from the liver of the bow fin was reduced. Similar treatment had no effect on substrate oxidation in liver mitochondria isolated from lake char. Oxidation of substrates by lake char red muscle mitochondria was stimulated by $T_3$ injection. Citrate synthase level was increased in red muscle suggesting that changes in enzyme activity may be in part responsible for short-term mitochondrial responses to $T_3$ injection.

Among the metabolic changes known to be associated with altered thyroid status in fish are those involving lipid metabolism (Plisetskaya *et al.* 1983). Tripathi and Verma (2003) showed that $T_4$ did not produce any significant effect on DNA content of the fish. The citrate synthase, RNA and protein contents of brain, liver and skeletal muscle of fish exposed to thiourea for 28 days decreased approximately 50-58% as compared to their levels in control individuals. Injection of $T_4$ to thiourea treated fish reduced about 3-fold increase in citrate synthase, RNA and protein. These macromolecular inductions by $T_4$ were blocked by actinomycine-D or cycloheximide. This suggests that $T_4$-induced *de novo* synthesis of macromolecules and stimulation of aerobic capacity. However, the activities of

G6-PDH and LDH in brain, liver and skeletal muscle of fish exposed to thiourea increased 2 times that of the activities in control individuals. Administration of $T_4$ to thiourea – exposed fish reduced LDH and G6-PDH activities by about 64-74% which reflected $T_4$-dependent inhibition in anaerobic power and selective anabolic activities of HMP pathway. Administration of thyroid hormone increased the activity of malate dehydrogenase in liver of the teleost *Mugil auratus* (Bonnet, 1970; Le Ray *et al.*, 1970). The effects of thyroid hormone on the activities of hepatic cytoplasm MDH, mitochondrial MDH and LDH of thiouracil treated fish (*Anabus testudineus*) have been demonstrated. They showed the thyroid hormone–dependent decrease in the activities of glucose 6-phosphate dehydrogenase and cytoplasmic and mitochondrial MDH.

Inui *et al.* (1995) elucidated the role of thyroid hormone in tissue development in metamorphosing flounder. Battershy *et al.* (1996) have studied the effect of $T_3$ on the enzymes of intermediary metabolism in dog shark. Hochachka (1962) found that long-term thyroid hormone administration causes an increase in the activity of hexose monophosphate pathway in the rainbow trout *Oncorhynchus mykiss*. Singh *et al.* (2008) showed that administration of thyroxine at low dose caused a significant increase in oxygen consumption and injection of thiourea to this fish significantly decreased oxygen consumption.

The $T_4$-induced increase in RNA and protein contents in fish has also been reported by other workers (Eales, 1980; Tripathi, 1998). However, both catabolic (Higgs *et al.*, 1979; Degani and Dosoretz, 1986) and anabolic (Singh, 1979) effects of thyroid hormones on protein metabolism in coho salmon, eels and freshwater catfish have been described.

### Effect of Ascorbic Acid

Ascorbic acid or vitamin C is a monosaccharide antioxidant found in both plants and animals. Ascorbic acid cannot be synthesized in humans and must be obtained from the diet (Smirnoff, 2001). It is due to absence of an active enzyme, I-gulonololactone oxidase in the liver (Burns, 1959). Most animals are able to produce this compound in their bodies and do not require it in their diets (Linster and Schaftingen, 2007). In the cells it is maintained in its reduced form by reaction with glutathione (Meister, 1994; Wells *et al.*, 1990). Ascorbic acid is a reducing agent and can reduce thereby neutralize reactive oxygen species such as hydrogen peroxide. Vitamin C is a water soluble antioxidant. It was first isolated in 1828 by Hungarian biochemist and Nobel prize winner Szent-Gyorgyl. Ascorbic acid helps in the metabolism of tyrosine, folic acid and tryptophan. It lowered the blood cholesterol and helps in the synthesis of carnitine and catecholamine that regulate the nervous system. It is needed for tissue growth and wound healing. It increases the absorption of iron in gut and helps in formation of neurotransmitters. Being an antioxidant it protects body from harmful effect of free radicals and pollution.

Vitamin C therapy is beneficial in the treatment of different infections and infectious diseases e.g., hepatitis, HIV, common cold, flu and influenza etc. Ascorbic acid is available in reduced form (L-ascorbic acid) and oxidized form (L-dehydro ascorbic acid). It is mainly found in citrus fruits. Fish and milk also contain small amount of Vitamin C. The absorption of Vitamin C, mainly occurs in the buccal mucosa, stomach and small intestine in human and guinea pigs. The buccal absorption of vitamin C is mediated by passive diffusion through the membranes of buccal cavity. While gastrointestinal absorption in through an efficient and active sodium-dependent, energy requiring and carrier mediated transport mechanism (Stevenson, 1974). About 80 to 90% ascorbic acid is absorbed in the gastrointestinal tract. The absorbed acid circulates freely in plasma, leucocytes and RBC and enter all the tissues (Olson and Hodges, 1987).

It has been shown that short-term supplementation of vitamin C lasting 2-4 weeks can significantly reduce the level of free radicals in the body (Naidoo and Lux, 1998). Dietary vitamin C is absorbed primarily by active transport in the small intestine, with absorption decreasing as intake increases. Approximately 70-90% of vitamin C is absorbed when dietary intake is between 30 and 180 mg/day. The kidney excrete dietary vitamin C in urine, but excrete virtually no vitamin C when intake of vitamin is very low (Wardlaw and Kessel, 2002). After absorption in small intestine, vitamin C is transported in the blood to cells in reduced form, ascorbic acid or ascorbate. The concentration of vitamin C varies in different tissues in the body. For instance, vitamin C concentrations are highest in the adrenal and pituitary glands, intermediate in liver, spleen, heart, kidney, lung, pancreas and white blood cell, and lowest in the muscle and red blood cells (Olson and Hodges, 1987). This vitamin may also possess some prooxidant. properties, meaning it can participate in oxidizing other molecules such as iron in blood stream (Alhadeff *et al.*, 1984).

Ascorbic acid is white crystalline sugar that naturally occurs in chemical forms of L-xylo ascorbic acid and D-xylo ascorbate. L-xylo is without vitamin activity (Thurnhan, 2000). The principle pathway of oxidation and turn over of ascorbic acid is believed to involve the removal of two electrons in succession and to yield first the ascorbate free radical (AFR) and then dehydroascorbate. Two molecules of AFR may react together to form one molecule of ascorbate and one dehydroascrobate. Alternatively, AFR may be reduced by a microsomal NADH-dependent enzyme, mono-dehydro-L ascorbate oxidoreductase to ascrobate

(Bates, 1981). Vitamin C also helps the body to absorb iron and to breakdown histamine, the inflammatory component of many allergic reactions (Gaby and Singh, 1991). Specially ascorbic acid works as co-enzyme to convert proline and lysine to hydroxy proline and hydroxy lysine, both important to the collagen structure. It also helps in production of thyroid hormones.

The primary antioxidant role of vitamin C is to neutralize free radicals. It can work both inside and outside the cells to combat free radical damages. As it has water soluble properties. Free radical seek out an electron pair to gain their stability. Vitamin C is an excellent source of electrons, therefore, it can donate electrons to free radicals such as hydroxyl and superoxide radicals and quench their reactivity (Bendich, 1990). It protects DNA of the cells from the damage caused by free radicals and mutagens. Besides this, vitamin C prevents free radical damages in lungs and even help to protect the central nervous system from damages (Kronhausen *et al.*, 1989). Vitamin C and vitamin E together can help to prevent blood from clotting, a condition that contributes to the risk of stroke (Kronhausen *et al.*, 1989). The synergistic combination of vitamin C and vitamin E may be further enhanced by addition of vitamin A. This combination is effective in enhancing the characteristics of enzyme and non-enzymatic antioxidants.

Vitamin C fights of widespread environmental pollutants including mono-oxide, hydrocarbons, pesticides and heavy metals. It stimulates the enzymes in the liver that detoxify the body. Ascorbic acid also protects us by preventing the development of nitrosamines, the cancer causing chemicals that stem from the nitrate contained in many foods (Gaby and Singh, 1991). It is suspected that vitamin C helps the body to reduce glycolysation, which is an abnormal attachment of sugars to protein. It also lowers accumulation of sugar sorbitol (Will and Tyres, 1996) which can damage eyes and kidneys. Vitamin C lowers blood pressure and cholesterol levels, helps in thinning the blood and protects it against oxidation and works in close synergism with vitamin E (Rath, 1993; Witakar, 1985; Trout, 1991). Atherosclerosis is a major contributor to heart disease; vitamin C may prevent this plaque formation by inhibiting the oxidative modification of low density lipoprotein (LDLs). LDLs may contribute to atherosclerotic process by scavenger receptor and influence on monocyte macrophage mortality. Ascorbic acid also helps to prevent atherosclerosis by strengthening the artery walls through its participation in synthesis of collagen and by preventing the undesirable addition of WBC to damaged arteries (Rath, 1993; Weber, 1996; Lehr, 1995). In addition vitamin C may increase the beneficial high-density lipoprotein (Gaby and Singh, 1991).

## Conclusions

Various pesticides are used in different parts of world to combat pest and diseases. They are harmful to non-target species. Pesticides are drained to the homes of fishes and pollute aquatic ecosystem. They hamper the health of fish and man too through ecological cycling and biological magnification. A number of adverse effects have been reported in fish in response to organophosphate and pyrethroid pesticides. Most of the effects are channelised through metabolism in fish. Some of the effects can be minimized via hormonal and antioxidant manipulation. Such strategy may be adopted to protect fish from harmful effects of pesticides.

## References

Ahmed, I., Hamid, T., Fatima, M., Chand, H.S., Jain, S.K., Athas, M. and Raisuddin, S. (2000). Induction of hepatic antioxidants in freshwater catfish *Channa punctatus* (Bloch) is a biomarker of paper mill effluent exposure. *Biochim. Biophy. Acta.*, 1523, 37-48.

Alhadeff, L., Gualtieri, C. and Lipton, M. (1984). Toxic effects of water soluble vitamins. *Am. J. Clin. Nutri.* 42, 33-40.

Amali, A.A., Kumar, L.C., Elizabeth Jayanthi, F.X. and Selvanayagan, M. (1996). Quinolphos – induced biochemical anomalies in *Cirrhinus mrigala* (Ham). *J. Environ. Bio.* 17, 121-124.

Anandhi, D.V. and Murthy, P.S. (1996). Malathion – induced biochemical changes in the liver of freshwater gobiid, *Glossogobius giuris* (Ham). Ind. *J. Environ. Toxicol.* 6, 31-39.

Arora, H.C. Shrivastava, S.K. and Seth, A.K. (1971a). Biossay studies of some organic insecticides. Part-1 and II. *Indian J. Environ. Hlth.* 13, 226-233.

Arora, H.C., Sharma, V.P. and Chattopadhya, S.N. (1971b). Bioassay studies of some commercial organic insecticides. Trials of Malathion with exotic and indigenous carps. *Indian J. Environ. Hlth.*, 13, 300-306.

Arunachalam, S., and Palnichmy (1980). Toxic and sublethal effects of carbaryl on a fresh water fish *M. vittatus* (Bloch). *Arch. Environ. Toxicol.* 9, 307-316.

Awasthi, M., Shah, P. Dubale, M.S. and Gohhia, P. (1984). Metabolic changes induced by organophosphates in the piscine organs. *Environ. Res.* 35, 320-325.

Ballantyne, J.S., John, T.M., Oommen, O.V. and Singer, T.D. (1992). Short-term effects of triiodothyronine on enzymes and mitochondria of holostean fish the bowfin (*Amia salva*) and the teleost (*Salvelinus namaycush*). *J. Exp. Zool.* 261, 105-109.

Banerjee, G., Rajendranath, T. and Dayakar, P. (1987). Acute and chronic toxicity of dichlorvos on the gills of *Clarias batrachus* 8th Ann. Session of AEB Jammu. Abstract 113, 81.

Bashamohideen, M.D., Obilesu, K. and Reddy, P.M. (1987). Behavioural changes induced by malathion and methyl parathion in the fresh water fish, *Tilapia mossambica. Environ Ecol.*, 5, 403-404.

Bates, C.J. (1981). The function and metabolism of vitamin-C in man. In vitamin-C (Ascorbic Acid), 1st ed., Counsell JN and Horing DH, Eds., Applied Science, London, pp, 1-22.

Battersbeg, B.J., Forlane, W.J., Ballantyne, J.S. (1996). Short term effects of 3,5,3-triiodothyroxine on the intermediary metabolism of the dogfish shark. *Squalus acanithias* : evidence from enzyme activities. *J. Exp. Zool.* 274, 151-162.

Begum, G. (2005). In vivo biochemical changes in liver and gill of *Clarias batrachus* during cypermethrin exposure and following cessation of exposure. 82, 185-196.

Begum, G. and Vijyaraghvan, S. (1999). Effects of acute exposure of organophosphate insecticide Rogor on some biochemical aspects of *C. batrachus* (Linnasus). *Environ. Res.* 80, 80-83.

Bendich, A. (1990). Antioxidant micronutrients and immune responses. In: Bendich, A and of Sciences, N.Y. pp, 175.

Benke, G.M., Cheever, F.E., Mirer and Murphy, S.D. (1974). Comparative toxicity, anticholinesterase action and metabolism of methyl parathion and parathion in sunfish and mice. *Toxicol. Appl. Pharmacol.* 28, 97-109.

Bonnet, B. (1970). Thyroide et milier interiens chez mugil sp. influenced dela nutrition *et al.* l environment. Ph.D. thesis, University D'Aix Mar-Seive.

Cappage, D.L., Mathew, E., Gary, H., Cook and Knight, J. (1975). Brain acetyl cholinesterase inhibition in fish as a diagnosis of environmental poisoning by malathion pesticide. *Biochem. Physiol.* 5, 536-542.

Chattopadhyay, D.P., Dighe, S.K., Nashikkar, A.B. and Dube, D.K. (1986). Species differences in the *in vitro* inhibition of brain acetylcholinesterase and carboxylesterase by mipafox, parasxon and sonan. *Pestic. Biochem. Physiol.* 26, 206-208.

Choudhary, D.K., Sadhu, A.K. and Mukhopadhyay, P.K. (1984). Toxicity of two organophosphorus insecticides malathion and phosphamidon to the fish, *Channa striatus. Environ. Ecol.* 2, 143-147.

Das, B.K. and Mukerjee, S.C. (2003). Toxicity of cypermethrin in *Labeo rohita* fingerlings : biochemical, enzymatic and haematological consequences. *Comp. Biochem. Physiol.* 184, 109-121.

David, M. Mushigeri, S.B. and Philip, G.H. (2003). Alterations in the levels of ions in tissues of freshwater fish, *Labeo rohita* exposed to fenvalerate, *Pollut. Res.* 22, 359-363.

Degani, G. and Dosoretz, C. (1986). The effects of 3,5,3-triiodo L-thyroxine and 17a methyl testosterone on growth and body composition of the glass stage of the eel (*Anguilla anguilla*). *Fish Physiol. Biochem.* 1, 145-152.

Devraj, P., Selvarajan, V.R. and Durairaj, S. (1991). Relationship between acetylcholine esterase and monoamine oxidase in brain regions of *Orechromus mossambicus* exposed to phoslane. *Ind. J. Exp. Biol.* 29, 790-792.

Dikshit, J.S.S., Raizada, R.B., Kumar, S.N., Shrivastava, M.K., Kulshrestha, S.K., and Adholia, U.N.(1990). Residues of DDT and HCH in major sources of drinking water in Bhopal, India. *Bull. Environ. Contam. Toxicol.* 45,389-393.

Durmaz, H., Sevgiler, Y. and Üner, N. (2006). Tissue-specific antioxidative and neurotoxic responses to diazinon in *O. niloticus*. *Pest. Biochem. Physiol.* 84, 215-226.

Eales, J.G. (1980). Thyroid hormone function in cyclostomes and fishes. In : Barrington, J.W. ed. Hormones and evolution, Vol. 1, New York : Academic Press Inc. p. 341-436.

Ferreira, J.T., Schoobee, H.J. and Smit, G.L. (1984). The uptake of the an aesthetic benzocaine – hydrochloride by the gills and skin of three freshwater fish species. *J. Fish. Biol.* 25, 35-41.

Gabi, S. K. and Singh, V. N. (1991). "Vitamin C,"- Vitamin intake and health:, A scientific review, Gaby, S. K., Bendich, A., Singh, V. N. and Machlin, L. (eds.). Marcel Dedder, N. Y.

Galow, A.A. and Godzi, T.A. (1994). Acute toxicity of deltamethrin and dieldrin to *O. niloticus* (Linn.) *Bull. Environ. Contam. Toxicol.* 52, 351-354.

Garg, S.K. (2007). Effect of oral administration of *l*-thyroxine ($T_4$) on growth performance, digestibility and nutrient retention in *C. punctatus* (Bloch) and *H. fossilis* (Bloch). *J. Fish. Physiol. Biochem.* 33, 347-358.

Ghosh, T.K. and Chatterjee, S.K. (1987). Histopathological Changes in liver and kidney of *Channa punctatus* during dichlorvos intoxication. $8^{th}$ Ann. Sess. of AEB and Sym. on *Environ. Poll. Pesti. Toxicol.* at Jammu, Abstract No. 22, pp. 15.

Gibson, J.R. and Ludke, J.R. (1973). Effect of SKF 525A on brain acetylcholinesterase inhibition by parathion in fishes. *Bull. Environ. Contam. Toxicol.* 9, 140-143.

Gupta, R. (1989). Effect of some pesticides on the brain and some tissues of fish. Ph. D. Thesis. Dr. H.S. Gour University, Sagar, India.

Hai, D.Q., Varga, S.I. and Matkovics, B. (1997). Organophosphate effects on antioxidant system of carp (*Cyprinus carpio*) and catfish (*Ictalurus nobulus*). *Comp. Biochem. Physiol.* 117, 83-88.

Harilal, (1989). Effect of some pesticides on the gonads of Fish. Ph.D. Thesis, Dr. H.S. Gour University, Sagar, India.

Higgs, D.A., Fagulund, U.H.M., McBride, J.R. (1979). Influence of orally administered *l*-thyroxine or 3,5,3 triiodo-1-thyroxine on growth, food consumption and food conversion of under yearling coho salmon (*Oncorhynchus kisutch*). *Can. J. Zool.* 57, 1974-1979.

Hochachka, P.W. (1962). Thyroidal effects on the pathways for carbohydrate metabolism in a teleost. *Gen. Comp. Endocrinol.* 2, 499-505.

Inui, Y., Yamano, K. and Miwa, S. (1995). The role of thyroid hormone in tissue development in metamorphosing flounder. *Aquaculture* 135, 87-98.

Jagadish, C.G. and Sahai, S. (1987). Effect of malathion on the pituitary gland of *Heteropneustes fossilis* (Bloch) *Environment and Ecotoxicology*, pp. 283-287. Published by : The academy of En. Bio., India 65715, Civil lines ( South), Muzaffarnagar, India.

Jarvinen, A.W., Nordhing, B.R. and Henry, M.E. (1983). Chronic toxicity of duraban (chlorpyriphos) to fathead minnow (*Pimephales promelas*) and the resultant acetylcholinesterase inhibition. *Ecotoxicol. Environ. Saf.* 7, 423-434.

Johnson, J.A. and Wallace, K.B. (1987). Species – related differences in the inhibition of acetyl cholinesterase by paraxon and malaxon. *Toxicol. Appl. Pharmacol.* 88, 234-241.

Joshi, U.M. and Desai, A.K. (1988). Biochemical changes in liver of fish, *T. mossambica* (Peters), during continuous exposure to monochrotophos. *Ecotoxicol. Environ. Saf.* 15, 272-276.

Khare, A., Singh, S. and Srivastava, K. (2000). Malathion induced biochemical changes in the kidney of freshwater fish *C. batrachus*. *J. Ecotoxicol. Environ. Monit.* 10, 11-14.

Konar, S.K. (1981). Pollution of water by pesticides and its influence on aquatic ecosystem. *Indian Rev. Life Sci.* 1,139-165.

Kronhausen, E., Kronhausen, P. and Demopoulos, H. B. (1989). Formula for Life, William Morrow and Co., New York, pp, 95-103.

Kumar, A., Sharma, B. and Pandey, R.S. (2007). Preliminary evaluation of the acute toxicity of cypermethrin and l-cyhalothrin to *C. punctatus*. *J. Bull. Environ. Contamin. Toxicol.* 79, 613-616.

Kumar, S. and Singh, M. (2000). Toxicity of dimethoate to a freshwater teleost *Catla catla* (Ham). *J. Exp. Zool.* 8, 83-88.

Kumarguru, A.K. (1995). Water pollution and fisheries. *Ecol. Env. Econo.* 1,140-150.

Kumble, G.B., Muley, D.V. and Deshpande, V.V. (1999). Enzymatic changes due to pesticidal stress in freshwater fish *Sarotherodeon mossambicus*. *Ind. J. Environ. Ecoplan.* 2, 27-31.

Lal, S. and Pandey, A.K. (1999). Ecotoxicological problems in freshwater bodies with particular reference to the fertilizer factory effluent in chilwa, lake, Gorakhpur. In: *Environmental Toxicology*. (Dwivedi, B.K. and Pandey G, eds.). Bioved Res. Soc., Allahabad pp.212-244.

Le Ray, C., Bonnet B., Febvre, A., Vallet, R. and Pic, R. (1970). Quelquos activities peripherique des hormones thyroidiennes. Observes Chez *Mugil auratus* L. (*Tileostien Mugilidae*). *Ann. Endocrinol.* 31, 567-572.

Lehr, Hans-Anton. (1995). Protection from oxidized LDL-induced leukocyte adhesion to microvascular and macrovascular endothelium in vivo by vitamin-C but not by vitamin-E. *Circulation* 91, 1525-1532.

Linster, C.L. and Schaftingen, V.E. (2007). Vitamin C biosynthesis, recycling and degradation in mammals. *Febs. J.* 274, 1-22.

Luthra, V. and Gupta, R.C. (2009). Haematological changes induced by monocrotophos and endosulfan in the blood of freshwater catfish *C. batrachus* (Bloch) during its annual reproductive cycle. *J. Exp. Zool. India.* 12, 75-77.

Lynshiang, D.S. and Gupta, B.B.P. (2000). Role of thyroidal and testicular hormone in regulation of tissue respiration in male air breathing fish, *Clarias batrachus. Ind. J. Exp. Biol.* 38, 705-712.

Massey, B.P. and Smith, C.L. (1968). The action of thyroxine on mitochondrial respiration and phosphorylation in the trout (*Salmo trutta Fario* L.). *Comp. Biochem. Physiol.* 25, 241-255.

Meister, A. (1994). Glutathione – ascorbic acid antioxidant system in animals. *J. Biol. Chem.* 269, 9397-9400.

Milam, C.D., Farris, J.L. and Wiluide, J.D. (2000). Evaluating mosquito control pesticides for effect on target and non-target organism. *Arch. Environ. Contam. Toxicol.* 39, 324-328.

Mishra, D.K., Tripathy, P. and Hata, A.K. (1991). Toxicity of Kilex carbaryl to a freshwater teleost *Channa punctatus* (B.). *J. Appl. Zool. Res.*, 2, 96-98.

Naidoo, D. and Lux, O. (1998). The effects of vitamin C and E supplementation on lipid and urate oxidation products in plasma. *Nutri. Res.* 18, 953-961.

Natarajan, G.M. (1984). Effects of lethal concentration ($LC_{50}$/48h) of metasystox on some selected enzyme system in the air breathing fish *C. striastus* (Bloch). *Comp. Physiol. Ecol.* 9, 29-32.

Neethirajan, K. and Mathavan, S. (2004). Sumidon – induced changes on some aspects of metabolism in the chosen tissue of freshwater catfish *Mystus vittatus* (Bloch). *J. Exp. Zool.* 7, 279-283.

Nemesok, J., Asztolos, B., Vig, E. and Orban, L. (1987). The effect of an organophosphorous pesticide on the enzymes of carp *Cyprinus carpio* (L.). *Acta Biol. Hung.* 38, 77-85.

Olsen, D.L. and Christenson, G.M. (1980). Effects of water pollutants and other chemicals on fish acetylcholinesterase (*in vitro*). *Environ. Res.* 21. 327-335.

Olsen, J. A. and Hodges, R. E. (1987). Recommended dietary intakes (RDI) of vitamin-C in humans. *An. J. Clin. Nate.* 45, 693-73.

Pandey, A.C., Gopal, K. and Pandey, A.K. (2000). Pollution and Fish Physiology: A review. *Aquacult* 1, 1-8.

Pandey, S., Ahmed, I., Parvez, S., Bin-Hafeez, B. Haque, R. and Raissuddin, S. (2006). Effect of endosulfan on antioxidants of freshwater fish *Channa punctatus* Bloch : 1. Protection against lipid peroxidation in liver by copper pre-exposure. *Arch. Environ. Contam. Toxicol.* 41, 345-52.

Pant, J.C. and Singh, T. (1983). Inducement of metabolic dysfunction by carbonate and organophosphorus compounds in fish, *Puntius conchonius*. *Pest. Biochem. Physiol.* 20, 294-298.

Peter, M.C.S. and Oommen, O.V. (1989). Oxidative metabolism in teleost, *Anabas testudineus* (Bloch). Effect of thyroid hormones on hepatic enzyme activities. *Gen. Comp. Endocrinol.* 73, 96-107.

Philip, G.H., Reddy, P.M. and Sridev, G. (1995). Cypermethrin induced *in vivo* alternations in the carbohydrate metabolism of freshwater fish *Labeo rohita*. *Ecotoxicol. Environ. Saf.* 31, 173-178.

Plisetskaya, E., Woo, N.Y.S. and Murat, J. (1983). Thyroid hormone in cyclostomes and fish and their role in regulation of intermediary metabolism. *Comp. Biochem. Physiol.* 74A, 179-187.

Ramakrishna, R. and Sivakumar, A.A. (1993). Effect of quinolphos on the fish *O. mossambicus*. *J. Ecol. Biol.* 5, 45-50.

Rao, J.V. (2006). Sublethal effects of an organophosphorous insecticide (RPR-II) on biochemical parameters of tilapia (*O. mossambicus*). *Comp. Biochem. Physiol.* 143, 492-498.

Rao, K.P.S. and Rao, K.V.R. (1979). Effect of sublethal concentration of methyl parathion on selected oxidative enzymes and organic constituents in the tissue of the freshwater fish, *Tilapia mossambica* (Peters). *Curr. Sci.* 48, 526-528.

Rao, K.S.P. and Rao, K.V.R. (1982). Toxicity of methyl parathion to freshwater teleost *Tilapia mossambica* (peters). *Comp. Physiol. Ecol.* 7, 271-274.

Rao, K.S.P., Rao, K.R.S.S., Prasad, K.S. and Rao, K.V.R. (1983). Methyl parathion impact on the esterases of a freshwater teleost *Tilapia mossambica* (Peters). *Ind. J. Fish.* 30, 148-154.

Rath, M., (1993). Eradicating heart disease. *Health Now.* San Francisco, CA.

Ray, A.K. and Medda, A.K. (1976). Effect of thyroid hormones and analogues on ammonia and urea excretion in lata fish (*O. punctatus*). *Gen. Comp. Endocrinol.* 29, 190-197.

Reddy, P.M. and Bashmohideen, M.A.D. (1995). Alternations in protein metabolism in skeletal muscle of fish, *C. carpio* during sublethal concentration of cypermethrin. *Environ. Monit. Assess.* 36, 183-190.

Sahib, K.A., Sailatha, P. and Rao, K.V.R. (1980). Impact of malathion on acetylcholinesterase in the tissue of fish *Tilapia mossambica* (Peters). A time course study. *J. Bio. Sci.* 2, 37-41.

Samuel, M. and Sastry, K.V. (1989). *In vivo* effect of monocrotophos in the carbohydrate metabolism of the freshwater snake headed fish *C. punctatus*. *Pestic. Biochem. Physiol.* 34, 1-8.

Sancho, E., Ferrando, M.D. and Andreu, E. (1997). Sublethal effects of an organophosphate insecticide on the European eel *Anguilla anguilla*. *Ecotoxicol. Environ. Saf.* 36, 57-65.

Saravanan, R., Revathi, K. and Murthy, P.B. (2009). Lambda cyhalothrin induced alternations in *Clarias batrachus*. *J. Environ. Biol.* 30, 265-270.

Sarkar, S.K. (1997). Herbicides and ecosystem. *Geobios.* 24, 293-308.

Sastry, K.V. and Malik, P.V. (1982). Acute and chronic effects of diazinon on the activities of three dehydrogenases in the digestive system of freshwater teleost fish *C. punctatus*. *Toxicol. Lett.* 10, 55-59.

Sastry, K.V. and Sharma, K. (1980). Diazinon effect on the activities of brain enzymes from *O. punctatus*. *Bull. Environ. Contam. Toxicol.* 24, 326-332.

Sastry, K.V. and Sharma, K. (1981). The effect of sublethal concentration (0.4 mg/l and 0.2 mg/l) of diazinon on brain enzymes of *Ophiocephalus* (*Channa punctatus*). *J. Environ. Res.* 11, 12-17.

Sastry, K.V. and Siddiqui, A.A. (1984). Some hematological, biochemical and enzymological parameters of a freshwater teleost fish *C. punctatus* exposed to sublethal concentration of quinolphos. *Pestic Biochem. Physiol.* 22, 8-13.

Sastry, K.V., Siddiqui, A.A. and Singh, S.K. (1982). Alternation in some biochemical and enzymological parameters in the snake head fish *C. punctatus* exposed chemically to quinolphos. *Chemosphere* 11, 1211-1216.

Saxena, P.K., Singh, V.P., Kondal, J.K. and Soni, G.L. (1989). Effect of some pesticides on *in vitro* lipid and protein synthesis by the liver of the freshwater fish teleost *C. punctatus* (Bloch). *Environ. Poll.* 58, 273-280.

Sayeed, I., Parvez, S., Pandey, S., Hafeez, B.B., Haque, R. and Raisuddin, S. (2003). Oxidative stress biomarkers of exposure to deltamethrin in freshwater fish *C. punctatus* (Bloch). 56, 295-301.

Saytadevan, S., Kumar, S. and Tembhre, M. (1993). Acetylcholinesterase activity and enzyme kinetics in the brain of common carp *Cyprinus carpio* subjected to sublethal exposure to dimethoate. *Biosci. Biotech. Biochem.* 57, 1566-1567.

Senger, M.R., Rico, E.P., Arizi, M.B., Rosenberg, D.B., Dias, R.D., Bogo, M.R. and Bonan, C.D. (2005). Carbofuran and malathion inhibit nucleotide hydrolysis in zebra fish (*Danio rerio*) brain membranes. *Toxicology.* 212, 107-115.

Shah, P.H. and Dubale, M.S. (1983). Biochemical changes induced by malation in the body organs of *Channa punctatus*. *J. Anim. Morphol. Physiol.* 30, 107-108.

Siddiqui, M.K.J., Rahman, M.F., Mahboob, M., Arjun, F. and Mustafa, M. (1988). Species differences in brain acetylcholinesterase and neuropathy target esterase response to monocrotophos. *J. Environ. Sci. Health.* Part B 23, 291-299.

Singh, A. and Srivastava, V.K. (1999). Toxic effect of synthetic pyrethroid permethrin on the enzyme system of the freshwater fish *C. striatus*. 39, 1951-1956.

Singh, A.K. (1979). Effects of thyroxine and thiorurea on liver and muscles energy stress in the freshwater catfish *H. fossilis*. *Biol. Anim. Biochem. Biophys.* 19, 1670-1976.

Singh, M.K., Nirala, M.K., Yasmin, R., Deo, D.P., Rajan, K.R. and Singh, M.P. (2008). Effect of L-thyroxine and thiouracil on oxygen consumption in *Mystus vittatus* (Bloch). *Proc. Zool. Soc.* India. 7, 119-122.

Singh, S. and Sahai, S. (1984). Histopathological changes in the gills of *Rasbora daniconius* induced by Y-BHC. *J. Environ. Biol.* 5. 65-69.

Singh, S. and Sahai, S. (1985) Identification malathion residues in some tissues of *Rasbora daniconius* (Ham.) teleost by thin layer chromatography. *Proc Sym. Assess. Environ. Pollut.* 159-162.

Smirnoff, N. (2001). L-ascorbic acid biosynthesis, *Vitam. Horm.* 61, 241-266.

Somnuek, C., Boonphakee, C., Cheevporn, V. and Tanaka, K. (2009). Gene expression of AChE in hybrid catfish (*C. gariepinus* and *C. macrocephalus*) exposed to chlorpyriphos and carbaryl. *J. Environ. Biol.* 30, 83-89.

Srabani, De, A.K., Ray, D.A.K. and Medda, A.K. (1989). Induction of NADP-dependent malic dehydrogenase activity in brain of singhi fish *H. fossilis* (Bloch) by 3,5,4 triiodothyronine. *Neurochem. Inter.* 14, 461-465.

Stevenson, N. R. (1974). Active transport of L-ascorbic acid in the human ileum. *Gastroenterology* 67, 952-958.

Tham, L.G., Perumal, N., Sayed, M.A., Shamaan, N.A. and Shukor, M.Y. (2009). Assessment of *C. batrachus* as a source of AChE for the detection of insecticides. J. Environ. Biol. 30, 135-138.

Thangaval, P., Ramaswamy, M., Sumathiral, K. and Amutha, K. (2005). Individual and combined effects of– ziram dimecron on the levels of serum prolactin and selected minerals of an edible freshwater fish, *O. mossambicus* (Peters). *Pest. Biochem. Physiol.* 81, 24-31.

Thurnham, D. I. (2000). Water-soluble vitamins (Vitamin-C and B Vitamins, thiamin, Riboflavin and Niacin). In "Human Nutrition and Dietetics". Garrow, J. S.; James, W.P.T.; Ralph, A. (Eds.). Churchill. Livingston Publishers, Edinburgh London, 249-257.

Tilak, K.S., Veeraiah, K. and Koteswara Rao, D. (2005). Biochemical changes induced by chlorpyriphos, an organophospate compound in sublethal concentration of the freshwater fish *Catla catla, Labeo rohita* and *Cirrhinus mrigala*. *J. Environ. Biol.* 26, 341-347.

Tripathi, G. (1998). Thyroid hormone – induced changes in cytoplasmic and mitochondrial protein of a teleost. *Z. Naturforsch* 53C, 120-124.

Tripathi, G. and Shukla, S.P. (1989). Induction of cytoplasmic and mitochondrial enzyme by thyroid hormone in a freshwater catfish. *Biochem. Int.* 19, 945-949.

Tripathi, G. and Shukla, S.P. (1990). Enzymatic and ultrastructural studies in a freshwater catfish. Impact of methyl parathion. *Biomed. Environ. Sci.* 3, 166-182.

Tripathi, G. and Verma, P. (2003). Differential effect of thyroxine in metabolic enzymes and other macromolecules in a freshwater teleost. *J. Expt. Zool.*, 296A, 117-124.

Tripathi, G., Chauhan, J.S., Shasmal, J. and Kachhawaha, N. (2007). Cypermethrin induced changes in protein content of brain, liver and skeletal muscle of a freshwater fish. *Biochem. Cell. Arch.* 7, 313-316.

Tripathi, G., Singh, H. and Shasmal, J. (2006). Effect of cypermethrin on protein content of gill, heart and kidney of a catfish *Clarias batrachus*. *Biochem. Cell. Arch.* 6, 123-127.

Trout, D. L. (1991). Vitamin-C and cardiovascular risks factors. *Am. J. Clin. Nutr.* 53, 322S-325S.

Upnishad, K., Singh, K.A., Nirala, M.K., Pallavi Nandan, G., Khan, A.A. and Kumar, D. (2008). Studies on calorific values, protein and water content in Channa punctatus (Bloch) as indicator for assessing water quality. *Proc. Zool. Soc. India.* 7, 105-109.

Varghese, S., Shameena, B. and Oommen, O.V. (2001). Thyroid hormone regulate lipid peroxidation and antioxidant enzyme activities in *A. testudineus* (Bloch). 128, 165-171.

Venkataramana, G.V., Sandhya Rani, P.N. and Murthy, P.S. (2006). Impact of malathion on the biochemical parameters of gobiid fish, *Glossogobius ginris* (Ham). *J. Environ. Biol.* 27, 119-122.

Venugopal, G., Swian, D. and George, J.P. (1988). Bioassay evaluation of tosicity of monocrotophos to a fresh water exotic crop, *Cyprinus carpio communis* (Lin.) Mortality and behaviour study, *J. Environ. Biol.* 9, 395-399.

Verma, S.R., Bansal, S.K. and Dalela, R.C. (1977). Bioassay trail with organic biocides in freshwater fish *Labeo rohita*. *Ind. J. Environ. Hlth.* 19, 107-115.

Verma, S.R., Bansal, S.K. and Dalela, R.C. (1978a). Toxicity of selected organic pesticides to a freshwater teleost fish, *Saccobranchus fossilis* and its applications in controlling water pollution. *Arch. Environ. Contam. Toxicol.* 7, 317-323.

Verma, S.R., Bhatnagar, M.C. and Dalela, R.C. (1978b). Biocides in relation to waer pollution Part- II Bioassay Studies to few biocides to a freshwater fish *Channa gachua*. *Acta. Hydrochem. Hydrobiol.* 6, 137-144.

Verma, S.R., Kumar, V., Chand, R. and Tonk, I.P. (1984). Use of static bioassay procedure for the evaluation of the relative toxicity of pesticides and relative resistance of fishes. *Proc. Sem. Eff. Pest. Ag. Fau.* 115-127.

Wang, C. and Murphy, S.D. (1982). Kinetic analogue of species differences in acetylcholinesterase sensitivity to organo-phosphorus insecticides. *Toxicol. Appl. Pharmacol.* 66, 409-419.

Wardlaw, M. and Kessel, M. (2002). Perspectives in Nutrition, Boston, McGraw Hill

Weber, C. (1996). Increased adhesiveness of isolated monocytes to endothelium is prevented by vitamin-C intake in smokers. *Circulation.*, 93, 1488-1492.

Well, W., Xu, D., Yang, Y. and Röeque, P. (1990). Mammalian thiol transferase (glutaredoxin) and protein disulfide isomerase have dehydroascorbate reductase activity. *J. Biol. Chem.* 265, 15361-15364.

Whitaker, M. J. (1985). Reversing Heart Disease. Wamer Books, New York, USA.

Will, J. C. and Tyers, T. (1996). Does diabetes mellitus increase the requirement for vitamin-C? *Nutr. Rev.* 54, 193-202.

Yadav, A., Gopesh, Pandey, R.S., Rai, D.K. and Sharma, B. (2007). Fertilizer industry effluent induced biochemical changes in freshwater teleost *C. striatus* (Bloch). *Toxicology.* 79, 588-595.

# 15

# Climate-associated Arthropod Borne Diseases of Human and Domestic Animals

*Dr. B.M. Sharma*

Climate-associated arthropod borne diseases have an enormous impact on human and domestic animals health and national economies. They fall into two categories. The first category includes pest specific diseases caused by stinging, biting and infestation or allergic reactions. Arthropods responsible for such diseases are ant, bee, wasp, horse fly, deer fly, bedbug and louse. A fundamental relationship between humanity and insects involves those insects that cause, or transmit, diseases of human, livestock, pest and poultry. Annoyance from mosquitoes and other biting or stinging insects are oftenly seen in some localities. Venoms of these insects may cause irritation, swelling, pain or even death of people who have become sensitized. Domestic animals are similarly affected and these insects, and others like face flies and bot flies which neither bite nor sting but greatly annoy cattle, may be direct cause of reduced yield of meat or milk (Hutchinson, 1976). The direct effects due to insect attack are considered disease only if the symptoms are systematic or tissue reaction occurs.

**Insects Biting**

Insects with sucking mouthparts usually inject saliva when they feed. This saliva may cause a general tissue reaction in the animals. Swelling and itching accompany bites of mosquitoes, horse flies, black flies, bed bugs and other biting insects. Apparently, proteins and other chemical constituents of the injected saliva trigger a defensive reaction by the victim's antibodies. Bites of most insects are not serious. Though the temporary itching caused by chiggers is maddening and black fly bites sometimes swell disproportionately to tiny insect that caused the bite. Insects that feed on blood are in most cases well adapted to taking only a small drink without eliciting a serious allergic reaction in the host. Ticks sometimes inject enough saliva in several days during which they feed on a host to cause a systematic reaction, tick paralysis, in both humans and domestic animals (particularly in cattle). A few insects, mainly eye gnats (Chloropidae : *Hippelates* sp.) and face flies, feed on liquid bodily secretions of humans and domestic animals. The eye gnats are most troublesome, for large numbers will irritate and annoy the host while feeding. It is suspected that these flies transmit bacteria causing conjunctivitis (pink eye). Other insects of several orders may accidentally cause eye irritation.

### Insects Stinging

In hymanopterans (ants, bees and wasps), the ovipositor has become modified into a sting for injecting poison produced by enlarged secondary glands. Most of such poisons are apparently complex chemical mixtures that disrupt transmission of nerve impulses and breakdown cells in the recipient. Most sever stinging incident results from accidental disturbance of honey bees or bumble bees and ants. A small number of persons (4-8%) are exceedingly sensitive to stings of hymenopterans and react to the small amount of protein in venom.

### Insects Infestation

Myiasis is a condition caused by infestation of humans or other animals by larval Diptera. Flies of several species oviposit on wild and domestic animals and occasionally, on people. The resulting maggots may feed within the digestive tract (horse bots) and the nasal passages (sheep, goat and deer bots). The human bot fly of tropical America does not select hosts directly but instead oviposits on a mosquito or other small arthropod. When the mosquito lands on a host, the bot fly eggs hatach and larvae make their way through the skin to feed (Horn, 1978). Another category includes vector specific diseases that transmit vectors through insect pests. Examples of pests in this category include the *Anopheles* sp. (vector of epidemic typhus), oriental rat flea, *Xynopsylla cheopis* (vector of plague), the tsetse fly, *Glossina* sp. (vector of trypanosomiasis), and black fly, *Simulium* sp. (vector of onchocerciasis). Certain insects may be both general pests and disease vectors, depending on the areas of infestation, for example, the house fly (*Musca domestica*) which is a vector of trachoma virus and bacterial dysenteries and *Blatella* sp. The later species is a nuisance everywhere but is also strongly incriminated in the spread of viral hepatitis. Transmission of pathogenic microorganisms is by far the most important climate-associated interrelationship between people, arthropods and disease. Disease-causing microorganisms transmitted by arthropods are probably originally associated with insect vectors exclusively and only later developed in hosts, livestock and humans. Cockroaches crawl directly to your food pantry from sewerage contaminated with pathogens of typhoid, cholera, hepatitis and polio. The human body louse is a proven vector of typhus. Transmission occurs mainly in two major ways :

### *Mechanical Transmission*

Many disease-causing microorganisms can be spread from diseased to healthy insects or people by mechanical means, merely by contamination of the insect vector as it moves around on or in a diseased organism. All insects are covered with bacterial and fungal spores, therefore, mechanical transmission can be demonstrated in laboratory. Most mechanically transmitted pathogens can be maintained and transmitted without insects. House flies (*Musca doemestica*) feed and breed on human and other excrement and are also given to walking about on food for human consumption. Cockroaches are partial to both sewers and pantries. Microorganisms causing dysentery, salmonella, typhoid, cholera, leprosy, tuberculosis and hepatitis, together with eggs of intestinal tapeworms and nematodes, all may occur in human excrement (James and Harwood, 1969). An important mechanism for mechanical transmission by insects is introduction of a pathogen during feeding.

### *Biological Transmission*

A number of pathogens develop within the arthropod vector before being transmitted to a vertebrate or plant host. Biological transmission may occur in three ways :

#### *Propagative pathogens*

Pathogens merely multiply within their arthropod vector. Epidemic typhus is an example of a propagative rickettsial disease of humans, the vector is the human body louse, *Pediculus humans*. Rickettsiae multiply within the louse, eventually biting the host. The pathogen can also introduced when a person crushes an affected louse or its feces in a cut or scratch. The disease is always debilitating and sometimes fatal in people and survivors may become reservoirs of the disease. An equally serious disease is plague, a propagative bacterial disease spread from rodents to humans by bites of fleas. The main source of disease is probably the black rat (*Rattus rattus*) and the major vector is the oriental rat flea (*Xenopsylla cheopis*). Plague is mainly a disease of rodents and is continuously present in many wild rodents. Plague bacilli ingested by a flea multiply in its digestive tract, blocking it so that the flea becomes hungrier, bites more frequently and requires association between people and rats.

*Cyclopropagative pathogens*

These pathogens not only multiply but also transform within the arthropod. The most important of cyclopropagative disease is malaria, a cyclic protozoan disease of mainly tropical region. Human malaria is caused by four *Plasmodium* sp. and transmitted by *Anopheles* sp. Man is the principal reservoir, but two *Plasmodium* sp. also infest chimpanzees. An Anopheles female ingests male and female plasmodia from a malarial victim. These cells fuse and migrate to the gut wall, where they multiply in a cyst. This bursts and the resulting sporozoites migrate to the salivary glands, from where they are injected into another person. This sexual cycle takes at least a week, during which the mosquito does not yet transmit the disease. Their cyclic reproduction in blood cells causes typical chills, high grade fever, fatigue, and weakness in the victim, sometimes with fatal results. Dengue, also commonly called as breakbone fever, and dengue haemorrhagic fever (DHF) have emerged as a major public health problem in India in recent years, with more and more new foci being formed in both urban and rural areas. The primary vector mosquito is *Aedes aegypti*, although *A. abbopictus* has also been recently incriminated in desert. Four viruses, dengue 1 to 4, classified in an antigenic complex of the Flavivirus genus, under family Flaviviridae, are the aetological agents of this entity. The four serotypes are antigenically very similar to each other but different enough to elicit only partial cross protections, after infection by one of them. In the urbanization process, which has left many without adequate water, sewer systems or waste management, and created new breeding grounds for the vector has hastened the spread of disease.

Trypanosomiasis, is as serve as malaria but is localized to tropical areas, where it spreads by various species of tsetse flies (*Glossina* sp.). Several protozoans of the genus *Trypanosoma* are responsible for the disease in humans and animals. There is an 18$^{th}$ day latent period while the protozoans multiply and move to the tsetse's salivary glands (James and Harwood, 1969). Thereafter, the insect can transmit the pathogens. In humans, dongs, horses, and camels, the trypanosomes move to the cerebrospinal fluid and cause progressive lethargy and often death. Cattle become extremely weak but usually survive. However, swine remains generally unaffected. Many of the large wild mammals of Africa are tolerant of *Trypanosoma* and serve as reservoirs, which greatly complicate management of sleeping sickness. Isolation and treatment of individual cases is important because people are the major reservoirs for human trypanosomiasis. Climate changes are expected to greatly influence the spread of diseases in human and animals.

*Cyclodevelopmental pathogens*

These pathogens transform, without multiplication, within the arthropod. Several insect species transmit roundworms that are parasites of people and domestic animals. Particularly in the tropics, biting dipterans transmit microscopic blood-infesting, microfilariae. Black flies transmit filariae that cause onchocerciasis, or river blindness, so called because the worm larvae invade the eyes and optic nerves. Black flies evidently ingest worm larvae from the host's skin, and these worms move through the stomach wall to develop further in the thoracic muscles. When the fly bites, the worms emerge from the fly and enter through the host's skin (DeLeon and Duke, 1966). Heartworm of dogs and cats is an increasingly important veterinary health problem in the United States. In this disease, filarial roundworms, transmitted by mosquitoes, concentrate in the heart and pulmonary arteries where they may block circulation to the lungs.

## Epidemiology

Epidemiology is the study of how diseases are maintained and spread, with emphasis on the conditions necessary for infecting many individuals. The epidemiology of any arthropod-borne disease has three foci : pathogen, vector and host. These are equally important, and changes in any one may be sufficient to commence or end an epidemic. It may depend on whether people are a primary or secondary host. Humans are secondary host for the oriental rat flea and primary for human body louse. For instance, in plague the pathogen is a bacterium, *Yersinia pestis*. Its usual host is rodent, both wild rodent and semi domestic rat. The principal vector is the oriental rat flea, which transmits infected *Yersenia pestis* from rodent to rodent. People become infected when bitten by an infected rat flea. Density and physical condition of the host are important cases of plague. Plague becomes more frequent when humans and rodents, especially, rats, are in close proximity. Persons who are unhealthy and under nourished generally have less resistance to disease. By over whelming the defences of its human host, plague may progress from bubonic form (transmitted by fleas) to pneumonic plague, transmitted by droplet infection through cough and sneeze.

## Vector-disease System

Understanding vector ecology and the intricate interrelationships of the vector with the pathogen and its human host, and animal reservoirs of the disease is a considerably more complex than the ecology of a typical insect pest of agricultural crops. Vectorial capacity represents the interaction of the physiological and biochemical factors that determine the susceptibility of an arthropod vector to become infective with the pathogen. Some ecological factors, such as population density and longevity, dispersal and flight range, and host preference and feeding patterns determine the probability of successful contact with the host. The dispersal of vectors from their breeding sites is of major importance. Such varying dispersal patterns complicate vector-control procedures. The behavioral patterns that determine the contact between vector and vertebrate host during the search for and indulgence in a blood meal are perhaps the most critical factors in the epidemiology of insect-borne diseases (Hamon *et. al.*, 1970). Their quantitative assessment is essential for system analysis and modeling of pest-disease management programs. The frequency of feeding is determined by the biting cycle or interval between blood meals, which is related to the gonotrophic cycle, but is also influence by temperature and humidity, and may be completely interrupted under unfavorable climatic conditions unless, as in *Anopheles atroparvus*, overwintering and diseases transmission continue in houses during winter.

Infection of the host represents the crucial interaction between vector and pathogen, and the host involved in disease transmission. The resting habits of adult mosquitoes or tsetse flies engorged with a host blood meal are good examples. Critical estimates of the size of the vector population and its age composition are essential to adequate modeling and system analysis. Relative estimates such as the number of larvae or pupae per unit area, the number of adults per house or room, the biting rate, or the adult emergence rate have been widely used in lieu of absolute estimates.

## Control Measures

Control of arthropod-borne diseases is usually a community-wide effort coordinated and applied by public agencies. This is necessary because of the wide range of some arthropod vectors, such as mosquitoes and black flies, and societal consequences of major disease epidemics. The control of arthropod-borne diseases is often a prerequisite to further economic development. Epidemiological strategies associated with climate change may provide the most useful and generally acceptable long-term solutions to the control of arthropod-borne diseases. The control methods used to implement of these tactics can be divided into three major tasks : (i) Source reduction techniques are of almost endless variety, and their employment alone and together with water conservation and land reclamation programs is limited only by the imagination and technical competence of the planner and by the facilities at his disposal (TVA, 1947; WHO, 1973). (ii) Insecticide application is important component of disease management programs (or) arthropod-borne diseases (WHO, 1971; Metcalf, 1989). (iii) Naturalistic control is destruction of insects or the prevention of their development by any natural forces, animate or inanimate (WHO, 1973; Metcalf and Metcalf, 1993).

Detailed knowledge of larval ecology of arthropod vectors, especially mosquitoes, was a basic component of control techniques a generation ago (TVA, 1947), was often ignored during the age of insecticides. It is being revived today under source reduction approaches to mosquito pest management (WHO, 1973). The reduction in arthropod population due to insecticide treatments is well known (Tripathi and Sharma, 2005). The spectacular successes resulting from the application of DDT (Tyagi, 1992) and other modern insecticides against the arthropod vectors of human and domestic animal diseases, particularly in the control of malaria and typhus, have obscured the necessity for understanding vector ecology and the especially complex interrelations between vector, pathogen, human victim and possible animal reservoir (Metcalf and Novak, 1994). Huffaker (1944) found the significant negative correlation between environmental temperature and developmental time from eclosion to emergence. Vector susceptibility to infection may involve genetic variations in the pathogen, i.e., virulence. These genetic controls must affect vital physiological factors involved in the nutrition, development and reproduction of the pathogen in the body of the host (Steiner *et. al.*, 1982).

Biological control methods, particularly the exploitation of predators like indigenous larvivorous fishes, bugs belonging to various families Napidae, Belostomatidae and Notonetidae have been widely used for mosquito control (Rajendra and Reuben, 1988). Sharma and Jain (1999) reported that the utility of aquatic plants like *Azolla* sp., blue green algae, *Nostoc* sp., *Oscillatoria* sp. as biological control agents for mosquito breeding habitats. Similar significant effects have been documented by other workers (Lin, 1983; Hobbs and Molina, 1983; Ansari and Sharma, 1991). Massive trapping with this animal kairomone mixture has become an important IPM tool for control of diseases Calvin and Gibson, 1992). Integrated efforts to replace the present and increasingly unsatisfactory reliance on the exclusively insecticidal control of insect vectors involve integrated control by combinations of chemical and biological methods. This will become a leading approach to pest management technology. A strong strategy may be developed in relation to climatic changes to control arthropod-borne diseases effectively.

## Conclusions

Insects impinge on us mainly through the diseases they can spread direct and transmit to humans and domestic animals. The interrelationship of the vector, with the pathogen and its human host, and animal reservoirs of disease is complex and to some extent environment dependent. In tropical and subtropicals regions, attentions have been paid to the role of insects in direct and transmitting protozoa, virus and nematode. Such pathogen are the causative agents of many important and wide spread human diseases, including malaria, plague, onchocerciasis, dengue and trypanosomiasis. Even more frequent than direct injury due to insects is their action as vectors, transmitting disease-inducing pathogens from one animal or human host to another. This transfer may be mechanical or biological means. Control methods mainly involve source reduction, insecticide application and naturalistic. Integrated efforts to reduce reliance on insecticidal control of insect vectors need on integrated control by combinations of chemical and biological methods. Epidemiological strategies coupled with environmental influences can provide long-term solutions to the control of arthropod-borne diseases.

## References

Ansari, M.A. and Sharma, V.P. (1991). Role of Azolla in controlling mosquito breeding in Ghaziabad district villages (U.P.). *Ind. J. Malariol.*, 28, 51-54.

Calvin, J. and Gibson, G. (1992). Host seeking behavior and management of tsetse. *Ann. Rev. Entomol.*, 37, 21-40.

DeLeon, J.R. and Duke, B.O.L. (1966). Experimental studies on the transmission of Guatemalan and West African strains of *Onchocerca volvulus* by *Smimulium ochraceum*, *S. metallicum* and *S. callidum*. *Trans. Roy. Soc. Trop. Med. Hyg.*, 60, 735-752.

Hamon, J., Mouchet, J. Brengues, J. and Chauvet, G. (1970). Problem facing anopheline vector control : vector ecology and behavior before, during and after application of control measures. *Misc. Publ. Entomol. Soc. Am.*, 7, 28-41.

Hobbs, J.H. and Molina, P.A. (1983). The influence of aquatic fern Salvinia auriculata on the breeding of Anopheles albimanus in costal Guatmala. *Mosq. News.*, 43, 456-459.

Horn, D.J. (Ed.) (1978). Insect and disease. In : *Biology of Insects*, pp. 376-402, W.B. Saunders Co., England.

Hutchinson, P. (1976). *How Animals Live*. Elsevier-Phaidon, Oxford.

Huffaker, D. (1944). The temperature relations of the immature stages of the malaria mosquito, Anopheles quadrimaculatus say, with a comparison of the developmental power of constant and variable temperatures in insect metabolism. *Ann. Entomol. Soc. Am.*, 37, 1-27.

James, M.T. and Harwood, R.F. (1969). *Herm's Medical Entomology*. McMillan Publishing Co., New York.

Lin, L.B. (1983). The effect of Azolla on mosquito breeding. *Parasitol. Today*, 4, 328-329.

Metcalf, R.L. (1989). Insect resistance to insecticides. *Pestic. Sci.*, 26, 333-358.

Metcalf, R.L. and Metcalf, R.A. (1993). *Destructive and useful insects : their habits and control.* McGraw-Hill Publication, New York.

Metcalf, R.L. and Novak, R.J. (1994) Pest management strategies for insects affecting humans and domestic animals. In: Introduction to Insect Pest Management (Eds., Metcalf, R.L. and Luckmann, W.H.), pp. 587-628, John Wiley & Sons, New York.

Rajendra, R. and Reuben, R. (1988). Laboratory evaluation of water fern Azolla pinnata for mosquito control. *J. Biol. Contl.*, 2, 114-116.

Sharma, M. and Jain, D. (1999). A role of Azolla in malaria vector control. *J. Appl. Zool. Res.*, 10, 183-185.

Steiner, W.W.M., Tebachnick, W.J., Rai, K.S. and Narang, S. (Eds.) (1982). *Recent developments in the genetics of insect disease vectors*. Stipes, Champaign, Ireland.

Tripathi, G. and Sharma, B.M. (2005) Effects of habitats and pesticides on aerobic capacity (citrate synthase activity) and survival of soil fauna. *Biomed. Environ. Sci.*, China, 18 : 169-175.

TVA (1947). *Malaria control of impounded waters*. USTVA, Washington, D.C.

Tyagi, B.K. (1992) Control of malaria vectors in India. *Ind. Rev. Life Sci.*, 12, 211-238.

WHO (1971). Vector control. *WHO Chron.*, 25, 201-242.

WHO (1973). *Manual on larval control operations in malaria programmes*. WHO, Geneva.

# 16

# Biological Impact of Disposal of Steel Industry Wastewater in the Lentic Water Bodies

*Sanyogita R. Verma, P.R. Chaudhari and S.R. Wate*

Pollution of river courses associated with industrial discharges is a global problem. In India, it is reported that 70% of the available water is polluted (Citizens Report, 1982). The source of pollution is identified to be industrial pollution constituting 8-16% of the wastewater and sewage comprises 84-92% (Chaudhari, 1982). A number of reports focus on the present status of pollution and its damaging effects (Chandra & Krishna, 1983; Mahadevan & Krishnaswamy, 1983; Jagdale et al, 1984; Geevarghese & Chandrasekharan, 1985; Chaudhary et al. 1985; Shankar et al. 1986; Ghose & Sharma, 1988; Shah, 1988; Singh & Singh, 1990; & Kumar 1999). But enough data on impact of wastewater on steel plant is not available; therefore the present work is carried out to evaluate the trophic status of Seonath River & Kharoon River.

The study area around Bhilai Steel Plant is largely a flat terrain with two rivers viz. Seonath River and Kharoon River, three streams / nallas viz. Kosa nalla, Samoda nalla, and Somni nalla. There are also a few numbers of small man-made ponds that are utilized for domestic use or for fishery purpose at different places and agricultural fields. The rivers, for the most part of their stretch, are clean. The nallas carry treated wastewater from Bhilai Steel Plant for a 4-5 kms distance before meeting with the rivers. They carry low pollution load and were observed to get self purified during their flow up to the river and then through the river, which is evident from visual observations as well as biological evaluation of the quality of the nalla and rivers.

The "Standard Methods for Water Analysis, 20th Edition (APHA, AWWA, WPCA, 1998)" was followed for the collection of water samples and analysis of phytoplankton, zooplankton, and benthos and macrophyte community. The water samples were collected from various places at each sampling station (Upstream, Downstream & confluence) and mixed together to prepare a composite water sample. The water bodies were visited and surveyed for general morphology, sources of pollution and the presence of macrophytes in them.

The three nallas flow for 2-3 km mostly through the forest area and then join with rivers. The bank of rivers has few villages in 10 km area around Bhilai Steel Plant. Therefore there is no other source of pollution to the river except some drains from some villages

The sampling station number and corresponding place of sampling are given in Table 1 & the diagrammatic sketch of study area is shown in Figure 1.

## Phytoplankton Community

### *Kosa nalla and Samoda nalla and the Combined nalla after their confluence*

The Kosa nalla and Samoda nalla are formed by the release of wastewater from Bhilai Steel Plant from Outlet "A" and Outlet "B" respectively. These two nallas join together before their discharge in Seonath River. The observations on phytoplankton flora in these two nallas and combined nalla after their joining are shown in Table 2.

The total phytoplankton count (number per ml) in water was observed to be 853, 1047 and 2471 in Kosa nalla, Samoda nalla and their combined nalla respectively. The community structure of phytoplankton indicate that

**Table 1**
**Sampling Stations**

| Sampling Stations | Place | Sampling Stations | Place |
|---|---|---|---|
| 1 | Kosa nalla | 6 | Confluence of Seonath and combined nalla |
| 2 | Samoda nalla | 7 | Somni nalla |
| 3 | Combined nalla | 8 | Kharoon River U/s |
| 4 | Seonath River U/s | 9 | Kharoon River D/s |
| 5 | Seonath River D/s | 10 | Confluence of Kharoon and Somni |

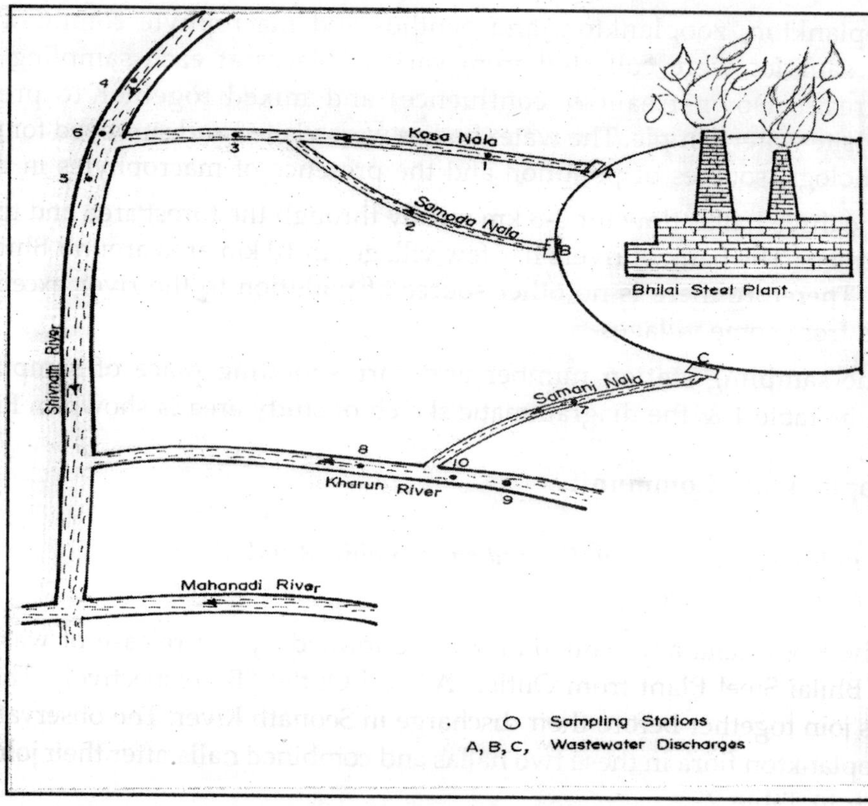


**Fig. 1: Diagrammatic Sketch of Study Area**

Bacillariophyceae (diatoms), the clean water phytoplankton indicator group, is dominant (53.8-57.6% by composition) in Kosa and Samoda nallas, while clean water indicator group (diatoms) and polluted water indicator groups (Chlorophyceae, Cyanophyceae and Euglenophyceae) are more or less equally dominant in combined nalla indicating â-mesotrophic level of nalla.

The diversity of phytoplankton is also found to be increased from 4 species each in Kosa nalla and Samoda nalla to 9 species in combined nalla indicating biodegradable nature of pollutants in the nalla and satisfactory self-purification of wastewater

*Seonath River*

The Seonath River receives the discharge of combined wastewater nalla. The total phytoplankton count (number per ml) was observed to be 1633, 2460 and 1829 respectively in upstream, downstream and confluence stretch of the river.

The phytoplankton count in upstream stretch is low. The downstream stretch of river is observed to be enriched due to discharge of nalla in the river but the count is still of much lower order and is comparable to slightly polluted river water.

The phytoplankton community structure (Table 2) showed the dominance of Chlorophyceae and Euglenophyceae in upstream river stretch, while the confluence point and downstream stretch of river showed the dominance of diatoms indicating satisfactory self purification process in the river, along with

**Table 2**
**Biological Characteristics of Phytoplankton Community (no/ml) in Different Sampling Station**

| Sr. No. | Algal Group | 1 | 2 | 3 | 4 | 5 | 6 | 7 | 8 | 9 | 10 |
|---|---|---|---|---|---|---|---|---|---|---|---|
| 1 | Cyanophyceae | 83 | 171 | 47 | 46 | -- | 37 | 760 | 148 | 210 | 378 |
| 2 | Chlrophyceae | 134 | 196 | 638 | 1058 | 984 | 597 | 105 | 296 | 294 | 244 |
| 3 | Bacillariophyceae | 489 | 563 | 115 | 161 | 660 | 828 | 70 | 2294 | 2016 | 1201 |
| 4 | Euglenophyceae | 147 | 117 | 611 | 368 | 436 | 367 | 70 | 74 | 42 | 35 |
| 5 | Pyrrophyceae | - | - | - | - | 324 | - | - | - | 84 | 35 |
| | **TOTAL COUNT** | **853** | **1047** | **2471** | **1633** | **2460** | **1829** | **1005** | **2812** | **2646** | **1893** |
| | **Number of Species** | **4** | **4** | **9** | **17** | **16** | **14** | **11** | **13** | **15** | **17** |

presence of diverse species of Chlorophyceae, Euglenophyceae and Pyrrhophyceae due to release of nutrients during self purification (degradation and stabilization of pollutants).

The downstream stretch of river showed slight reduction in clean water indicator species and appearance of diverse organic pollution indicator species. But the total count is still comparable to oligotrophic enriched water and show satisfactory self-purification process in the river.

These observations showed that the Seonath River is slightly polluted in upstream stretch, while downstream stretch of river, after receiving combined wastewater nalla, showed only slight impact on river water quality indicating good self purification capacity of Seonath River.

### *Somni nalla*

Total phytoplankton count no/l in Somni nalla was observed to be 1005 (Table 2), which again indicate low standing crop of phytoplankton which may be due to limiting concentration of some major nutrients. However, higher diversity indicated presence of diverse nutrients in Somni nalla, though may be at low concentration. The community structure of phytoplankton indicated that Cyanophyceae (75.5%) was dominant, problem due to presence of organic contamination in Somni nalla. However, presence of Bacillariophyceae at low concentration indicated that water quality was not very bad.

Thus Somni nalla is â-mesotrophic in nature with diverse but low concentration of mineral nutrients and showed slight organic contamination.

### *Kharoon River*

Kharoon River at upstream station showed phytoplankton count (no/l) of 2701 and dominance of Bacillariophyceae (81.6%) indicating nutrient rich river water with marginal organic pollution. The river is large and deep with ample flow of water. It dilutes the discharge of polluted water from Somni nalla considerably, resulting in no appreciable change in the phytoplankton flora at confluence point and at down stream point.

Kharoon River appeared to be more enriched than Seonath River but is β-mesotrophic in nature with some organic contamination. Because of large flow of river and relatively good quality of river water, only slight impact due to confluence with Somni nalla was observed on the river water quality and the pollutants in Somni nalla were observed to be self purified in the stretch of the river.

## Zooplankton Community

### *Kosa nalla, Samoda Nalla and the Combined Nalla after their confluence*

The observations on zooplankton are presented in Table 3. Kosa nalla, Samoda nalla and combined nalla formed by the union of these two nallas have shown very low count (no/l) of zooplankton viz. 4, 6 and 8 respectively. Only *Brachionus* sp. was recorded from both the nallas and combined nalla. These observations showed that physico-chemical quality of these nallas was good except some organic contamination as indicated by phytoplankton indicators of organic pollution.

### *Seonath River*

Seonath River showed the presence of only one zooplankter species from Rotifera (*Brachionus* sp.), which is the indicator of water pollution. The zooplankton count was found to be increased from 6 no/l in upstream region to 12 no/l in downstream region and 15 no/l at the confluence of wastewater stream with

**Table 3**
**Biological Characteristics Zooplankton Community (no./l) in Different Sampling Stations**

| Sr. No. | Zooplankton Group | 1 | 2 | 3 | 4 | 5 | 6 | 7 | 8 | 9 | 10 |
|---|---|---|---|---|---|---|---|---|---|---|---|
| 1 | Rotifera | 4 | 6 | 8 | 6 | 12 | 15 | 8 | 9 | 9 | 12 |
| 2 | Copepoda | - | - | - | - | - | - | 10 | 4 | 5 | 3 |
| 3 | Cladocera | - | - | - | - | - | - | - | - | 4 | - |
| | **Total count** | **4** | **6** | **8** | **6** | **12** | **15** | **18** | **13** | **18** | **15** |
| | **Number of Species** | **1** | **1** | **1** | **1** | **1** | **1** | **2** | **5** | **6** | **4** |

river (Table 3). The diversity of zooplankton is less due to less amount of phytoplankton count in river water. Presence of *Brachionus* sp. indicated presence of slight organic pollution in river water.

### *Somni nalla*

Somni nalla showed low zooplankton count viz. 18 no/l (Table 3). Copepoda (*Eubrachipus* sp.) (indicator of clean water) and Rotifera (*Keratella* sp.) (indicator of organic pollution) were observed to be dominant in Somni nalla indicating slightly polluted water quality. However, Somni nalla was observed to be more enriched than Kosa and Samoda nallas.

### *Kharoon River*

The observations on zooplankton in Kharoon River are shown in Table 3. The zooplankton count at upstream point was found to be low viz. 6 no/l indicating good water but dominance of Rotifera in the water sample indicated presence of organic pollution in the river. At confluence point, the zooplankton was observed to be marginally increased i.e 7 no/l due to dilution of wastewater stream. The downstream point of river showed the increase in zooplankton count due to enrichment of river water by the discharge of Somni nalla.

### Benthos Community

The observations on benthos community are given in Table 4. Kosa and Samoda and combined nallas showed the presence of *Branchiura* sp. (Diptera)

**Table 4**
**Biological Characteristics of Benthos Community (no./m²) in Different Sampling Stations**

| Sr. No. | Zooplankton Group | 1 | 2 | 3 | 4 | 5 | 6 | 7 | 8 | 9 | 10 |
|---|---|---|---|---|---|---|---|---|---|---|---|
| 1 | Oligochaeta | - | - | - | 6 | 4 | 4 | - | 6 | 3 | 2 |
| 2 | Diptera | 5 | 4 | 4 | - | 6 | 3 | 10 | - | 8 | 5 |
| 3 | Mollusca | - | - | - | 3 | 5 | 3 | - | - | - | - |
| | **Total count** | **5** | **4** | **4** | **9** | **15** | **10** | **10** | **6** | **11** | **7** |
| | **Number of Species** | **1** | **1** | **1** | **2** | **3** | **3** | **2** | **1** | **2** | **2** |

with density of 5 organism/m$^2$, 4 organism/m$^2$ and 4 organism/m$^2$ respectively, which is indicator of oligotrophic nature of water quality, however with slight organic pollution as indicated by phytoplankton.

Seonath River showed the dominance of Oligochaeta and Mollusca in upstream region with 9 organism/m$^2$ showing oligotrophic nature of river water. At downstream point, the river showed higher count of benthos (15 specimem/m$^2$) with dominance of Diptera and Mollusca due to enrichment of river water showing β-mesotrophic water quality. At confluence point, a count of 10 organism/m$^2$ was observed.

Somni nalla showed the dominance of Diptera with two genera namely *Branchiura* sp. and *Tubifex* sp. with higher count i.e. 10 specimem/m$^2$ showing enriched water quality. Therefore, the count of benthos in Kharoon River was found to be increased after its confluence with Somni nalla in downstream region from 6 organism/m$^2$ in upstream stretch to 11 specimem/m$^2$ in downstream stretch showing slightly polluted or β-mesotrophic water quality. The number of benthos species was again reduced to 7 organism/m$^2$ in downstream zone of self purification.

## Macrophytes

The study area have ample amount of available water due to presence of rivers, reservoirs and small ponds. This has created moist marshy areas at various places and along the shores of water bodies which are found to be covered by macrophytes (Table 5) such as *Typha* sp. and *Ipomoea* sp. The shallow water bodies formed due to soil excavation for brick-making along the Kharoon River and at other places are densely infested by submerged macrophytes such as *Hydrilla* sp. and filamentous macro algae mostly *Pithophora* sp. These water

**Table 5**
**Macrophytes Observed in Study Area**

| Sr. No. | Macrophytes | Abundance |
|---|---|---|
| 1 | *Ipomoea* sp. | D |
| 2 | *Typha anguatata* | C |
| 3 | *Typha latifolia* | C |
| 4 | *Eichhornia* sp. | R |
| 5 | *Hydrilla* sp. | C |
| 6 | *Lemna* sp. | R |
| 7 | Filamentous algae *Pithophora* sp. | C |

**D = Dominant; C = Common; R = Rare**

bodies are also covered by *Lemna* sp. and *Eichhornia* sp. (water hyacinth) at some places. The macrophytes are more common in shallow shore portions of reservoirs which may form the source of eutrophication of lake after their death and decay in annual growth cycle.

## Primary Productivity

The rate of production of phytoplankton biomass per day in water bodies is called primary productivity and is the indicator of their trophic status. Rate of primary production is high when nutrients (organic and inorganic) are available in plenty in water bodies and is low when nutrients are low or limiting or environmental conditions are not suitable for the growth of phytoplankton.

The primary productivities in different water bodies in study area were assessed and are shown in Table 6. The primary productivity in combined nalla after confluence of Kosa and Samoda nallas was observed to be 0.27 $gCm^{-2}d^{-1}$ and in Somni nalla 0.24 $gCm^{-2}d^{-1}$ which is very low due to lower level of nutrients or limiting nutrients in these nallas showing oligotrophic nature of these nallas, however, with some organic contamination as indicated by phytoplankton.

The primary productivity in Seonath River was also low (1.29 $gCm^{-2}d^{-1}$) (oligotrophic), which was marginally increased to 1.92 $gCm^{-2}d^{-1}$ (mesotrophic) after its confluence with wastewater stream of Kosa and Samoda nallas, showing that the wastewater stream was less polluted and self purification process of Seonath River was able to take care of it.

**Table 6**
**Primary Productivity in Water Bodies**

| Sr. No. | Water bodies | Primary Productivity ($gC/m^2/d$) |
|---|---|---|
| 1 | Waste nalla 500 m D/s Confluence of Kosa & Samoda nalla | 0.27 |
| 2 | Seonath River 500 m U/s Disposal | 1.29 |
| 3 | Seonath River 500 m D/s Disposal | 1.92 |
| 4 | Confluence of Seonath River & Wastewater Stream (A+B) | 1.50 |
| 5 | Somni nalla | 0.24 |
| 6 | Kharoon River U/s Disposal | 5.12 |
| 7 | Kharoon River D/s Disposal | 7.20 |
| 8 | Confluence Point of Kharoon River and Somni nalla containing wastewater from Outlet 'C' | 6.70 |

0-1.5 $gC/m^2/d$ : Oligotrophic water body
1.5-5.0 $gC/m^2/d$ : Mesotrophic water body
>5.0 $gC/m^2/d$ : Eutrophic water body

The primary productivity of Kharoon River was relatively high (5.12 $gCm^{-2}d^{-1}$) which was marginally increased to 7.2 $gCm^{-2}d^{-1}$ at downstream due to the discharge from Somni nalla in the river. Thus, the Kharoon River is able to self purify its water course even after necessary discharged pollutants from Somni nalla. This is mainly due to large flow of water in river and high dilution of Somni nalla water after its discharge.

## Conclusions

The Seonath River is less polluted than Kharoon River; however both the rivers have capability to assimilate the discharged pollutants through wastewater nallas. The wastewater nalla are not highly polluted except low level of nutrients and organic contamination which were easily degraded in river water. Macrophytes were found to infest the shallow shores of river, were also responsible for additional eutrophication through their death and decay due to annual cycle of growth. The primary productivity of all the three nallas and Seonath River was found to be low when compared with Kharoon River.

Present study has been carried out to evaluate the impact of disposal of wastewater from Bhilai Steel Plant on the aquatic community in receiving stream and river. Tropic status of nalla and river before and after disposal of wastewater was successfully assessed on the basis of phytoplankton, zooplankton, benthos and macrophytes as well as primary productivity. Three man made wastewater streams/nallas carry wastewater from Steel Plant for a distance of 4-5 kms. Two nallas viz. Kosa and Samoda showed low density of plankton. These nallas unite with each other and then meet the Seonath River. The Seonath River showed algal bloom at downstream point indicating that wastewater was not toxic and got stabilized during their course of flow in the nalla. Therefore, Seonath River got enriched at downstream point. The river water was found to be further reclaimed before its confluence with Kharoon River. Kharoon River quality was not observed to be changed due to confluence with third nalla i.e. Somni coming from steel plant. The biodiversity of both the rivers was observed to be improved due to discharge of wastewater nalla. The same observations were reflected in the data on primary productivity. The study indicated that wastewater from steel plant is suitably treated and disposed off in 4-5 km long nallas so that it is stabilized in the nallas made specially for this purpose with beneficial impact on river ecosystem.

**Acknowledgment**

The authors wish to thank Director, NEERI, Nagpur for providing facilities to carryout this work

## Reference

APHA (American Public Health Association) (1998). *Standard Methods for Examination of Water & Wastewater,* 20th edition, APHA, AWWA Washington, U.S.A.

Chandra, S. & Krishna, G. (1983). Effect of tannery waste disposal on the quality of Ganga at Kanpur. *Poll. Res.,* 2: 63-64.

Chaudhari, N. (1982). Water & Air Quality Control.:The Indian Content Central Board for the Prevention & Control of Water Pollution, New Delhi, India.

Chaudhary, V.K. and Ojha, C.S P. (1985). Environmental impact assessment of the river Ganges at Varanasi. *Proc Int. Sem. Environ. Imp. Assessment. Wat. Res.* Proj. Univ. of Roorkee, 12-14 Dec; pp. 760-771.

Citizens Report (1982). The State of Indian's environmental centre for Science & Environment, New Delhi, India.

Geevarghese, C. & Chandrasekharan, M.P. (1985). Studies on ecological impact of a newsprint factory waste on the river ecosystem. *Poll. Res.*, 4; 21-24.

Ghose, M.C & Sharma, C.B. (1988). Effect of drain water on the physico-chemical & bacteriological characters of the river Ganga at Patna, Bihar. In *"Ecology & Pollution of Indian Rivers"* (R.K. Trivedy ed). pp 255-269. Ashish Publising House, New Delhi, India.

Jagdale, M.H., Salunkhe, M.M. Vibhute. Y.B., Khodaskar, S.N & Ghuge, K.N. (1984). Pollution of Godavari river water at Nanded. *Poll. Res.*, 3: 83-84.

Kumar, A. (1999). Comparative studies of phytoplankton in two ecologically different lentic freshwater ecosystems. In: *Modern Trends in Environmental pollution & Ecoplanning* (Ed. A. Kumar). ABD Publisers. Jaipur.

Mahadevan, A., & Krishnaswamy, S. (1983). A quality profile of Vaigai (S. India). *Indian J. Environ. Hlth.*, 25: 288-299.

Shah, A.R. (1988). Physico-chemcial aspects of pollution in river Jhelum(Kashmir) 1981-1983. In *"Ecology & Pollution of Indian Rivers"* (R.K. Trivedy ed). pp 163-207. Ashish Publising House, New Delhi, India.

Shankar, V., Samgu, R.P.S., & Joshi, G.C. (1986). Impact of distillery effluents on the water quality & ecosystem of river Reh in Doon Valley. *Poll. Res.*, 5: 137-142.

Singh, O.K. & Singh, C.P. (1990). Pollution studies on river Subamarekha around industrial belt at Ranchi (Bihar). *Indian J. Environ. Hlth.* 32; 26-33.

# 17

# Impact of Cooling Water Discharges on Primary Producers of a Reservoir

*T.K. Ghosh and Santosh Zargar*

## Preamble

The alternations induced in the ambient temperature can lead to wide manifestations in species distribution and community structure (Countant and Suffern 1979, Manush *et al.* 2004). In general, elevated water temperature causes changes in species composition, species dominance, standing crop and productivity of biota including phytoplankton communities in any aquatic ecosystem. Towie (2005) reported that in North Sea, UK 15 species of fish had shifted as much as 400 kilometers in to cooler waters and 6 species of fish had moved in to deeper water in their search of cooler living conditions. Thus warm water discharges from power plants into the receiving water bodies may adversely affect aquatic ecology. A small change in temperature above 15°C can cause substantial release of phosphorus because of the exponential increase in conversion rates with increasing temperature (Kaushik 1981).

In India, all the operating power plants, which employ once-through systems for cooling, are designed for a temperature rise across the condenser of 8-10°C. The Environmental Protection Rules stipulated that the heated water discharged from a power plant to the receiving water should have a temperature not more than 7°C higher than the intake temperature. Power engineers have pointed out that if the rule is implemented indiciously the cooling water requirement of a power plant would increase substantially and major design modifications have also to be incorporated leading to escalation in the unit cost of power.

In the absence of exhaustive data on the response of aquatic organisms and ecosystems in the tropics to elevated temperatures, the only option is to draw inferences, from the experiences in the subtropical and temperate areas. Since, sufficient data on similar line are not available in tropical environment; present paper delineates phytoplankton population dynamics and various physico-chemical characteristics of the Kadra reservoir where cooling water is discharged by Kaiga Nuclear Power Plant.

## Experimental

Two units of 220 MWe each Pressurized Heavy Water Reactor (PHWR) type nuclear power plants were commissioned in the year 1999 at Kaiga site. The site is located on the left bank of the freshwater Kadra reservoir, formed due to construction of a dam on Kali river at about 58 km (by road) east of Karwar in the State of Karnataka. The dam is constructed at about 11 km downstream of the site. While the water spread area of the reservoir at FRL is 37.75 $km^2$, its dead storage capacity is 179.86 M Cum. The height of the dam above deepest foundation level at overflow section is 48 m. The once through condenser cooling water for the nuclear reactor is drawn at the rate of 1,25,205 $m^3$/hr from the Kadra reservoir and discharged back to the reservoir through a 1.6 km long open channel.

The water samples were collected from three different layers, viz. surface, 3-m depth and bottom of the water column with the help of a boat from the Kadra reservoir at nine locations (Fig. 1), namely, intake point (1), discharge point (DP: 2), 50 m from DP (3), 100 m from DP (4), 500 m from DP (5), 1000 m from DP (6), 100 m from right side (90°) of 100 m point (7), 100 m from right side (90°) of 500 m point (8) and dam site point (12 km from DP: 9) in each month from January 2002 to December 2003 for analyzing various abiotic/biotic (phytoplankton) parameters which would subsequently help in understanding the water quality and substrate suitability for aquatic organisms.. Different abiotic (dissolved oxygen, pH, turbidity, conductivity, ortho- phosphates, total phosphate, nitrates, sulphate, heavy metals etc. and biotic (phytoplankton) of water samples were estimated following standard methods (1999). A number of indices have been developed for the assessment of qualities of water & biota, and these have been reviewed by Hellawell (1978), Lagler (1956), Hanes (1980) Ghosh (1998) and Zargar and Ghosh (2006). In the present study *Shannon Wiener, Palmer Pollution, Species Evenness, Similarity* and *Habitat* indices were applied to the data, generated from phytoplankton samples of the reservoir.

Impact of temperature individually (26, 31, 33, 36°C and above) and in combination with chlorine (10, 5, 2, 1, 0.5, 0.25, 0.125 $mgl^{-1}$ and below) on select phytoplankton, isolated from Kadra reservoir water samples and cultured in their respective/ developed media, was assessed in the laboratory. In another experiment, the expression of Heat Shock Protein (HSP) 70 was assessed at 26, 31, 33, 36, and 39° C in select phytoplankton.

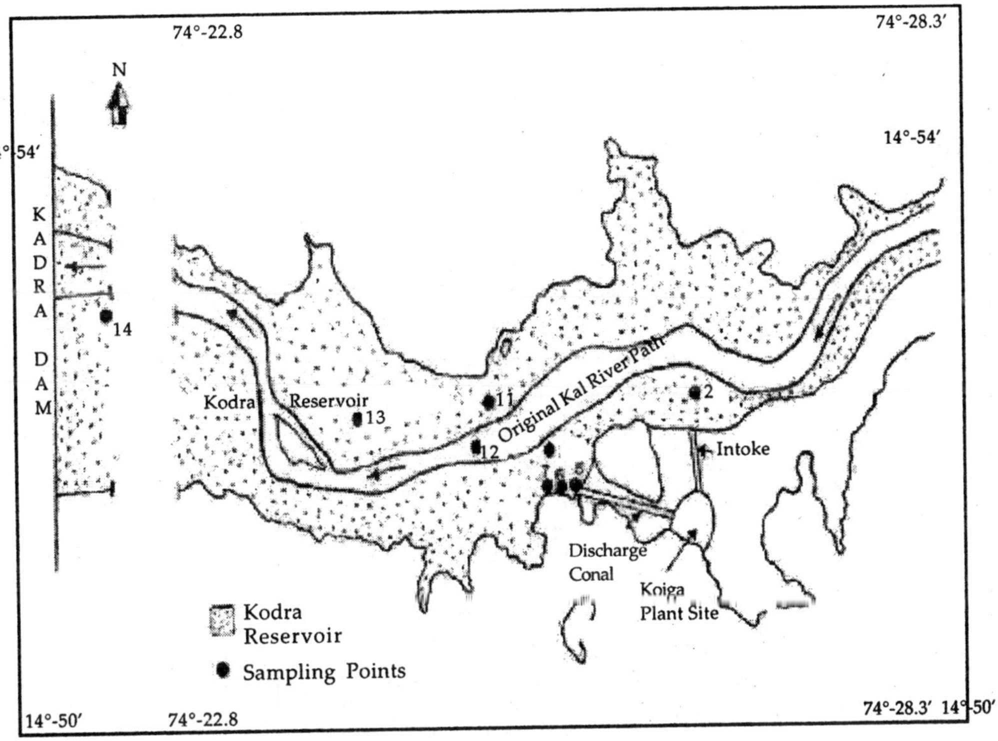


**Fig. 1: Map indicating sampling points at Kadra reservoir**

## Impact of Temperature on Abiotic Characteristics

Water temperature is one the most important abiotic parameters controlling the survival and propagation of living components of an aquatic medium. The compiled data for respective seasons at select stations (1, 2, 4 and 5) have been presented in Table1. DO was always found to be more in surface followed by middle and bottom waters during the study period. DO levels near discharge point areas, where waters had comparatively higher temperature, were observed to be marginally decreased till 100 m from right side (90°) of 100 m point. pH values of surface and 3 m depth waters were always lower irrespective of season around discharge point area. Nutrient load, in terms of phosphate, orthophosphate and nitrate, was maximum during monsoon and summer seasons as compared to winter season. However, conductivity was not found to be correlated with temperature irrespective of seasons. Total hardness, suspended solid, alkalinity and TDS did not show much difference in different seasons. Levels of heavy metals were found within the permissible limits of drinking water standard as specified in Bureau of Indian Standards-10500 – 1991. It may be assumed that heavy metals were not in alarming state in the reservoir. Although chlorine is

**Table 1**
**Physico- chemical parameters of Kadra reservoir waters at different depths during four seasons in 2002-03**

| Parameter | Location | | Winter | Summer | Monsoon | Post Monsoon |
|---|---|---|---|---|---|---|
| pH | Surface | 1 | 7.4 | 7.0 | 7.1 | 7.2 |
| | | 2 | 7.0 | 6.5 | 6.8 | 6.8 |
| | | 4 | 7.1 | 6.5 | 6.8 | 6.8 |
| | | 5 | 7.1 | 6.7 | 7.0 | 7.0 |
| | Middle | 1 | 7.4 | 7.1 | 7.2 | 7.3 |
| | | 2 | 7.3 | 6.6 | 7.0 | 7.1 |
| | | 4 | 7.4 | 6.7 | 7.1 | 7.3 |
| | | 5 | 7.4 | 6.9 | 7.2 | 7.2 |
| | Bottom | 1 | 7.6 | 7.4 | 7.6 | 7.5 |
| | | 2 | 7.5 | 6.9 | 7.3 | 7.4 |
| | | 4 | 7.4 | 6.9 | 7.3 | 7.5 |
| | | 5 | 7.4 | 7.3 | 7.5 | 7.1 |
| Dissolved Oxygen (mg/l) | Surface | 1 | 7.2 | 6.9 | 7.7 | 6.6 |
| | | 2 | 6.3 | 5.8 | 5.8 | 5.1 |
| | | 4 | 6.6 | 5.9 | 6.6 | 6.9 |
| | | 5 | 6.0 | 5.9 | 7.0 | 7.1 |
| | Middle | 1 | 6.4 | 5.7 | 7.4 | 6.3 |
| | | 2 | 5.7 | 5.0 | 5.5 | 4.9 |
| | | 4 | 6.2 | 5.1 | 6.1 | 6.8 |
| | | 5 | 6.0 | 5.2 | 6.8 | 6.6 |
| | Bottom | 1 | 5.2 | 5.2 | 5.6 | 4.8 |
| | | 2 | 5.0 | 5.0 | 4.9 | 4.9 |
| | | 4 | 4.9 | 4.8 | 5.4 | 5.5 |
| | | 5 | 4.5 | 3.8 | 6.0 | 3.8 |
| TDS (mg/l) | Surface | 1 | 58.5 | 58.5 | 58.0 | 58.0 |
| | | 2 | 54.0 | 58.3 | 55.5 | 56.0 |
| | | 4 | 53.0 | 54.3 | 54.3 | 62.5 |
| | | 5 | 55.0 | 54.5 | 52.3 | 56.5 |
| | Middle | 1 | 62.0 | 55.3 | 49.5 | 57.0 |
| | | 2 | 57.0 | 57.3 | 52.0 | 56.5 |
| | | 4 | 59.0 | 54.3 | 55.5 | 59.5 |
| | | 5 | 61.0 | 56.0 | 52.8 | 58.0 |
| | Bottom | 1 | 60.0 | 53.8 | 56.5 | 57.0 |
| | | 2 | 58.5 | 53.0 | 51.8 | 60.0 |
| | | 4 | 60.0 | 54.8 | 50.8 | 61.0 |
| | | 5 | 59.5 | 58.0 | 51.0 | 58.0 |

*Table Contd...*

| Parameter | Location | | Winter | Summer | Monsoon | Post Monsoon |
|---|---|---|---|---|---|---|
| O-phosphate (mg/l) | Surface | 1 | 0.002 | 0.007 | 0.014 | 0.006 |
| | | 2 | 0.002 | 0.015 | 0.024 | 0.005 |
| | | 4 | 0.004 | 0.006 | 0.008 | 0.005 |
| | | 5 | 0.004 | 0.017 | 0.009 | 0.009 |
| | Middle | 1 | 0.005 | 0.010 | 0.012 | 0.006 |
| | | 2 | 0.011 | 0.013 | 0.024 | 0.004 |
| | | 4 | 0.004 | 0.015 | 0.012 | 0.008 |
| | | 5 | 0.013 | 0.013 | 0.014 | 0.006 |
| | Bottom | 1 | 0.003 | 0.021 | 0.010 | 0.006 |
| | | 2 | 0.010 | 0.026 | 0.013 | 0.013 |
| | | 4 | 0.005 | 0.010 | 0.015 | 0.008 |
| | | 5 | 0.004 | 0.022 | 0.014 | 0.011 |
| Total Phosphate (mg/l) | Surface | 1 | 0.04 | 0.04 | 0.05 | 0.04 |
| | | 2 | 0.05 | 0.06 | 0.05 | 0.03 |
| | | 4 | 0.05 | 0.08 | 0.04 | 0.08 |
| | | 5 | 0.05 | 0.05 | 0.07 | 0.03 |
| | Middle | 1 | 0.08 | 0.10 | 0.07 | 0.03 |
| | | 2 | 0.07 | 0.09 | 0.06 | 0.07 |
| | | 4 | 0.05 | 0.03 | 0.04 | 0.03 |
| | | 5 | 0.05 | 0.05 | 0.05 | 0.05 |
| | Bottom | 1 | 0.07 | 0.05 | 0.06 | 0.05 |
| | | 2 | 0.06 | 0.08 | 0.05 | 0.05 |
| | | 4 | 0.05 | 0.07 | 0.06 | 0.04 |
| | | 5 | 0.05 | 0.08 | 0.06 | 0.05 |
| Nitrate (mg/l) | Surface | 1 | 0.27 | 0.50 | 0.28 | 0.47 |
| | | 2 | 0.35 | 0.44 | 0.49 | 0.44 |
| | | 4 | 0.39 | 0.53 | 0.52 | 0.53 |
| | | 5 | 0.43 | 0.53 | 0.56 | 0.43 |
| | Middle | 1 | 0.54 | 0.50 | 0.28 | 0.48 |
| | | 2 | 0.35 | 0.44 | 0.36 | 0.37 |
| | | 4 | 0.32 | 0.49 | 0.67 | 0.53 |
| | | 5 | 0.19 | 0.71 | 0.43 | 0.47 |
| | Bottom | 1 | 0.34 | 0.56 | 0.32 | 0.50 |
| | | 2 | 0.27 | 0.45 | 0.40 | 0.39 |
| | | 4 | 0.37 | 0.57 | 0.51 | 0.70 |

*Table Contd...*

| Parameter | Location | | Winter | Summer | Monsoon | Post Monsoon |
|---|---|---|---|---|---|---|
| Alkalinity (mg/l) | Surface | 1 | 34.8 | 36.3 | 33.0 | 36.4 |
| | | 2 | 32.8 | 36.1 | 33.6 | 34.5 |
| | | 4 | 34.5 | 33.2 | 33.4 | 38.5 |
| | | 5 | 35.0 | 33.8 | 30.9 | 37.2 |
| | Middle | 1 | 37.1 | 32.8 | 31.4 | 35.0 |
| | | 2 | 37.3 | 35.6 | 31.0 | 30.7 |
| | | 4 | 35.5 | 33.2 | 32.0 | 37.5 |
| | | 5 | 35.0 | 34.0 | 29.8 | 36.0 |
| | Bottom | 1 | 37.5 | 34.6 | 30.8 | 34.5 |
| | | 2 | 38.7 | 34.0 | 29.9 | 36.3 |
| | | 4 | 37.3 | 34.5 | 31.7 | 36.0 |
| | | 5 | 37.3 | 36.2 | 30.5 | 36.5 |
| Conductivity (mSi) | Surface | 1 | 108.5 | 114.7 | 107.3 | 107.0 |
| | | 2 | 104.9 | 106.0 | 102.7 | 107.0 |
| | | 4 | 114.1 | 105.1 | 105.5 | 115.0 |
| | | 5 | 112.5 | 101.9 | 100.2 | 114.0 |
| | Middle | 1 | 101.8 | 115.2 | 106.3 | 106.5 |
| | | 2 | 117.7 | 95.5 | 93.1 | 109.5 |
| | | 4 | 119.0 | 95.6 | 103.0 | 110.5 |
| | | 5 | 116.6 | 105.0 | 97.2 | 114.5 |
| | Bottom | 1 | 104.9 | 106.0 | 102.7 | 107.0 |
| | | 2 | 119.0 | 107.8 | 103.0 | 110.5 |
| | | 4 | 111.5 | 113.1 | 97.9 | 115.0 |
| | | 5 | 116.6 | 105.0 | 97.2 | 114.5 |
| Suspended Solid (mg/l) | Surface | 1 | 3.5 | 7.9 | 5.1 | 5.9 |
| | | 2 | 8.0 | 5.8 | 7.7 | 8.9 |
| | | 4 | 6.5 | 4.8 | 6.5 | 7.9 |
| | | 5 | 3.5 | 5.1 | 7.2 | 6.4 |
| | Middle | 1 | 8.5 | 13.8 | 8.1 | 7.7 |
| | | 2 | 13.0 | 5.4 | 9.3 | 10.7 |
| | | 4 | 7.5 | 7.0 | 8.9 | 8.5 |
| | | 5 | 5.0 | 11.5 | 10.5 | 9.5 |
| | Bottom | 1 | 27.5 | 33.0 | 19.9 | 13.0 |
| | | 2 | 14.0 | 12.6 | 13.7 | 22.7 |
| | | 4 | 16.5 | 12.0 | 14.5 | 10.8 |
| | | 5 | 7.1 | 14.2 | 16.2 | 18.0 |

*Table Contd...*

| Parameter | Location | | Winter | Summer | Monsoon | Post Monsoon |
|---|---|---|---|---|---|---|
| Total Hardness (mg/l) | Surface | 1 | 46 | 43.5 | 39.5 | 42 |
| | | 2 | 42 | 39.5 | 34.6 | 41 |
| | | 4 | 43 | 40 | 36.5 | 43 |
| | | 5 | 42 | 39.5 | 37 | 40 |
| | Middle | 1 | 42 | 41.5 | 39 | 39 |
| | | 2 | 46 | 39 | 33.85 | 39 |
| | | 4 | 44 | 40.5 | 36 | 41 |
| | | 5 | 42 | 42 | 35.5 | 42 |
| | Bottom | 1 | 42 | 40.5 | 38 | 42 |
| | | 2 | 44 | 40.5 | 36.1 | 39 |
| | | 4 | 44 | 40 | 36.6 | 39 |
| | | 5 | 42 | 41 | 35 | 44 |
| Temperature (°C) | Surface | 1 | 27.8 | 31.3 | 29.6 | 28.6 |
| | | 2 | 31.0 | 36.2 | 33.5 | 33.1 |
| | | 4 | 31.0 | 35.3 | 33.3 | 32.7 |
| | | 5 | 29.8 | 33.6 | 31.8 | 32.6 |
| | Middle | 1 | 27.8 | 30.2 | 29.3 | 26.3 |
| | | 2 | 28.0 | 34.6 | 31.2 | 29.4 |
| | | 4 | 29.8 | 33.1 | 30.0 | 27.7 |
| | | 5 | 28.0 | 30.8 | 29.3 | 30.0 |
| | Bottom | 1 | 26.0 | 27.6 | 25.8 | 24.4 |
| | | 2 | 27.3 | 31.9 | 28.3 | 26.5 |
| | | 4 | 28.0 | 31.2 | 28.4 | 26.1 |
| | | 5 | 27.0 | 28.6 | 27.2 | 22.0 |

normally applied to prevent biological fouling in pipes and culverts of the power station cooling systems, it could not be detected at discharge point. Apparently, enhanced water temperature, due to discharge of cooling water in the reservoir, did not cause remarkable adverse impact on abiotic characteristics beyond 100 m from the discharge point (DP) of the Kadra reservoir water.

## Effects of Temperature on Phytoplankton Population

In the present investigation, the pattern of distribution, and density of phytoplankton with respect to temperature were carried out in each month for two years for evaluating potential productiveness at different sampling locations.

Variability of plankton distribution is reflected by prevailing environmental parameters, which cause spatial difference in the abundance and diversity of plankton community. Thus spatial and temporal changes in phytoplankton communities depend on physical and biological conditions of the ecosystem. During the study period, 49 genera of phytoplankton were identified from water samples of Kadra reservoir. Four major groups, viz. bacillariophyceae, cyanophyceae, chlorophyceae and pyrohphyceae were recorded.

### *Temporal Variation*

During the period of monitoring, the groups cyanophyceae, chlorophyceae and bacillariophyceae dominated the other groups in all the seasons. Phytoplankton counts were found to be minimum in the month of August 2002 at bottom water (4 no.ml$^{-1}$) and maximum in March 2003 at surface water (868 no.ml$^{-1}$).Counts were maximum at surface waters as compared to middle and bottom layers in all the seasons. In general, peaks of phytoplankton were found in winter (2002-03) and summer (2003) seasons in all the layers of select four sampling points. However, a similar trend was not found in 2002. Phytoplankton were less at surface, middle and bottom waters of all the nine sampling points during monsoon as compared to other seasons. A DT of 7.3°C at discharge point triggered 67 percent reduction of phytoplankton in winter-2003, while only 45 percent was reduced at 3.4°C DT in monsoon-2003. Goldman and Home (1983) reported that seasonal appearance of algae is predominantly regulated by temperature and light intensity with the other physical factors. In the present investigation, the members of chlorophyceae and cyanophyceae outnumbered the other groups, viz. bacillariophyceae, pyrohphyceae and euglenophyceae in surface and middle waters irrespective of seasons. According to Venkateswarlu (1981), occurrence of low number of diatoms may be due to the dominance of cyanophyceae which generally do not allow other algae to grow simultaneously. In bottom waters, diatoms were the major components of phytoplankton. Zhang and Prepas (1996) found dominance of diatoms in lower water columns in lakes of Alberta, Canada, while Zafar (1964) had observed lesser number of diatoms in ponds with excess of organic matter. Kadra reservoir, from where samples were collected, is known to possess poor organic matter at bottom (Madhyastha *et al.* 2002).

### Spatial Variation

Phytoplankton counts were gradually increased in straight line starting from DP in all the seasons of surface and middle waters, whereas in the bottom waters, the same pattern was not found. In general, phytoplankton count was less within 1000 m from DP as compared to intake and dam site points. DP exhibited comparatively less count of phytoplankton amongst all the sampling points during the study period. Beyond discharge point, phytoplankton populations were increased gradually and were restored to nearly 90 and 100 percent of normal (intake) levels at a distance of 500m and 1km respectively from the DP. Zargar and Ghosh (2005) reported that the chlorophyll-a values were found to be increased from DP towards dam site, however, the rate of carbon fixation by algae was found to be low upto 500 m as compared to intake point. Student's t-test showed significant differences of densities at DP, 50 m from DP, 100 m from DP and 100 m from right side of 100 m points ($P < 0.05$) while comparing with intake point. Although chlorophyceae and bacillariophyceae were the major groups as found near discharge point area, filamentous members, i.e. *Anabaena* sp., *Oscillatoria* sp., *Phormidium* sp., *Spirulina* sp. etc. of group cyanophyceae were also found in the same area. Amongst the cyanophyceae group, the densities of filamentous members were more near discharge point area as compared to other areas. This resulted in comparatively lesser density of cyanophyceae members near discharge point with reference to intake point. Literature reveals that blue green algae are generally most tolerant to thermal discharges, though these are dominant when temperature exceeds 32°C for long periods (Langford 1983). Further, Krenkel and Parker (1969) reported that when the temperature exceeded 34.5°C, there was an increase in blue green algal species in the areas near the discharge and 200 m downstream. However, Turoboyski (1973) reported a few species of *Navicula*, belonging to group bacillariophyceae, were most favored in heated waters (27-30°C).

## Correlation of Phytoplankton with Temperature and Other Parameters

The correlation coefficient between phytoplankton with temperature and other parameters were carried out at 5 percent significance levels. While significant positive correlations of total phytoplankton were observed with conductivity, TDS, and dissolved oxygen; significant negative correlation was found with temperature($R=-0.27$).Earlier, Saran and Adoni (1985) and Tongeren *et al.* (1992) found negative correlation of phytoplankton with increase of temperature. The groups cyanophyceae and bacillariophyceae had positive correlations with conductivity, TDS, alkalinity and total hardness. Direct correlation ($R=0.34$) was observed between pyrohphyceae and dissolved oxygen. Group chlorophyceae was found to be positively correlated with conductivity ($R=0.20$), TDS ($R=0.24$) and dissolved oxygen ($R=0.31$). While phytoplankton and phosphorus did not show any significant correlation, the importance of phosphorus in controlling the standing stock of fresh water phytoplankton has been well documented in many studies (Pearsall 1930, Sommer *et al.* 1990, Laxminarayan and Somshekhar *et al.* 2002, Water Quality Glossary; www.ew.govt.nz).In the present studies, temperature is the principle factor for reduced phytoplankton density within 500 m from DP. This resulted in non-correlation of phytoplankton population and nutrients within the study area, where majority of the sampling points are within 500 m from DP. Total phytoplankton has strong correlation with their groups chlorophyceae (R=0.82) followed by bacillariophyceae (R=0.72) and pyrrhophyceae (R=0.55). Group cyanophyceae was found to be positively correlated with the other groups, viz. chlorophyceae (R=0.25), bacillariophyceae (R=0.20) and pyrrhophyceae (R=0.19).However, bacillariophyceae had positive correlations with chlorophyceae (R=0.42) and pyrrhophyceae (R=0.32).

**Diversity Indices**

Shannon Wiener diversity values were generally higher in winter and summer seasons as compared to monsoon season (Fig. 2). Minimum diversity of phytoplankton was found at 20 m from DP in monsoon when the temperature was found to be 34°C. Martin Creek Power station on the Delaware showed distinct reductions in species-diversity when temperature in the discharge point had reached 37°C (Langford 1983). The alterations in the phytoplankton community were only partially reflected in a slight decrease in species diversity at the heated water area (DP and 100m from DP, points) as depicted in Fig 2. Several studies have shown that species diversity of phytoplankton was decreased in areas consistently heated over 30°C (Patric 1969, Trembley 1960, Simmons 1974, Langford 1983). In contrast, Carpenter (1973) and McMahon and Docherty (1975) had reported that phytoplankton diversity might increase with small

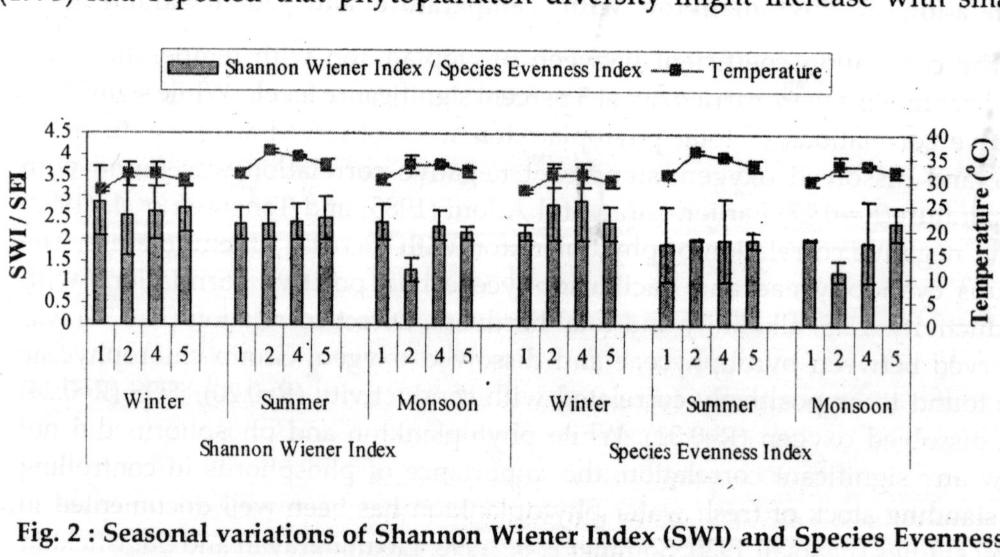


**Fig. 2 : Seasonal variations of Shannon Wiener Index (SWI) and Species Evenness Index (SEI) of phytoplankton at select points of Kadra reservoir**

temperature increments (5 – 10°C) provided that the maximum temperature did not exceed 32 – 35°C. It is widely accepted that species diversity decreases in aquatic communities under stress (Odum 1971, Wetzel 1975). This may be applicable to cases of more severe pollution, but in the present study the species diversity index did not alter remarkably due to alterations in phytoplankton structure and dynamics. Since pollution tolerant species replaces the susceptible species, Green and Vascoto (1978) had earlier opined that the relation between high diversity and high environment quality was not always valid. Species Evenness Index (SEI) values, as depicted in Fig. 2, varied from 1.27 (monsoon, DP, 33.5°C) to 2.91 (winter, 100 m from DP, 31.5°C). The values indicate that distribution of phytoplankton genera was most uneven at DP as compared to 100 m from DP, where the genera were most evenly distributed. Palmer Pollution Index (PPI) value (0) support the oligotrophic nature of the lake water at all sampling points throughout the study period. The PPI value was 'zero' since none of the indicator phytoplankton genera was found above 50 no. ml$^{-1}$ in any of the water sample collected from different sampling points of Kadra reservoir. Similarity Index values indicating availability of similar types of genera amongst certain points like intake & DP, DP & 500 m etc. were poor during most of the seasons (Fig 3). The decrease in Sorenson's similarity index between these stations in this study indicated less floral overlap and predictability of community composition. The decrease in predictability of community composition could be correlated with increase in water temperature due to discharge of warm water in the reservoir. However, increased (>0.6) similarity indices were recorded in stations between intake & 500 m, intake & 1000 m, intake & dam site, DP & 50 m, DP & 100 m, 1000 m & dam site and 1000 m & 100 m from right side (90° angle) of 500 m. Habitat Index of each genus of plankton was evaluated in two zones of the reservoir (Table 2), namely, Zone 1 comprising of sampling points within 500 m from DP (DP, 50 m from DP,100 m from DP,100 m from right side (90$^0$ angle) of 100 m point,500 m from DP and100 m from right side (90$^0$ angle) of 500 m point) and Zone 2 covering the points beyond 500 m from DP (intake, 1000 m from DP and dam site). The index values reveal that *Melosira* sp. and *Anacystis* sp. were the most preferred phytoplankton at zone1 and zone 2 respectively. Kumar (1990) reported that the extra load of certain algae, such as, *Melosira* sp., *Anacystis* sp. etc. is an indication of nutrient enrichment of the water body. However, in Kadra reservoir, phytoplankton were not found to be correlated with nutrients significantly. Although, the average temperature values at zone 1 were found to be 33°C, zone 2 had 30°C. Thus, it may be inferred that the temperature 33°C and 30°C may be favorable for *Chlorella* sp. and *Anacystis* sp. respectively. According to Hickmann (1974), temperature ranges 27-36°C are

**Table 2**
**Ranking of phytoplankton, based on Habitat Index in two zones of the Kadra reservoir**

| S No. | Phytoplankton | Zone 1 (within 500m from discharge point) | | Zone 2 (beyond 500m from discharge point) | |
|---|---|---|---|---|---|
| | | Habitat Index | Ranking | Habitat Index | Ranking |
| 1 | *Melosira* sp. | 10.15 | 1 | 10.17 | 2 |
| 2 | *Chlorella* sp. | 9.83 | 2 | 9.31 | 3 |
| 3 | *Anacystis* sp. | 8.73 | 3 | 10.52 | 1 |
| 4 | *Cyclotella* sp. | 5.62 | 4 | 4.14 | 7 |
| 5 | *Ceratium* sp. | 5.47 | 5 | 2.22 | 14 |
| 6 | *Oscillatoria* sp. | 5.03 | 6 | 3.4 | 11 |
| 7 | *Navicula* sp. | 4.53 | 7 | 1.27 | 22 |
| 8 | *Scenedesmus* sp. | 4.45 | 8 | 4.7 | 6 |
| 9 | *Staurastrum* sp. | 4.23 | 9 | 6.15 | 4 |
| 10 | *Chlorococcum* sp. | 4.16 | 10 | 4.75 | 5 |
| 11 | *Fragillaria* sp. | 3.11 | 11 | 3.61 | 9 |
| 12 | *Euastrum* sp. | 2.65 | 12 | 1.97 | 15 |
| 13 | *Ankistrodesmus* sp. | 2.49 | 13 | 2.95 | 12 |
| 14 | *Nitzschia* sp. | 2.46 | 14 | 0.66 | 29 |
| 15 | *Synechocystis* sp. | 2.42 | 15 | 1.83 | 17 |
| 16 | *Stauroneis* sp. | 2.13 | 16 | 0.19 | 40 |
| 17 | *Pediastreum* sp. | 2.12 | 17 | 0.79 | 27 |
| 18 | *Desmedium* sp. | 2.1 | 18 | 1.54 | 21 |
| 19 | *Chroococus* sp. | 1.96 | 19 | 3.99 | 8 |
| 20 | *Rhizosolenia* sp. | 1.71 | 20 | 1.04 | 24 |
| 21 | *Microspora* sp. | 1.58 | 21 | 0.55 | 30 |
| 22 | *Dictylococcus* sp. | 1.49 | 22 | 1.95 | 16 |
| 23 | *Merismopedia* sp. | 1.38 | 23 | 1.54 | 21 |
| 24 | *Phormedium* sp. | 1.25 | 24 | 0.86 | 26 |
| 25 | *Synedra* sp. | 1.22 | 25 | 2.22 | 14 |
| 26 | *Arthrodesmus* sp. | 1 | 26 | 3.41 | 10 |
| 27 | *Ulothrix* sp. | 0.98 | 27 | 2.49 | 13 |
| 28 | *Closteridium* sp. | 0.8 | 28 | 1.15 | 23 |
| 29 | *Frustulia* sp. | 0.79 | 29 | Absent | NA |

| | | | | | |
|---|---|---|---|---|---|
| 30 | *Raphidiopsis* sp. | 0.63 | 30 | 1.56 | 20 |
| 31 | *Scleriopsis* sp. | 0.52 | 31 | 0.37 | 35 |
| 32 | *Apimyxa* sp. | 0.44 | 32 | 0.97 | 25 |
| 33 | *Mougeotia* sp. | 0.44 | 33 | 0.41 | 33 |
| 34 | *Tetraedron* sp. | 0.4 | 34 | 0.22 | 38 |
| 35 | *Asterionella* sp. | 0.29 | 35 | 0.47 | 31 |
| 36 | *Scytonema* sp. | 0.24 | 36 | Absent | NA |
| 37 | *Closteriopsis* sp. | 0.23 | 37 | 0.23 | 37 |
| 38 | *Pinnularia* sp. | 0.21 | 38 | 0.35 | 36 |
| 39 | *Oocystis* sp. | 0.16 | 39 | Absent | NA |
| 40 | *Selenastrum* sp. | 0.16 | 39 | 0.09 | 41 |
| 41 | *Mastogloia* sp. | 0.14 | 40 | Absent | NA |
| 42 | *Aphynomyzone* sp. | 0.12 | 41 | 0.45 | 32 |
| 43 | *Phacus* sp. | 0.1 | 42 | 0.4 | 34 |
| 44 | *Cerasterius* sp. | 0.08 | 43 | 0.41 | 33 |
| 45 | *Cocconeis* sp. | Absent | NA | 0.2 | 39 |
| 46 | *Cosmarium* sp. | Absent | NA | 1.58 | 19 |
| 47 | *Gomphonema* sp. | Absent | NA | 0.55 | 30 |
| 48 | *Spirulina* sp. | Absent | NA | 1.64 | 18 |
| 49 | *Zygnema* sp. | Absent | NA | 0.73 | 28 |

NA: Not Applicable

favorable to *Chlorella* sp. Although *Zygnema* sp., *Spirulina* sp., *Gomphonema* sp. etc. were not recorded at zone 1, these were found to be present at zone 2. It, thus, clearly indicates that these phytoplankters avoid higher temperature, and prefer zone 2 possessing comparatively less temperature (30°C).Wallace (1955) reported optimum growth temperature of *Gomphonema* sp. between 22-26 $^0$C, which is much lesser than zone 1. Lower rankings of *Cerasterias* sp., *Selenastrum* sp. etc. indicated that these genera did not prefer the Kadra reservoir habitat irrespective of temperature.

## Laboratory Studies

Laboratory studies revealed that during the experiment, optimum growth temperature for the select algae, viz. *Chlorella vulgaris*, *Scenedesmus quadricauda* and *Chroococcus prescotii* were estimated as 26°C but it showed considerable growth up to 36°C, which was found to be the limit of thermotolerence of the select algae. However, 0.25 mgl$^{-1}$ of chlorine adversely affected the growth at 36°C (Zargar and Ghosh 2007). On the other hand, Zargar *et al.* (2006$_{a,b}$) detected HSP 70 at 39°C in select algae. It may be added that the expression of HSP 70

**Winter**

| Sampling point | 1 | 2 | 3 | 4 | 5 | 6 | 7 | 8 | 9 |
|---|---|---|---|---|---|---|---|---|---|
| 1 | 1 | 0.47 | 0.47 | 0.53 | 0.61 | 0.72 | 0.46 | 0.63 | 0.77 |
| 2 | | 1 | 0.81 | 0.79 | 0.59 | 0.66 | 0.58 | 0.51 | 0.4 |
| 3 | | | 1 | 0.64 | 0.62 | 0.64 | 0.45 | 0.59 | 0.6 |
| 4 | | | | 1 | 0.59 | 0.6 | 0.76 | 0.6 | 0.59 |
| 5 | | | | | 1 | 0.61 | 0.42 | 0.53 | 0.73 |
| 6 | | | | | | 1 | 0.42 | 0.68 | 0.67 |
| 7 | | | | | | | 1 | 0.3 | 0.46 |
| 8 | | | | | | | | 1 | 0.55 |
| 9 | | | | | | | | | 1 |

**Summer**

| Sampling point | 1 | 2 | 3 | 4 | 5 | 6 | 7 | 8 | 9 |
|---|---|---|---|---|---|---|---|---|---|
| 1 | 1 | 0.36 | 0.59 | 0.35 | 0.63 | 0.77 | 0.49 | 0.74 | 0.66 |
| 2 | | 1 | 0.83 | 0.88 | 0.5 | 0.37 | 0.68 | 0.43 | 0.46 |
| 3 | | | 1 | 0.8 | 0.49 | 0.6 | 0.71 | 0.41 | 0.37 |
| 4 | | | | 1 | 0.38 | 0.37 | 0.69 | 0.4 | 0.55 |
| 5 | | | | | 1 | 0.75 | 0.48 | 0.74 | 0.66 |
| 6 | | | | | | 1 | 0.53 | 0.81 | 0.64 |
| 7 | | | | | | | 1 | 0.54 | 0.54 |
| 8 | | | | | | | | 1 | 0.69 |
| 9 | | | | | | | | | 1 |

**Monsoon**

| Sampling point | 1 | 2 | 3 | 4 | 5 | 6 | 7 | 8 | 9 |
|---|---|---|---|---|---|---|---|---|---|
| 1 | 1 | 0.36 | 0.59 | 0.35 | 0.63 | 0.77 | 0.49 | 0.74 | 0.75 |
| 2 | | 1 | 0.83 | 0.88 | 0.5 | 0.37 | 0.68 | 0.43 | 0.4 |
| 3 | | | 1 | 0.8 | 0.49 | 0.6 | 0.71 | 0.41 | 0.36 |
| 4 | | | | 1 | 0.38 | 0.37 | 0.69 | 0.4 | 0.28 |
| 5 | | | | | 1 | 0.75 | 0.48 | 0.74 | 0.58 |
| 6 | | | | | | 1 | 0.53 | 0.81 | 0.82 |
| 7 | | | | | | | 1 | 0.54 | 0.53 |
| 8 | | | | | | | | 1 | 0.76 |
| 9 | | | | | | | | | 1 |

**Fig. 3 : Seasonal variation of Similarity Index of phytoplankton among different points of Kadra reservoir**

was assessed at 26, 31, 33, 36, and 39° C in *Chlorella Vulgaris* and the higher level of expression was observed only at 39° C in comparison with control (Fig. 4). Further, population of plankton in Kadra reservoir, at a distance beyond 500 m from discharge point, was comparable to upstream (intake) water. The area upto 500 m from discharge point (average surface temperature; 31.8 to 34.2°C) may be treated as mixing zone. Investigation revealed that the average temperature of the water beyond this zone varied between 29.4 and 31.6°C, which satisfy the laboratory determined MATL (maximum allowable temperature limit for the alga: 36°C).

Studies revealed that the heated effluents from Kaiga Nuclear Power plant caused changes in dissolved oxygen and pH of water, species composition, species dominance, standing crop and productivity of phytoplankton communities within 500 m from End of Discharge Canal point of Kadra reservoir when two units were running in full capacity. The outcome of this research suggests that temperature 36° C may be considered as the maximum allowable temperature limit for the alga. Since applications of stress protein to environmental monitoring are still limited, the present study provides validation for the development of HSP as a biomarker of exposure and effect. The data base generated will serve as a reference source for initiation of review/reaffirm the thermal water quality standards and also would be useful for the management of thermally stressed fresh water reservoirs.

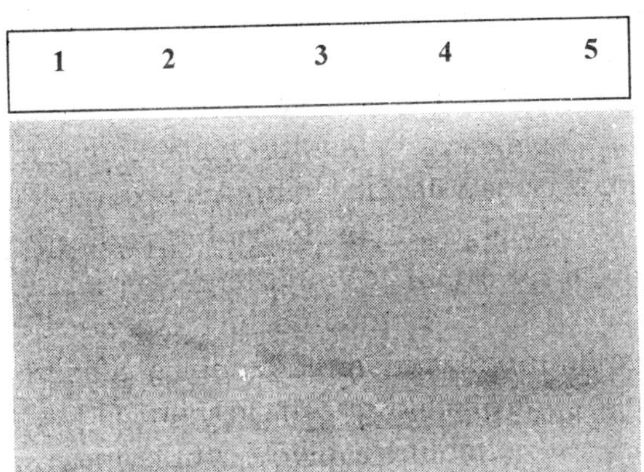


1: Control; 2: 0 hour; 3: 1 hour; 4: 3 hour; 5: 6 hour
(Lanes 2-5 indicate HSP70 expressions after exposure of the alga at 39° C; however, Lane 1 indicates the same of the alga at 26° C)

**Fig. 4. Western blots (HSP70) of freshwater alga, Chlorella vulgaris at different hours of exposure at 39° C**

## Conclusions

The aberrations in the ambient temperature can lead to wide manifestations in species compositions of biota. In general, elevated water temperature causes changes in species diversity, dominance, standing crop and productivity of biota including phytoplankton communities in any aquatic ecosystem. In view of this, warm water discharges from power plants into the receiving water bodies may adversely affect aquatic ecology. In the absence of exhaustive data on the response of aquatic organisms and ecosystems in the tropics to elevated temperatures, the only option was to draw inferences from the experiences in the subtropical and temperate areas. Since sufficient data on similar line are not available in tropical environment; present paper delineates phytoplankton population dynamics and various physicochemical characteristics of the Kadra reservoir where cooling water is discharged. The study suggests the heated effluents from Kaiga Nuclear Power plant caused changes in dissolved oxygen and pH of water, species composition, species dominance, standing crop and productivity of biota including phytoplankton communities within 500 m from End of Discharge Canal point of Kadra reservoir when two units were in operation at full capacity.

Laboratory studies revealed that optimum growth temperature for select algae, viz. *Chlorella vulgaris*, *Scenedesmus quadricauda* and *Chroococcus prescotii* was 26°C but it showed considerable growth up to 36°C, which was found to be the limit of thermotolerance of the algae. However, 0.25 $mgl^{-1}$ of chlorine adversely affected the growth at 36°C. On the other hand, HSP 70 was detected at 39°C.

## References

Coutant, C.C. & Suffern, J.S. (1979). Temperature influences on growth of aquatic organisms. *in*: Lee S.S., Sengupta, (ed) Waste heat management and utilization. Hemisphere Publ. Corp., Washington, D.C., pp. 113-124.

Ghosh, T.K. (1998). Application of different indices for assessment of aquatic biota in EIA studies; *in*: K.D. Mishra (ed) Ecology of Polluted Waters and Toxicology. Technoscience Publications, Jaipur, India, pp. 1-18.

Goldman, C.R. and Home A.J. (1983). Limnology. Pub. Mc Graw Hill Inc. Japan.

Green, R.H. and Vascoto, G.L. (1978). A method for the analysis of environmental factors controlling patterns of species compositions in aquatic communities. *Wat. Res.* 12: 583-590.

Hanes, T. (1980). Vegetation and wildlife impact analysis *in*: Environmental Impact Analysis Handbook (Eds: J.G. Rrau and D.C. Wooten). McGraw-Hill Book Comp., New York, pp. 7.1-7.59.

Hellawell, J.M. (1980). Biological surveillance of rivers. Water Research Centre, Medmenham.

Hickman, M. and Klarer, D.M. (1974). The growth of some epiphytic algal in a lake receiving thermal effluent. *Arch. Hydrobiol* 74 : 403-426.

Kaushik, N. K. (1981) Water Pollution and Management Reviews. (Varshney, C. K. C. ed.). South Asian Publishers Pvt. Ltd., New Delhi.

Krenkel, P. A. and Parker F. L. (1969). Biological aspects of thermal pollution, Vanderbilt University press.

Lagler, K.F. (1956). Freshwater fishery biology. WMC Brown Co., Dubuque, Iowa .

Langford, T.E. (1983). Electricity generation and the ecology of natural waters. Liverpool University Press, Liverpool.

Madhyastha, M.N., Sridhar K.P., Tanuja K. and Arun A.B. (2002). Studies on benthic ecology and themed tolerance of some selected species from Kaiga Environment. p. 114-121 *in* N.B. Venkatramani and N. Sukumaran *(eds)* Proc. Nat. Symposium on Thermal ecology. Tirunelveli, India.

Manush, S.M., Pal, A.K., Chatterjee, N., Das, T. & Mukherjee, S.C. (2004). Thermal tolerance and oxygen consumption of Macrobrachium rosenbergii acclimated to three temperatures. J. Therm. Biol. 29: 15-19.

Mcmahon, J.W. and Docherty, A.E. (1975). Effects of heat enrichment on species succession and primary production in fresh water plankton. p 529 - 45 *in* IAEA. Environmental effects of cooling systems at Nuclear Power Stations, Vol. 1, Vienna: International Atomic Energy Agency.

Odum, E.P. (1971). Fundamentals of ecology. W.B. Samunders co. Philadelphia London, Toronto.

Patrick, R. (1969). Some effects of temperature on freshwater algae. p 161-185. *in* F. L. Parker and P. A. Krenkel *eds*. Biological aspects of thermal pollution. Vanderbilt University Press, USA.

Pearsall, W.H. (1930). Phytoplankton in the English lakes. The proportions in the water of some dissolved substances of biological importance. *Ecol.* 18: 306.

Saran, H. M. and Adoni, A.D. (1985). Limnological studies on seasonal variation in phytoplanktonic primary productivity in Sagar lake. p 71-76 *in* A. D. Adoni *ed.* Proc. Nat. Symp. Pure & Appl. Limnology. Bot. Soc., Sagar.

Simmons, C.N. (1974). An ecological extraction of heated water discharge on phytoplankton blooms in the Potomac River. *Hydrobiologia,* 45 (4): 441-466.

Standard methods for the examination of water and wastewater (1999). 20th edition, APHA, AWWA and WPCF, Washington, D.C.

Tongeren, U.F.R. van and Liers, L. Van, Gulati, R.D., Potema, G., Boesewinkel, P.J. and Bruyn, De (1992). Multivariate analysis of the plankton communities in the Loosdercht lakes : relationship with the chemical and physical environment. *Hydrobiologia* 233: 105-117.

Towie N (2005). Fish get hooked on cooler towers. http://www.bioedonline.org/news

Trembley, F.J. (1960). Research projects on effects of condenser discharge water on Aquatic life Progress Report 1956-1959. Bethlehom, Pennoylvania; Institute of Research, Lehigh University.

Turoboyski, L. (1973). Investigation on influence of heated waters from the power station at Skawina on Skawina and Vistula rivers, Poland. *Arch Hydrobiologia* 20(3): 443-460.

Venkateswarlu, V. (1981). Algal as indicators of river water quality and pollution. *in* Biological Indicators of Environmental Pollution. Proc of the WHO workshop, Osmania University, Hyderabad, India.

Vijayalakshmi R., Nair, S.N., Gajbhiye and Sayeed, F.H. (1983). Organic carbon content of zooplankton from the nearshore waters Bombay. *Ind. J. Mar. Sci.*12:183-185.

Wallace, N. (1955). The effect of temperature on the growth of some freshwater diatoms. *Not. Nat. Acad. Nat. Sci. Philadelphia* 280:11.

Water Quality Glossary www.ew.govt.nz

Wetzel, R.G. (1975). Limnology. W.B. Samunders Co., Philadelphia, London, Toronto.

Zafar, A.R. (1964).On the ecology of algae in certain fish ponds of Hyderabad. *Hydrobiologia* 23:176-196.

Zargar, S. and Ghosh, T. K. (2005). Impact of cooling water discharges from Kaiga Nuclear Power Plant on chlorophyll-a and primary productivity of the Kadra reservoir. *J. Inland Fisheries Soc. India* 37 (1), 1-8

Zargar, S. and Ghosh, T. K. (2006). Influence of cooling water discharges from Kaiga Nuclear Power Plant on selected indices applied to plankton population of Kadra Reservoir. *J. Environ.Biol.* 27 (1/2),191-198

Zargar, S. and Ghosh, T. K. (2007). Thermal and biocidal (chlorine) effects on select freshwater plankton. *Archieves of Environmental Contamination and Toxicology,* 53, 191-197.

Zargar, S., Krishnamurthi, K., Sarvanadevi, S., Ghosh, T. K. and Chakrabarti, T. ($2006_a$). Temperature induced stress on growth and expression of hsp in freshwater alga *Scenedesmus quadricauda.* J. *Biomedical and Environmental Sciences* 19 (6): 414-421.

Zargar, S., Krishnamurthi, K., Sarvanadevi, S., Ghosh,T. K. and Chakrabarti, T. ($2006_b$). Growth response and heat shock protein (hsp) expression in *Chlorella vulgaris* exposure to elevated temperatures. *Asian Fisheries Science* 19 (1/2): 69-74.

Zhang, Y., and Prepas, E. E. (1996). Regulations of the dominance of planktonic diatoms and cyanobacteria in four eutrophic hard water lakes by nutrients, water column stability and temperature. *Can. J. Fish. Aquat. Sci.* 53: 621.

# Index